# 173
# Advances in Biochemical Engineering/Biotechnology

## Aims and Scope

This book series reviews current trends in modern biotechnology and biochemical engineering. Its aim is to cover all aspects of these interdisciplinary disciplines, where knowledge, methods and expertise are required from chemistry, biochemistry, microbiology, molecular biology, chemical engineering and computer science.

Volumes are organized topically and provide a comprehensive discussion of developments in the field over the past 3–5 years. The series also discusses new discoveries and applications. Special volumes are dedicated to selected topics which focus on new biotechnological products and new processes for their synthesis and purification.

In general, volumes are edited by well-known guest editors. The series editor and publisher will, however, always be pleased to receive suggestions and supplementary information. Manuscripts are accepted in English.

In references, Advances in Biochemical Engineering/Biotechnology is abbreviated as *Adv. Biochem. Engin./Biotechnol.* and cited as a journal.

More information about this series at http://www.springer.com/series/10

Magnus Fröhling • Michael Hiete
Editors

# Sustainability and Life Cycle Assessment in Industrial Biotechnology

With contributions by

S. Albrecht · L. Asveld · T. Beck · U. Bos · C. Chen · P. L. Chu · G. Festel · M. Fröhling · F. Glombitza · W. M. Griffin · M. Hiete · R. Kermer · K. Kümmerer · J. P. Lindner · H. L. MacLean · C. Macombe · P. Osseweijer · D. Pleissner · J. A. Posada · I. D. Posen · S. Reichel · G. Reniers · P. Saling · B. A. Saville · K. Schürrle · A. Venkatesh

Springer

*Editors*
Magnus Fröhling
TUM Campus Straubing for
Biotechnology and Sustainability
Technical University of Munich (TUM)
Straubing, Germany

Michael Hiete
Department of Business Chemistry
Ulm University
Ulm, Germany

ISSN 0724-6145 ISSN 1616-8542 (electronic)
Advances in Biochemical Engineering/Biotechnology
ISBN 978-3-030-47065-4 ISBN 978-3-030-47066-1 (eBook)
https://doi.org/10.1007/978-3-030-47066-1



This Springer imprint is published by the registered company Springer Nature Switzerland AG.
The registered company address is: Gewerbestrasse 11, 6330 Cham, Switzerland

# Preface

Industrial biotechnology (IB) is often referred to as a key technology to contribute to respecting planetary boundaries and achieving sustainable development goals. In particular, it is expected to deliver contributions to the mitigation of greenhouse gas emissions, a secure resource supply and positive economic impacts, especially for rural areas. These positive effects have, however, to be proven and secured. At the same time, negative impacts regarding other sustainability criteria are possible and need to be avoided, and the acceptance of this new technology by stakeholders is not necessarily given.

Against the sketched background, sustainability and life cycle assessment of IB and its consequences may take an important role in the further development and dissemination. These assessments can accompany and focus on the development of products and processes towards sustainable solutions, support investment decisions and policy making and serve as an objective basis in stakeholder communication. Thus, it is the aim of this book to explore this field. Over 12 chapters, experts in the individual fields define the scope and characteristics of IB. They identify sustainability aspects and demonstrate the applicability and value of various assessment methods. They also examine the status quo, challenges and prospects of such assessments. Thus, we want to give students, researchers and practitioners an introduction and overview about basic concepts, current research and future developments.

We owe thanks to the editors of this book series for their idea on the topic of this volume and approaching us to edit it and fruitful discussions about the contents. Our deep thanks go to all the authors of the chapters, who took up our initial ideas and developed them to the most valuable contributions presented. Finally, we thank the team of Springer Nature and their contractors for accompanying and supporting us through the process of elaborating, editing and publishing this book.

Straubing, Germany — Magnus Fröhling
Ulm, Germany — Michael Hiete
March 2020

# Contents

# Part I
# Introduction

# Part I
# Introduction

Adv Biochem Eng Biotechnol (2020) 173: 3–10
DOI: 10.1007/10_2020_123

Published online: 31 March 2020

# The Sustainability and Life Cycle Assessments of Industrial Biotechnology: An Introduction

**Magnus Fröhling and Michael Hiete**

## Contents



**Abstract** Industrial biotechnology (IB) uses biological and biochemical processes in industrial production and is often regarded as an emerging key technology revolutionizing the production of many products while protecting resources and the environment and fostering economic development. This contribution describes the background and sketches the content of the volume 'Sustainability and Life Cycle Assessment of Industrial Biotechnology' in the Springer series 'Advances in Biochemical Engineering/Biotechnology'. The field of IB is introduced from different perspectives (milestones in IB history, economics of biotechnology industry, environmental and social as well as ethical issues and impacts, green chemistry) and in several applications fields (production of chemicals, geobiotechnology in mining).

M. Fröhling (✉)
Technical University of Munich (TUM), TUM Campus Straubing for Biotechnology and Sustainability, Straubing, Germany
e-mail: magnus.froehling@tum.de

M. Hiete
Ulm University, Department of Business Chemistry, Ulm, Germany
e-mail: michael.hiete@uni-ulm.de

**Graphical Abstract**

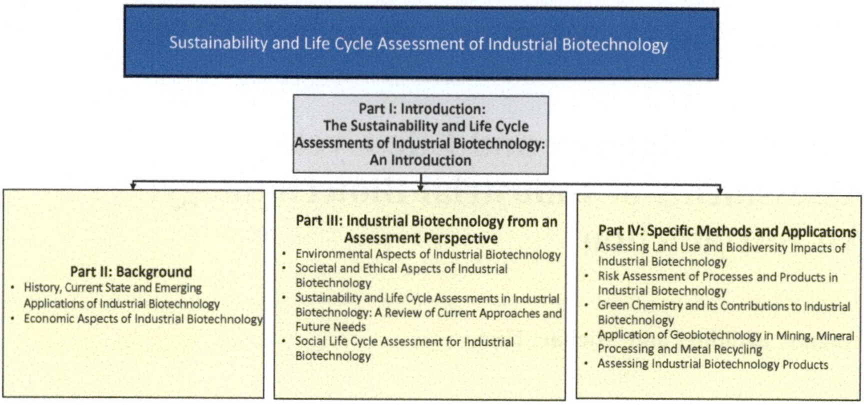


**Keywords** Application, Economy, History, Industrial biotechnology, LCA, Outline, Sustainability asessment

## 1 Introduction

Industrial biotechnology (IB) is a young but highly promising field. Using biological and biochemical processes in industrial production, IB offers more resource- and energy-efficient, less emission intensive methods of production that remain economical. IB is often seen as a key to the protection of non-renewable resources and climate change mitigation, and it is also seen as an enabling technology for economic development, especially in rural areas (cf., e.g. [1, 2]).

IB is widely applicable to our energy and fuel supplies [3–5], renewable and non-renewable material production [6, 7], food and feed provision [8, 9] and the production of pharmaceuticals and cosmetics [10, 11]. Furthermore, compared to traditional industrial processes, IB has potential in the context of smaller production capacities [12] that stimulate innovation and new business opportunities, particularly for start-ups.

However, the intended economic, ecological and social benefits of many new IB applications still need to be proven. Newly developed processes compete with those that have been optimised over decades [13], which are symbiotically integrated into the overall production system. The functional benefits of biotechnological processes and their products need to be demonstrated and assessed in a comparative manner. To that end, Peter Saling [14] in Chapter 12, for example, performed an Eco-Efficiency Analysis to compare biotechnologically produced astaxanthin, astaxanthin from fermentation and chemically produced Lucantin® Pink, all of which are used as pigment in salmon, bream and shrimp feed. Cutting-edge processes often come with higher production costs than those of existing processes. These higher costs may be reduced over time with further development and higher production volumes [15], and the production of biotechnological atorvastatin is an example of this process optimization. However, considerable research and development efforts are needed to further develop these processes, and success is not guaranteed. Environmental benefits in some areas, such as climate change mitigation, may be achieved at the expense of deficits in others, such as land competition or the loss of biodiversity.

In this context, sustainability and life cycle assessments can provide a valuable set of tools and methodologies to support research and development, decision-making, policy development and communication with stakeholders. Accordingly, the field's fast-emerging literature covers specific approaches in green chemistry, including the methodological toolset in general and that of IB and its applications in particular.

Therefore, the aim of this book is to explore the field of sustainability and life cycle assessments for IB. Over the 12 chapters, experts in the individual fields define the scope and characteristics of IB. They identify sustainability aspects and demonstrate the applicability and value of various assessment methods. They also examine the status quo, challenges and prospects of such assessments.

## 2 Volume Overview

Following the first introductory part with this first outlining chapter, the part II of the book explores IB's historical and economic background. In the second chapter 'History, Current State, and Emerging Applications of Industrial Biotechnology' [16], Karsten Schürrle gives a fascinating and comprehensive overview of IB's development over the past 150 years. He describes how the biotechnological discoveries since the sixteenth century laid the foundation for later industrial uses. In considerable detail, he traces the developments that led to today's state of affairs, and he gives a glimpse of a potential future in which these technologies transform industries, societies and individuals.

In chapter 'Economic Aspects of Industrial Biotechnology' [17], Gunter Festel explores the economic development and dissemination of this technology. He shows how IB's currently small market share is expanding at a rapid pace. He analyses the companies in the field and describes their roles. While multinational enterprises are influential in terms of commercial development, technological development is mainly driven by small- and medium-sized enterprises. Small and nascent spin-offs from universities and research institutions also have an important role in putting research into practice. For a further diffusion of IB, Festel underscores the importance of collaborations between all types of enterprises; he also highlights the importance of venture capital, more entrepreneurial thinking in research and development and more and better incentives to attract investors and staff. In addition, he states that funding schemes should better consider the needs and particulars of nascent high-tech companies.

In part III of the book, IB is viewed from an assessment perspective.

In chapter 'Environmental Aspects of Biotechnology' [13], Venkatesh et al. briefly present IB's key environmental aspects. Then, they analyse the direct and indirect environmental impacts that arise over the life cycle of bio-based products, from feedstock supply and production to product use and end of life. This is followed by a brief review of environmental impacts in industries that use biotechnological products, such as enzymes, in food production. The geographic and temporal aspects of environmental impacts are also covered, as are multi-output allocations and strategies for dealing with uncertainty, variability and trade-offs. The findings are further detailed and illustrated in three case studies that centre on widely discussed IB products: lignocellulosic ethanol, bio-based plastics and enzymes used in the detergent industry. The case studies underline the notion that the generalisations of assessment findings are delicate. Finally, the emerging topics of genetic engineering, the shift from fossil to bio-based feedstock and IB's maturity-driven improvements are analysed.

In chapter 'Societal and Ethical Issues in Industrial Biotechnology' [18], Asveld et al. examine IB controversies that involve the production of synthetic artemisinin, vanillin and algae-based oil as raw material for surfactant production. The authors identify and discuss five societal topics relevant to IB: (1) the definition and measurement of what is considered sustainable, (2) the definition of what is considered natural, (3) the management of emerging IB risks, (4) the innovation trajectories of IB development, and (5) the question of who benefits from IB in terms of economic justice. The Responsible Research and Innovation framework is presented as a guideline for learning from different actors to achieve 'societally robust innovations', and the authors discuss in detail the framework's dimensions anticipation, reflexivity, inclusion and responsiveness with respect to IB.

In the sixth chapter 'Sustainability and Life Cycle Assessment of Industrial Biotechnology' [19], Fröhling and Hiete deal with the current approaches to and future needs of IB sustainability and life cycle assessments. In a comprehensive review, the authors take a methodological and application-based perspective to explore and categorise various methodologies, methods and tools. In so doing, they describe IB from an assessment perspective. They set out challenges and solutions, and they sketch possible avenues of future research, thereby identifying important research fields and referencing specific works in this book as well as important external literature, which can be pursued by readers in their own works.

In chapter 'Social Life Cycle Assessment for Industrial Biotechnology' [20], Catherine Macombe accounts for the potential positive and negative social impacts of products, processes and methods. In her comprehensive article, she offers an in-depth analysis of the literature on existing approaches of social life cycle assessment (SLCA), which she divides into two main types. Type 1 SLCAs assess current states of products or systems mostly regarding corporate social responsibility criteria. Type 2 SLCAs follow a consequential approach and try to assess the social effects of a change by, for example, using methods such as input-output tables. Macombe reviews 58 articles largely related to biofuel and investigates the motivations behind the studies as well as their results as they relate to IB's social impacts. It becomes evident that no commonly agreed-upon framework exists; however, researchers often used a combination of several methods. On that basis, Macombe discusses what SLCA can and should achieve for IB by analysing its limitations and motivations and proposing rules for conducting a reliable SLCA. In her conclusion, she stresses that IB may lead to larger system changes that need to be considered in SLCAs, as the current approaches often consider only marginal changes.

In part IV of the book, IB's current and emerging applications and assessments are covered, and the provision of substrates from agriculture and forestry is among its most relevant environmental impacts. In the eighth chapter 'Assessing Land Use and Biodiversity Impacts of Industrial Biotechnology' [21], Lindner et al. concisely provide a general overview of land use issues in agriculture and forestry, with an emphasis on how life cycle assessment considers biodiversity loss and land competition. Specifically, the authors focus on the Land Use Indicator Value Calculation Tool, which allows for the assessing of impacts on soil quality and has its origin in the authors' Department of Life Cycle Engineering at the University of Stuttgart.

In chapter 'Risk Assessment of Processes and Products in Industrial Biotechnology' [22], Chen and Reniers explore IB's hazards and risks and the approaches of assessing and managing them. After demonstrating that IB shares many of the risks of chemical technology – as well as other risks – Chen and Reniers briefly present the risk assessment methods commonly used in the chemical industry, including HAZOP, fault tree analyses and the risk matrix. Then, IB's biological and traditional hazards are analysed in greater detail, and a checklist for each is presented in order to identify the possible hazards in concrete problems. The authors also describe how risk analyses for occupational health and the environment can be conducted in the context of IB. Finally, they present a risk analysis of bioenergy-production accidents in biogas plants, highlighting the value of such analyses, which include fishbone diagrams of the accident causes.

The exploration of green chemistry's possible contributions to IB and vice versa is the topic of the tenth chapter 'Green Chemistry and Its Contribution to Industrial Biotechnology', which is written by Pleissner and Kümmerer [23]. After setting the scene of IB and green chemistry, the authors point to sustainability weak points in current IB processes. Their examples demonstrate how the principles of green chemistry could make biomass production as well as upstream and downstream processing more sustainable by, for example, efficiently utilizing the entire biomass and the implementation of circular processes. Both the potential and challenges are further illustrated using a case study on biotechnological production of adipic acid from lignin.

IB is mostly discussed in the context of using biogenic feedstock. However, promising applications exist also in further contexts. In chapter 'Application Potentials of Geobiotechnology in Mining, Mineral Processing and Metal Recycling' [24], Glombitza et al. explain how IB can be used in the named fields. They detail the role of microorganisms in the geosphere and sketch the development of geobiotechnology, which can be seen as part of IB. The authors show how chemical oxidation and reduction processes can be supported by microbes through (1) mobilisation by means of bioleaching, (2) metal extraction by immobilisation and (3) the microbial-mediated formation of metal organic compounds. Further, they deal with metal extraction through the direct participation of microorganisms, specifically (1) bioaccumulation through a storage of metals in living cells and (2) biosorption, which is the metal storage on the cell surface or envelope. Finally, they briefly sketch how $CO_2$ could be utilised in biomass formation using iron ions or $H_2$. Throughout, the authors give overviews of the relevant reactions and necessary conditions, and they show what an application might look like and which technologies already exist.

In chapter 'Assessing Industrial Biotechnology Products with LCA and Eco-Efficiency' [14], Peter Saling from BASF SE takes the perspective of a practitioner and a tool developer. As a practitioner, he provides several case studies in which Eco-Efficiency Analyses are applied to compare chemical and biotechnological processes for astaxanthin and vitamin $B_2$ production as well as laccase and sodium iodate as a reaction agent for C-N coupling. Results are mixed, which highlights the need to perform thorough assessments to identify the best option for a given task. As a tool developer, he presents new developments in the Eco-Efficiency Analysis approach and the new social analysis implemented in the SEEbalance® method.

With this book, we give students, researchers and industrial and political practitioners an up-to-date overview of the important and fast-emerging topics of IB's sustainability and life cycle assessments, including references to relevant literature and guidance for further work.

## References

1. German Federal Government (2020) Nationale Bioökonomiestrategie, Berlin
2. European Commission (2011) High level expert group on key enabling technologies, Brussels
3. Weiland P (2010) Biogas production: current state and perspectives. Appl Microbiol Biotechnol 85(4):849–860. https://doi.org/10.1007/s00253-009-2246-7
4. Hahn-Hägerdal B, Galbe M, Gorwa-Grauslund MF et al (2006) Bio-ethanol – the fuel of tomorrow from the residues of today. Trends Biotechnol 24(12):549–556. https://doi.org/10.1016/j.tibtech.2006.10.004
5. Gouveia L, Oliveira AC (2009) Microalgae as a raw material for biofuels production. J Ind Microbiol Biotechnol 36(2):269–274. https://doi.org/10.1007/s10295-008-0495-6
6. Laure S, Leschinsky M, Fröhling M et al (2017) Assessment of an organosolv lignocellulose biorefinery concept based on a material flow analysis of a pilot plant. Cellul Chem Technol 48 (9–10):793–798
7. Haitz F, Radloff S, Rupp S et al (2018) Chemo-enzymatic epoxidation of Lallemantia Iberica Seed Oil: process development and economic-ecological evaluation. Appl Biochem Biotechnol 185(1):13–33. https://doi.org/10.1007/s12010-017-2630-1
8. Karlovsky P, Suman M, Berthiller F et al (2016) Impact of food processing and detoxification treatments on mycotoxin contamination. Mycotoxin Res 32(4):179–205. https://doi.org/10.1007/s12550-016-0257-7
9. Petersen A, Wang C, Crocoll C et al (2018) Biotechnological approaches in glucosinolate production. J Integr Plant Biol 60(12):1231–1248. https://doi.org/10.1111/jipb.12705

10. Tyler B, Gullotti D, Mangraviti A et al (2016) Polylactic acid (PLA) controlled delivery carriers for biomedical applications. Adv Drug Deliv Rev 107:163–175. https://doi.org/10.1016/j.addr.2016.06.018
11. Banat IM, Makkar RS, Cameotra SS (2000) Potential commercial applications of microbial surfactants. Appl Microbiol Biotechnol 53(5):495–508. https://doi.org/10.1007/s002530051648
12. Falcone PM, Hiete M (2019) Exploring green and sustainable chemistry in the context of sustainability transition: the role of visions and policy. Curr Opin Green Sustain Chem 19:66–75. https://doi.org/10.1016/j.cogsc.2019.08.002
13. Venkatesh A, Posen ID, MacLean HL et al (2019) Environmental aspects of biotechnology. Adv Biochem Eng Biotechnol. https://doi.org/10.1007/10_2019_98
14. Saling P (2019) Assessing industrial biotechnology products with LCA and eco-efficiency. Adv Biochem Eng Biotechnol. https://doi.org/10.1007/10_2019_102
15. Ma SK, Gruber J, Davis C et al (2010) A green-by-design biocatalytic process for atorvastatin intermediate. Green Chem 12(1):81–86. https://doi.org/10.1039/b919115c
16. Schürrle K (2019) History, current state, and emerging applications of industrial biotechnology. Adv Biochem Eng Biotechnol. https://doi.org/10.1007/10_2018_81
17. Festel G (2018) Economic aspects of industrial biotechnology. Adv Biochem Eng Biotechnol. https://doi.org/10.1007/10_2018_70
18. Asveld L, Osseweijer P, Posada JA (2019) Societal and ethical issues in industrial biotechnology. Adv Biochem Eng Biotechnol. https://doi.org/10.1007/10_2019_100
19. Fröhling M, Hiete M (2020) Sustainability and life cycle assessment in industrial biotechnology: a review of current approaches and future needs. Adv Biochem Eng Biotechnol. https://doi.org/10.1007/10_2020_122
20. Macombe C (2019) Social life cycle assessment for industrial biotechnology. Adv Biochem Eng Biotechnol. https://doi.org/10.1007/10_2019_99
21. Lindner JP, Beck T, Bos U et al (2019) Assessing land use and biodiversity impacts of industrial biotechnology. Adv Biochem Eng Biotechnol. https://doi.org/10.1007/10_2019_114
22. Chen C, Reniers G (2018) Risk assessment of processes and products in industrial biotechnology. Adv Biochem Eng Biotechnol. https://doi.org/10.1007/10_2018_74
23. Pleissner D, Kümmerer K (2018) Green chemistry and its contribution to industrial biotechnology. Adv Biochem Eng Biotechnol. https://doi.org/10.1007/10_2018_73
24. Glombitza F, Kermer R, Reichel S (2019) Application potentials of geobiotechnology in mining, mineral processing, and metal recycling. Adv Biochem Eng Biotechnol. https://doi.org/10.1007/10_2018_82

# Part II
# Background

Adv Biochem Eng Biotechnol (2020) 173: 13–52
DOI: 10.1007/10_2018_81

Published online: 23 January 2019

# History, Current State, and Emerging Applications of Industrial Biotechnology

**Karsten Schürrle**

## Contents



**Abstract** The past 150 years have seen remarkable discoveries, rapidly growing biological knowledge, and giant technological leaps providing biotechnological solutions for healthcare, food production, and other societal needs. Genetic engineering, miniaturization, and ever-increasing computing power, in particular, have been key technological drivers for the past few decades. Looking ahead, the eventual

K. Schürrle (✉)
DECHEMA - Gesellschaft für Chemische Technik und Biotechnologie e.V, Frankfurt am Main, Germany
e-mail: karsten.schuerrle@dechema.de

transition from fossil resources to biomass and $CO_2$ demands a shift toward a 'bio-economy' based on novel production processes and engineered organisms.

**Graphical Abstract**

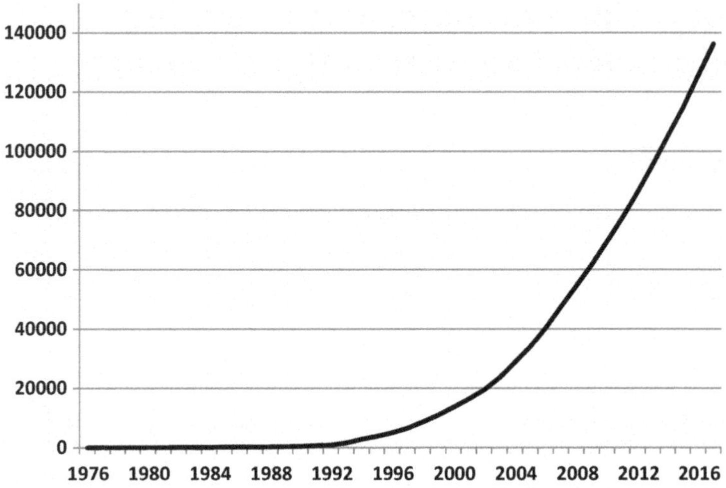


**Keywords** Chemicals, Food, Genetic engineering, Industrial biotechnology, Pharmaceuticals, Quantitative biology, Recombinant proteins, Synthetic biology

## 1 Millennia of Experience

Employing microorganisms for the production, processing, and preservation of food dates back to long before recorded history. Beer, wine, vinegar, bread, yoghurt, soy products, and cheese have been part of the human diet for thousands of years, as documented by archaeological finds from Mesopotamia, Egypt, China, and Central Europe. For millennia biotechnology has helped mankind conquer the planet. For example, preserving feed by the fermentation of grass, herbs, and leaves has enabled livestock breeding in colder regions for some thousand years [1]. In the eighteenth century, European sailors loaded barrels of fermented cabbage on board their vessels to prevent scurvy during their extended exploration voyages.

The use of enzymes can also be traced back to the early days of civilization. Cheese-making, already known in the Neolithic era [2], has depended on chymosin, obtained from the stomachs of calves and sheep, until recently when genetic engineering provided access to the microbially produced recombinant enzyme. In East Asia molds have been used for making fermented grains and soybeans, which provide a source of some 50 enzymes used in subsequent fermentations; in much the same way, the enzymes of malt (steeped and sprouted barley or other cereal grains) have served to make alcoholic beverages in the West [3]. The retting of flax and hemp is also a centuries-old biotechnological process, which makes use of consortia of either anaerobic bacteria or filamentous fungi and yeasts. Canine and avian excrements had been used to treat skins and hides for leather-making until the early twentieth century, when manufactured pancreatic proteases degrading the non-collagenous proteins replaced these odorous agents.

In the first long phase of biotechnology, although the number and variety of products grew considerably, from a technological point of view, progress was rather slow. Yeasts, Acetobacter, lactobacilli, consortia of bacteria, and some enzymes had been the invisible helpers for millennia, and the production processes – aerobic and anaerobic fermentations, biotransformations – did not significantly change. Biotechnology, essentially, was precious know-how generated from experience, while practitioners did not know what actually performed the conversions. However, sound knowledge of the ingredients, their sources and quality, optimum compositions, and the choice of suitable vessels (bioreactors), as well as the careful elaboration of optimum processing conditions, have been critical for the success of bioprocesses until today.

## 2 The Age of Discoveries

The invention of the microscope in the seventeenth century, credited to Zacharias Jansen (1588–1631), Christiaan Huygens (1629–1695), and Antoni van Leeuwenhoek (1632–1723), laid the foundation of modern biological research and the scientific investigation of cellular processes. Remarkably, the discovery of cells and microbes did not entail deeper insights into the agents behind bio(techno)logical processes before the nineteenth century. Chemistry was already flourishing then and microorganisms were still believed to emerge spontaneously, although earlier experiments with bacterial cultures had made Lazzaro Spallanzani (1729–1799) conclude that life is not an inherent feature of inorganic matter. He found that living bacteria, when treated with high temperatures in sealed tubes, never re-appeared in these tubes. In 1783, Spallanzani also observed that gastric juice could digest meat in vitro. It was the first report on reactions catalyzed by an active substance, later named pepsin. In 1828, Friedrich Wöhler's (1800–1882) synthesis of urea from cyanates and ammonia shattered the age-old belief that only organisms can produce the molecules of life. Three years later, Robert Brown (1773–1858) discovered that the nuclei of cells were involved in the embryogenesis of plants. His observations inspired Matthias Schleiden (1804–1881) and Theodor Schwann (1810–1882) to postulate that cells were the fundamental subunits of life, forming the tissues of animals and plants. Around 1840, Schwann, Charles Cagniard de la Tour (1777–1859), and Friedrich Traugott Kützing (1807–1893) independently reported that microorganisms, namely yeasts, accounted for the fermentative production of ethanol. However, the leading chemists Jöns Berzelius (1779–1848), Justus von Liebig (1803–1873), and Friedrich Wöhler (1800–1882) strongly opposed this view and insisted that fermentation was a chemical reaction due to decaying organic matter. It was Louis Pasteur (1822–1895) who, in 1857, provided clear evidence of the decisive role of yeast cells in fermentation. A few years later, he eventually refuted the age-old 'spontaneous generation' hypothesis by demonstrating that microorganisms travel through the air and regenerate and grow after landing on nutrient media. Bacteria were soon identified as the culprits when fermentations turned sour. Louis Pasteur, who investigated these processes in the 1880s, found that mildly heating the broth could prevent the undesired effect. 'Pasteurization' would become a very important technology for food production.

Fifty years after Spallanzani's observation of enzymatic activity, Anselme Payen (1795–1871) isolated an enzymatic complex from malt that catalyzed the breakdown of starch into glucose. In 1858, Moritz Traube (1826–1894) published his theory of ferments, 'enzymes' in today's terminology, which he correctly identified as proteins catalyzing biochemical reactions. His conclusions, drawn from experiments with plant extracts that converted guaiacum into colored products, contradicted Pasteur's view that *living* cells were indispensable for fermentations. It took almost 40 years until Eduard Buchner's (1860–1917) seminal experiments demonstrated that cell-free yeast extracts could catalyze the multi-step reaction sequence from sucrose to ethanol. Buchner's discoveries are widely regarded as the beginning of biochemistry and won him the Nobel Prize in chemistry of 1907. Around that time, Emil Fischer (1852–1919), who studied enzymatic conversions of saccharides, first explained the specificities of enzymatic reactions by a 'key-lock' principle. In 1905, Arthur Harden (1865–1940) and William John Young (1878–1942) discovered coenzymes, the small molecules that are essential for the functioning of many enzymes. The first industrial application of microbial enzymes started in 1894 in the United States, where Jokichi Takamine (1854–1922) prepared mixtures of amylases from *Aspergillus oryzae* fermentations to be used for the production of alcoholic beverages from grains.

The latter half of the nineteenth century also saw unprecedented progress in biomedical research and therapy. In 1876, Robert Koch (1843–1910) confirmed that 'pathogenic' bacteria were responsible for infections and that different bacteria caused different diseases. His discoveries well explained the remarkable improvements in healthcare achieved since Ignaz Semmelweis (1818–1865) had campaigned for the use of bactericidal chemicals some 30 years earlier. Koch's results laid the foundations of medicinal microbiology and eventually led to the development of antimicrobial chemotherapies in the twentieth century. In 1880, Ilya Ilyich Mechnikov (1845–1916) discovered phagocytes, which, as he correctly suspected, were an organism's primary line of defense (innate immunity), capable of engulfing and digesting any foreign agent they encountered. During the 1880s, Louis Pasteur successfully developed live attenuated microbes into vaccines against chicken cholera, anthrax, and rabies – almost eight decades after Edward Jenner (1749–1823) had produced the first vaccine by applying cowpox to prevent smallpox in humans. In 1890, Emil von Behring (1854–1917) and Shibasaburo Kitasato (1853–1931) discovered that sera prepared from the blood of previously infected mammals could protect non-infected animals and humans against diphtheria and tetanus ('passive immunization'). In the following years, Paul Ehrlich (1854–1915) identified the 'antitoxins' of the sera as mixtures of polyclonal antibodies, 'side-chains' in his words, which, as he correctly assumed, were selected against antigens, thus conferring immunity. Ehrlich and von Behring helped to establish the large-scale industrial production of sera, which started in 1894 at Farbwerke Hoechst in Germany, and von Behring later spent a large sum of his Nobel Prize money to found Behring-Werke in Marburg, a factory that produced sera against diphtheria, tetanus, and bovine tuberculosis.

At the turn of the twentieth century, Carl Correns (1864–1933), Erich von Tschermak (1871–1962), and Hugo DeVries (1848–1935) independently rediscovered the work of Gregor Johann Mendel (1822–1884), who – about 40 (!) years earlier, and largely unnoticed – had established the rules of heredity from breeding experiments with peas. Referring to Mendel's 'factors', Wilhelm Johannsen (1857–1927) introduced the terms 'gene', 'genotype', and 'phenotype'. In 1910, Thomas Hunt Morgan (1866–1945) identified chromosomes, which Walter Flemming (1843–1905) had described 30 years earlier, as the carriers of genes. Morgan also confirmed that certain traits are sex-specifically inherited, and by 1913 his collaborator Alfred Henry Sturtevant (1891–1970) had succeeded in deriving the first genetic linkage map of the fruit fly (*Drosophila melanogaster*) from carefully analyzing crossing-over events. These tedious experiments mark the beginning of modern genetics. However, it still remained unknown which biomolecules – proteins, sugars, or nucleic acids – actually stored genetic information. It should be noted that Friedrich Miescher (1844–1895), who had discovered nucleic acids in 1869, was also the first to use centrifugation to isolate nuclei and other organelles – a method that would become indispensable for biological research.

## 3 Industrial Products and More Discoveries

The nineteenth century's discoveries paved the way for non-food biotechnology products. Enzymatic bating, invented and marketed by Otto Röhm (1876–1939), was one of the first non-food applications of technical enzymes. Röhm is also credited with introducing pancreatic trypsin ('Burnus') as a commercial 'washing enzyme' in 1914 and commercializing pectinase (polygalacturonase) preparations for the production of fruit juices. Pasteur's observation, made in 1861, that butanol was a major product of the anaerobic cultivation of *Clostridia* was amended at the turn of the century when Franz Schardinger (1853–1917) found that these bacteria were also able to produce acetone. Building on these findings, Auguste Fernbach (1860–1939) established a butanol fermentation process based on potato starch in 1910. Four years later, World War I raged. The shortage of acetone – needed for the production of cordite propellant – and glycerol – needed for the production of nitroglycerin – urged both sides to tap into the potential of microorganisms. Chaim Weizmann (1874–1952), who later became Israel's first president, is acclaimed for developing the first large-scale biotechnological production of chemicals by applying *Clostridium acetobutylicum* to produce acetone and butanol. Both these chemicals have become important feedstocks for the chemical industry, serving as solvents and as precursors of butadiene and isoprene, the monomers of synthetic rubber. Over five decades, acetone-butanol ethanol (ABE) fermentation was the primary source of butanol until it eventually gave way to the more competitive petrochemical production in the 1960s. On the German side, Carl Neuberg (1877–1956) developed a biotechnological process providing glycerol from a 'shunt' alcoholic fermentation. Instead of reducing acetaldehyde, which was trapped by hydrogen sulfide, the yeast cells converted dihydroxyacetone phosphate into glycerol. James Currie's (1883-unknown) wartime work on *Aspergillus niger* strains secreting large amounts of citric acid into the medium laid the foundation for the industrial production of this metabolite, which started in Belgium in 1919 [4]. Tadeusz Reichstein's (1897–1996) synthesis of L-ascorbic acid (vitamin C), developed in 1933, is a milestone of industrial biotechnology, as it elegantly combined chemical synthesis and microbial biotransformation. *Gluconobacter oxydans*, endowed with sorbitol dehydrogenase, was employed for the conversion of D-sorbitol to L-sorbose. The industrial process, achieving yields of 60% from glucose, dominated the vitamin C market for over 60 years. In the 1960s, Chinese manufacturers successfully replaced the subsequent chemical oxidation step from L-sorbose to 2-keto-L-gulonic acid by mixed-culture fermentation with *Ketogulonigenium vulgare* and *Bacillus megaterium*. The two-step process increased the overall process yield to 90% from glucose [5]. Another early example of industrial microbial transformation is the synthesis of a precursor of L-ephedrine, established by German BASF/Knoll AG in the 1930s. *Saccharomyces cerevisiae* and *Candida utilis* were used to catalyze the condensation of pyruvic acid and benzaldehyde, yielding the chiral hydroxy ketone.

The interwar period saw important discoveries and technological progress [6]. Between 1909 and 1929 Phoebus A.T. Levene (1863–1940) identified the chemical components of nucleic acids. He also confirmed the phosphate-sugar-base structure of the nucleotide units; however, he suggested a false cyclic tetramer structure of DNA. In 1929, Herman Muller (1890–1967), a former collaborator of Morgan, reported on mutations induced by X-rays, supporting the theory of natural selection and the hypothesis that genes are individual molecular units. Induced mutations have accelerated strain development and breeding since then. The 1930s were a prolific era of biochemistry. Despite the short lifetimes and low steady-state concentrations of intermediates, Fritz Meyerhof (1884–1951), Gustav Embden (1874–1933), and other biochemists were able to accomplish a detailed step-by-step outline of the glycolysis pathway by the end of the decade. By 1937, building on prior work by Albert Szent-Györgyi von Nagyrápolt (1893–1986), who had identified key components, Hans Adolf Krebs (1900–1981) was able to delineate the citric acid cycle, the central hub of the metabolism of all organisms. In the following years, these findings were the origin of the elucidation of the biosynthetic pathways to important biotechnology products. The electron microscope was another major breakthrough for biological research. Invented in 1931 by Ernst Ruska (1906–1988) and Max Knoll (1897–1969), it allowed for the investigation of nanometer-sized intracellular structures, viruses, and larger biomolecules and has been generating invaluable insights for decades.

# 4 Antibiotics and Other Drugs

The era of chemotherapy began in 1909, when Paul Ehrlich and Hata Sahachirō (1883–1938), systematically searching for 'magic bullets', stumbled upon an arsenic compound (arsphenamine) that was able to kill both the syphilis bacterium and trypanosoma without harming the mammalian hosts. Named Salvarsan, it was the first non-natural, man-made drug. Another triumph of chemical-pharmaceutical research followed in 1932, when sulfamidochrysoidine (prontosil) was enlisted for the war against infections. Developed by Gerhard J.P. Domagk (1895–1964), it was the first in a series of sulfonamides that set off the antibiotic revolution in medicine. Alexander Fleming's (1881–1955) accidental discovery of penicillin in 1927 is probably the most prominent example of serendipity. The drug, produced by *Penicillium* molds, meant a giant leap for chemotherapy. Although initially eclipsed by the success of the sulfonamides, penicillin heralded the era of highly effective antibiotics produced by industrial biotechnology. However, it had been a long way from discovery to industrial production. Due to their beta-lactam ring structure – which Dorothy Crowfoot Hodgkin (1910–1994) was able to elucidate by X-ray crystallography in 1945 – penicillins are notoriously unstable molecules that had long frustrated chemists seeking to isolate them. It had taken 12 years of research, mainly performed by Howard W. Florey (1898–1968) and Ernst B. Chain (1906–1979), until a sufficiently stable variety, produced by *Penicillium notatum*, was available and enough material could be accumulated for clinical trials. The outbreak of World War II accelerated the development of large-scale industrial production, which was eventually established in the United States, where government, universities, and pharmaceutical companies had joined their forces. Capitalizing on *Penicillium chrysogenum*'s ability to accept synthetic acyl compounds as side chain precursors, novel penicillin variants with superior pharmacological profiles became available during the 1950s. Starting from biotechnologically produced 6-aminopenicillanic acid (6-APA), the common molecular core unit of penicillins, the 1960s saw the arrival of novel semi-synthetic analogues with improved properties. However, a continuous supply of chemical diversity was needed, since pathogenic *Staphylococcus aureus* strains inevitably became resistant, mainly due to adapting to the beta-lactamase. In 1945, Giuseppe Brotzu (1895–1976) discovered cephalosporin C, a beta-lactam antibiotic from *Acremonium chrysogenum*. It was the first drug of a new class of antibiotics, with a 7-aminocephalosporanic acid (7-ACA) core unit. More stable than penicillins to acidic conditions and beta-lactamases, 7-ACA is the origin of many semisynthetic derivatives still used today. Further important classes of antibiotics have been discovered since 1943, when Albert Schatz (1920–2005) and Selman Abraham Waksman (1888–1976) isolated streptomycin, the first antibiotic cure for tuberculosis. Tapping into the amazing chemical diversity of the secondary metabolism of actinomycetes proved extremely rewarding. These filamentous bacteria served as sources of many potent drugs such as (glyco-) peptides, aminoglycosides, tetracyclines, macrolides, and beta-lactams. Two thirds of the ~6,000 antibiotics known today were found with actinomycetes species, of which individual strains like *Streptomyces hygroscopicus* can produce up to 200 different antibiotics. Today, about 160 antibiotics are on the market, most of them semi-synthetic derivatives of natural bacterial or fungal products.

The availability of life-saving antibiotics, mainly due to the accelerated evolution of strains during the postwar decades, is one of the greatest achievements of industrial biotechnology. Because secondary metabolites are naturally produced in extremely small, hardly detectable amounts, productivities had to be increased and immense scale-up of the processes was necessary. Since the 1940s, strain selection, induced mutations, special nutrients, submerged culture technology, fed-batch fermentation, and efficient work-up technologies, e.g., membrane filters, have improved the productivities by factors of thousands. Modern *penicillium* production strains achieve titers of up to 90 g/L and global production capacities amount to several thousand metric tons per year. Between 1940 and 1962, more than 20 classes of antibiotics reached the market. However, only two new classes have been added since then. The supply of new antibiotics had been sufficient to keep up with the emergence of resistant bacteria until the 1990s, when multi-resistant strains emerged [7, 8].

Natural products, mostly first identified as antibiotics, have also been developed into important drugs for indications other than infections. A detailed analysis of new medicines approved by the United States Food and Drug Administration (FDA) between 1981 and 2010 revealed that 34% of all small molecule drugs were either natural products or direct derivatives of natural products [9]. Including natural-product-inspired total syntheses, the share is close to 50%. Microbially derived statins (e.g., compactin, lovastatin, pravastatin), immunosuppressants (e.g., cyclosporin A, rapamycin), and mycotoxins, as well as insecticides, herbicides, and antiparasitics, account for large sales today. Sixty percent of the antineoplastic compounds used for tumor therapy are natural products and derivatives.

Steroid drugs and hormones constitute a very important class of drugs, and so far, seven Nobel Prizes have been awarded for steroid research. Steroid drugs and hormones are widely used as anti-inflammatory, diuretic, anabolic, contraceptive, antiandrogenic, progestational, and anticancer agents. Cortisone, a steroid hormone from the adrenal cortex, was first isolated in the 1930s by Edward Kendall (1866–1972) and Tadeusz Reichstein, the inventor of the elegant industrial synthesis of vitamin C (see Sect. 3). When cortisone proved effective as a pain-relieving drug for patients with rheumatoid arthritis, in the 1940s, the demand went up by orders of magnitude. Merck & Co soon (Kenilworth, NJ, USA) developed a 37-step chemical synthesis starting from bovine bile acid (deoxycholic acid), yielding a rather expensive product. In 1952, biotechnology made a difference when the Upjohn Company (Kalamazoo, MI, USA) patented the microbial 11α-hydroxylation of progesterone, an early intermediate in the synthesis of cortisone, by *Rhizopus arrhizus*. The biotransformation step significantly cut the synthesis by 26 steps, reducing costs and depressing the sales price by a factor of 20. Since then, biotransformation steps have been incorporated into numerous partial syntheses of new steroids. It is now possible to hydroxylate the steroid nucleus with defined stereochemistry at virtually any position via microbial transformation. Industrial-scale 11α-, 11β-, and 16-α-hydroxylations are exclusively achieved by microbial transformations, and the microbial catabolism of phytosterol side chains yields C-19 steroids, C-22 steroids, and 17-ketosteroids, the precursors of adrenocortical hormones and contraceptives. Until 1975, diosgenin, a sapogenin prepared from Mexican Wild Yam (*Dioscorea villosa*), almost exclusively served as starting material for the hemi-synthesis of steroids. About 2,000 metric tons were processed annually. When the Mexican producers decided to raise the price tenfold, the pharmaceutical companies looked into alternatives and turned to byproducts of soybean and canola oil production [10]. The production processes have been continuously improved, e.g., by nonaqueous solvents with increased solubility of substrates, immobilization of enzymes and whole cells, and by developing continuous processes for economic product recovery. Genetic engineering of steroid-transforming microorganisms significantly enhanced productivities. However, a fermentative laboratory-scale production of hydrocortisone (= cortisol, the active metabolite produced in the liver) was realized as recently as 2003, when a French team successfully established the complex biosynthetic pathway in baker's yeast, involving eight mammalian proteins and a plant protein [11].

## 5 Primary Metabolites

Knowledge gained from the large-scale fermentation of antibiotics also boosted the industrial production of primary metabolites. Numerous biotechnological routes to amino acids, organic acids, alcohols, vitamins, and flavors were developed after WW II [12].

Amino acid production is the prime example: Microbially produced amino acids have been available since 1955, when Kyowa Hakko Kogyo Company in Japan optimized *Corynebacterium glutamicum* for the production of L-glutamic acid from glucose and ammonia. Monosodium glutamate, widely used as a flavor enhancer, is still the major amino acid in terms of tonnage, exceeding 1.6 million metric tons per year (2007). Nine of the 20 proteinogenic amino acids are essential; they have to be taken up with food because our body cannot synthesize them at a sufficient level. Since the 1960s industrial biotechnology has been the major supply: various species of the genera *Corynebacterium* and *Brevibacterium* have been developed for the industrial production of lysine, threonine, phenylalanine, and tryptophan, which are marketed as feed additives. Racemic methionine, supplied with poultry feed, is still produced by chemical synthesis. Bypassing feedback regulation is key to all biotechnological (over) production. In the case of amino acid production, this has been achieved by selecting auxotrophic and regulatory mutants. The fermentation of L-lysine from glucose or sugar, established in Japan during the 1960s, was steadily improved in this way, with yields exceeding 50% and reaching product titers of 170 g/L. Current high-performance production strains have been constructed by metabolic engineering (see Sect. 13) [13, 14]. This strategy comprises the genetic engineering of biosynthetic pathways, central metabolism, cofactor-regeneration systems, uptake and export systems, energy metabolism, global regulation, and stress responses. Advancements in genome analysis have revolutionized strain improvement, allowing for the reengineering of more efficient producers, building on knowledge that has been accumulated over years of industrial strain development [15]. Systems biology information has become essential for rational metabolic design, predicting targets to be engineered and metabolic states that will yield maximum production [16].

Citric acid is a versatile chemical produced by biotechnology [4]. Seventy percent of the global production of 1.5 million metric tons (2016) is used in the food and beverage industry. The pharmaceutical industry needs one fifth of the production for use as an antioxidant, effervescent agent, pH corrector, blood preservative, and iron carrier, as well as for ointments and cosmetic preparations. Moreover, citrates serves as a foaming agent for the softening and treatment of textiles, as a substitute for phosphate in commercial detergents, and to delay the hardening of cement. Citrates are also used in metallurgy to deliver metal ions. Taking on earlier studies by Currie (see Sect. 3), the conditions for citric acid fermentation were established during the 1930s and 1940s. Specific *A. niger* strains that are able to overproduce citric acid in different types of fermentation processes have been developed. Today, 80% of citric acid is obtained from submerged cultures, which have the advantage of lower investment and maintenance costs compared with surface cultures. During the 1960s, when crude oil was cheap, citric acid was also produced industrially by different *Candida* spp. fed on alkanes and carbohydrates.

The success story of biotechnological vitamin production (see Sect. 3), was enhanced when *Eremotheticum ashbyi* and *Ashbyia gossypii*, two fungal overproducers of riboflavin (vitamin $B_2$), were discovered in the 1930s and 1940s. The latter was optimized as an industrial producer that supplied the markets until the 1990s, when it was substituted by a recombinant *Bacillus subtilis* strain. The genome of the recombinant *B. subtilis* contained two copies of the *rib* operon controlled by strong promoters and two purine analogue-resistance mutations that deregulated the entire pathway. Vitamin $B_{12}$ (cyanocobalamin) is the largest and structurally most complicated vitamin. This coenzyme is essential for the functioning of isomerases and methyltransferases involved in DNA synthesis, and in fatty acid and amino acid metabolism. Since only bacteria and archaea are able to synthesize the molecule, biotechnology had to provide a solution: In the late 1940s several bacteria were developed into industrial producers, of which *Propionibacterium shermanii* and *P. denitrificans* turned out to be the most suited strains. Today, annual global production of cyanocobalamin is about 25 metric tons [17].

## 6 Structural Insights

Inspired by Max von Laue's (1879–1960) discovery of X-ray diffraction in 1912, William Henry Bragg (1862–1942) and his son William Lawrence (1890–1971) invented X-ray crystallography a year later. This invaluable method has been key to elucidating the structures of large biomolecules [18]. In 1926, James Batcheller Sumner (1887–1955) first crystallized an enzyme (urease). Besides definitely proving that enzymes are catalytic proteins, his result suggested that proteins could be subjected to X-ray crystallography. However, it took decades until the first enzyme, lysozyme, was imaged, in 1965. Organic chemistry and X-ray crystallography turned out to be a powerful combination for the structural analysis of complex molecules: Around 1940 Alexander Todd (1907–1997) determined the structures of nucleotides from chemical degradation, and later he also succeeded in re-synthesizing them. Dorothy Crowfoot Hodgkin (1910–1994) soon confirmed Todd's results by X-ray crystallography. In 1945, she and colleagues solved the structure of penicillin, containing a β-lactam ring, and by 1956 she eventually elucidated the three-dimensional structure of vitamin $B_{12}$. Structural biology saw another heyday in 1958, when Max Perutz (1914–2002) and John Kendrew (1917–1997) succeeded in determining the tertiary structure of myoglobin, a 17-kDa protein consisting of 153 amino acids. The high beam intensity of a synchrotron was first used in 1970 for studying an insect muscle protein. These X-ray sources allow for studying the structures of challenging proteins that are difficult or impossible to grow into large crystals. In the 1990s the resolution of crystallographic images of large proteins passed a critical threshold for discriminating single atoms. Structural analysis by nuclear magnetic resonance (NMR) spectroscopy originated from the pioneering work of Felix Bloch (1905–1983) and Edward Mills Purcell (1912–1997) in the 1940s. Kurt Wüthrich's (1938-) studies in the 1970s demonstrated that NMR was a powerful tool for providing structural information on biomolecules in solution.

Electron microscopy has been further improved for studying the structures of large biomolecules and complexes. Performed at the temperatures of liquid nitrogen or helium, the latest 'cryo' electron microscopy techniques allow the observation of specimens in their native environment without the need for staining or fixation in a lattice [19].

By March 2018, about 140,000 structures of proteins and other large biomolecules had been filed with the Protein Data Bank, at a steadily accelerating rate (Fig. 1).

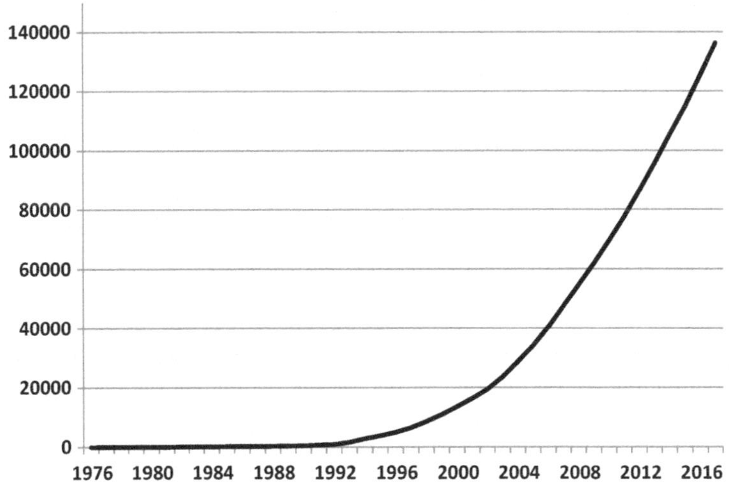


**Fig. 1** Structures filed with the Protein Data Bank; source: [20], with permission from the author

## 7 Molecular Biology and Genetics

Enhancing the productivity of fermentations by mutagenesis required neither complete understanding of the underlying biochemical pathways nor knowledge of the genes encoding the constituent enzymes. Sometimes even partial understanding of pathway regulation helped researchers to select improved strains. For example, lysine inhibits aspartate kinase, an enzyme upstream in the pathway of its synthesis. Feedback-resistant mutants producing more lysine than the original strain could be selected by using *S*-(2-aminoethyl) cysteine, a synthetic lysine analogue and potent inhibitor of the enzyme. Although industrial biotechnology had been thriving for decades, in the early 1940s little was known about the molecular machinery of bacteria and cells. This was about to change with a series of seminal experiments beginning in 1941 when George W. Beadle (1903–1989) and Edward L. Tatum (1909–1975) found that enzymes were affected by mutations. They concluded that 'a given enzyme will usually have its final specificity set by one and only one gene' [21], a correct and useful hypothesis that later needed slight modification when alternative splicing was discovered with eukaryotic cells. 'Genes', the units of inheritance, were believed to be made up of protein, since DNA was widely considered to be too simple a molecule to be the substance of genetic information. So it came as a big surprise in 1944 when Oswald Theodore Avery (1877–1955), Colin M. MacLeod (1909–1972), and Maclyn McCarty (1911–2005) proved that DNA was the '*stuff of life*'. They showed that pure DNA, isolated from an infectious pneumococcus strain, transferred the heritable property of virulence to a non-infectious bacterium. Observing the transforming activity being destroyed by DNA-digesting enzymes further supported their conclusions. Today regarded as a ground-breaking experiment that ushered in the age of molecular genetics, their discovery was initially greeted with scepticism and they were never awarded the Nobel Prize. Two years later, Tatum and Joshua Lederberg (1925–2008) discovered that bacteria could exchange genetic material, and in 1950 Alfred Hershey (1908–1997) and Martha Chase (1927–2003) impressively confirmed Avery's results. Using radio-labeled nucleotides and amino acids, they showed that bacteriophages only transferred DNA into the host cells while the viral proteins remained outside. In 1952, Lederberg also found that bacteriophages were shuttling genetic information between bacteria, thus conferring resistance to antibiotics. The same year saw a major technological breakthrough when Frederick Sanger (1918–2013) published the amino acid sequences (primary structures) of bovine insulin A and B, which he had determined by the chromatographic analysis of the chemically labeled peptidic fragments. The method was soon superseded by Pehr Edman's (1916–1977) degradation method, which was better suited for automation. The two primary structures first proved that proteins have a defined sequential chemical composition, suggesting that the corresponding information is analogously encoded in the DNA.

However, the structure of DNA still remained a mystery. Around 1950, Erwin Chargaff (1905–2005) found that in purified DNA the ratios of adenine/thymine and cytosine/guanine were always close to unity, while the nucleobase composition could vary significantly from one species to another.

The three-dimensional structure of DNA was eventually elucidated in 1953 when James D. Watson (1928-) and Francis Crick (1916–2004) presented the famous double helix model of B-DNA featuring two antiparallel strands of DNA held together by hydrogen bonds between thymine-adenine and cytosine-guanine base pairs. Their seminal work, which capitalized on the careful X-ray diffraction experiments performed by Rosalind Franklin (1920–1958) and Maurice Wilkins (1916–2004), elegantly explained Chargaff's rules and suggested a molecular mechanism of heredity [22, 23].

The iconic structure is rightfully regarded as the origin of a scientific revolution, and the following years would become the heyday of molecular biology when many puzzles of heredity could be solved.

An important step was made in 1955 when Arthur Kornberg (1918–2007) first isolated bacterial DNA polymerase I and demonstrated DNA polymerization in vitro. DNA polymerases would become an invaluable tool of biomolecular research, culminating in the invention of the polymerase chain reaction (PCR) 30 years after Kornberg's discovery (see Sect. 10). In 1958, Matthew Meselson (1930-) and Franklin Stahl (1929-) used nitrogen isotopes to elegantly prove that DNA replication is semiconservative, meaning that each of the two new helices consists of one strand from the original helix and one newly synthesized strand.

A year later, Jacques Monod (1910–1976) and François Jacob (1920–2013) proposed a model of the control of protein synthesis distilled from their studies of the lactose (*lac*) operon of *Escherichia coli*. The model suggested that the synthesis of proteins encoded in the operon is suppressed when a repressor protein, encoded by a regulatory gene, binds to a specific site on the DNA, the operator, next to the genes encoding the proteins. Monod and Jacob suspected that RNA molecules were the messengers of genetic information. The operon model was an important step toward understanding the regulation of protein synthesis and was a milestone in biotechnology.

Breaking the genetic code was top of the agenda in the early 1960s. Sophisticated genetic analysis of mutations in the *lac* operon and bacteriophage T4 helped elucidate the nature of the messenger and the genetic code. The concept of messenger RNA is associated with Francis Crick's 'Central Dogma of Molecular Biology', published in 1958 [24], which asserts that DNA is transcribed to a messenger molecule that serves as a template for the synthesis of proteins. Studying phage-infected bacteria, Sydney Brenner (1927-), Jacob, and Meselson could confirm in 1961 that messenger RNA (mRNA) was the carrier of genetic information from DNA to ribosomes.

In the same year, by careful analysis of frame-shift mutations of the T4 bacteriophage gene rIIB, Francis Crick and Brenner concluded that triplets of nucleotides code for each amino acid, a result which agreed well with earlier theoretical considerations. This finding prompted Marshall Nirenberg (1927–2010) and Heinrich Matthaei (1929-) to investigate protein synthesis by cell-free experiments with isolated ribosomes, natural amino acids, and synthetic RNA oligomers. When they added poly-uracil (U) oligomers they found that only poly-phenylalanine peptides were synthesized, thus proving that UUU was a triplet coding for phenylalanine. In 1965, Robert Holley (1922–1993), who earlier had succeeded in isolating and sequencing alanine transfer RNA (tRNA), was able to work out its tertiary structure, this being the first elucidated chemical structure of a biologically active polynucleic acid. Transfer RNA turned out to be the missing molecular 'adaptor' that chemically correlates RNA codons and amino acids in protein synthesis. Mainly through work by Nirenberg, Philip Leder (1934-), and Har Gobind Khorana (1922–2011), who developed a synthetic route to oligonucleotides (see Sect. 14), the genetic code was completely decrypted by 1966.

## 8 Genetic Engineering

Within a few years of the mid-1960s, a stream of important discoveries paved the way for recombinant DNA technology. By 1967, three teams of researchers had independently reported on DNA joining enzymes, so-called ligases, which catalyzed the covalent closure of single-strand breaks in duplex DNA. In the same year, Matthew Meselson (1930-) and Werner Arber (1929-) first described so-called type I restriction enzymes that recognized specific nucleotide sequences of DNA and cleaved double strands. Since the cleavage occurred at random distances from the recognition sites, these type I restriction enzymes were of limited use as reagents for genetic studies. It was a significant breakthrough when Hamilton O. Smith (1931-) prepared *Hind II* and *Hind III* endonucleases from the bacterium *Haemophilus influenzae* in 1970. Specifically cleaving DNA at the recognition sites, these type II restriction enzymes proved useful for laboratory work, e.g., for mapping DNA. The resulting fragments could be separated by agarose gel and polyacrylamide gel electrophoresis, two important analytical methods that had been introduced a few years earlier by several teams from virus and plant research. Today, more than 3,500 different type II restriction enzymes are known [25]. In the same year as Hamilton O. Smith's breakthrough, David Baltimore (1938-) and Howard Temin (1934–1994) independently reported on the discovery of reverse transcriptase. This enzyme, found in RNA retrovirus, violated the 'Central Dogma of Molecular Biology', since it transcribed genetic information from RNA to DNA to be integrated into the host genome and replicated along with it. Synthesizing DNA copies of genes from mRNA using reverse transcriptase would become an invaluable method for genetic engineering.

The next logical step was the cloning of foreign DNA [26]. Paul Berg (1926-) was the pioneer who, in 1972, used the available enzymatic tools (five laboratories had provided him with enzymes) to first insert foreign DNA into a plasmid. The *E. coli* plasmid he used for the cut-and-paste experiment contained the galactose operon, and the linearized SV40 monkey virus genome served as the insert. A year later, Stanley Cohen (1935-) and Herbert Boyer (1936-) first demonstrated that eukaryotic DNA could be transcribed from a recombinant bacterial plasmid. They inserted frog

DNA, coding for ribosomal RNA, which had been prepared using a restriction enzyme that left fragments with ‘sticky ends’. These two genetic engineering experiments mark the beginning of the era of recombinant DNA technology and are generally considered a watershed of biotechnology. The first widely used genetic engineering vector was pBR322, constructed and sequenced in 1977–1978; parts of it are still present in modern vectors.

## 9 Recombinant Products

In 1976, Herbert Boyer and Robert A. Swanson (1947–1999), a businessman experienced in venture capital financing, co-founded Genentech, the first biotechnology company seeking to commercialize recombinant DNA technology. As a proof-of-concept, in the following year Genentech researchers successfully expressed a synthetic human somatostatin (14 amino acids) gene in *E. coli*. However, human insulin would be the ultimate target promising the highest rewards. By September 1978, Genentech’s researchers had won the race for cloning and expressing the synthetic genes for chain A (77 bp) and chain B (104 bp) of human insulin. They later also succeeded in re-engineering a synthetic proinsulin gene containing both segments that suited commercial production, as developed by Eli Lilly, Genentech’s pharmaceutical partner. In 1982, insulin became the first recombinant pharmaceutical approved for clinical use [26].

The production of recombinant pharmaceuticals is *the* success story of modern biotechnology. Several companies were founded in the late 1970s, and some rapidly grew to become large biopharmaceutical companies: Biogen S.A., headquartered in Geneva, Switzerland, developed Interferon A and recombinant vaccines against hepatitis B. Applied Molecular Genetics, later Amgen, succeeded in producing human erythropoetin (EPO) and granulocyte colony-stimulating factor (G-CSF), and Chiron Corporation specialized in producing recombinant vaccines. Large pharmaceutical companies, initially reluctant to enter the field, soon launched research and development (R&D) programs and forged alliances with start-ups to co-develop recombinant pharmaceuticals.

Today, about 400 recombinant biopharmaceuticals are on the market, and biopharmaceuticals account for about one quarter of all drug approvals every year. In terms of revenue, pharmaceutical biotechnology is by far the largest sector of biotechnology: In 2016, the global biopharmaceutical market was valued at US$163 billion, claiming a 23% share of the pharmaceuticals market. A total of 42 biologic therapeutics, including 25 antibodies, had reached blockbuster status, with sales above the US$1 billion threshold, eight of them generating revenue of more than US $5 billion. Six companies posted combined biologics sales above US$10 billion [27].

Innovative start-ups and small- or medium- sized enterprises (SMEs), mostly founded as spin-offs from universities, have been instrumental, if not indispensable, for filling the pipelines of the pharmaceutical industry with novel drug candidates.

Since the early 1990s roughly 3,000 SMEs have been founded on both sides of the Atlantic, and 670 biotech companies were listed on major stock markets in 2016.

Since the market launch of recombinant insulin, *E. coli* and baker's yeast (*Saccharomyces cerevisiae*) have served as production organisms for recombinant biopharmaceuticals. However, in order to work properly, some therapeutic proteins require human-like glycosylation, which microbial producers cannot achieve. In 1986, Genentech's human tissue plasminogen activator (tPA) was the first recombinant therapeutic protein produced by mammalian cells to get market approval. Due to their capacity for proper protein folding, assembly, and post-translational modification, immortalized Chinese hamster ovary (CHO) cells have become the dominant system for the production of recombinant therapeutic proteins. The productivity of mammalian cells cultivated in bioreactors has reached the gram-per-liter range, a more than 100-fold yield improvement over titers seen for similar processes in the mid-1980s. This increase in volumetric productivity has mainly resulted from improvements in media composition and process control.

In the 1990s, the demand for therapeutic antibodies boosted cell culture technology. A sufficient supply of monoclonal antibodies (mAbs) specifically targeting an antigen had been the dream of medicine since Paul Ehrlich imagined 'magic bullets'. In 1975, Georges Köhler (1946–1995) and César Milstein (1927–2002) presented hybridoma technology. Hybridoma cell lines resulted from the fusion of antibody-producing B-cells with myeloma cells that conferred 'immortality' on the cancer cells. These cell lines allowed for the production of (theoretically) unlimited quantities of mAbs, which have found numerous applications as bioanalytical reagents, diagnostics, and therapeutics. Muromonab-CD3 (orthoclone OKT3®, Ortho Pharmaceutical Corporation, Raritan, NJ, USA), approved in 1986 for the prevention of kidney transplant rejection, was the first therapeutic mAb on the market. Limitations due to the murine origin of mAbs prompted researchers to develop 'chimeric' and 'humanized' antibodies; the latter consisting of human polypeptide chains, except for the antigen-binding sites. In the early 1990s, Greg Winter (1951-) employed in vitro molecular evolution (see Sect. 10) to generate antibody fragments comprising specific antigen-binding regions. This technology allowed for the generation of 'fully humanized' antibodies, bypassing hybridoma technology and immunization. In 2002, adalimumab (Humira®, Abbott Laboratories, Chicago, IL, USA), raised against tumor necrosis factor (TNF)-alpha, was the first fully humanized mAb granted marketing approval for treating rheumatoid arthritis. As of August 2017, 67 monoclonal antibody products had been approved in the United States and the European Union, with combined world-wide sales of nearly US$85 billion in 2015 [28].

The impressive success stories of the past four decades may likely eclipse the fact that the genetic engineering of eukaryotic cells was not invented by humans. About 400 million years ago, when plants had gone ashore, the soil microorganism *Rhizobium radiobacter* (former name *Agrobacterium tumefaciens*) became capable of transferring plasmid DNA into host plants, inducing gall tumors by activating a set of genes (T-DNA) located on the tumor-inducing (Ti) plasmid. The tumors are associated with the biosynthesis of plant growth hormones and are suspected to serve as a nutrient source for the bacterium. In 1983, Jozef Schell (1935–2003) and Marc Van Montagu (1933-) engineered the Ti plasmid into a vector that allowed for transferring foreign genes into plant cells by replacing the tumor-inducing genes with exogenous DNA. Since then *Rhizobium radiobacter* has become a powerful tool for modern plant genetics, breeding, and the production of transgenic plants. Insecticidal crystalline (Cry) proteins and vegetative insecticidal proteins (Vips) from *Bacillus thuringiensis* (Bt), which was discovered in 1901 by Shigetane Ishiwata (1868-1941), are known to be very effective against some devastating insect pests while being safe for beneficial insects, wildlife, and people. Crops genetically engineered to produce Bt proteins were introduced in the 1990s and have become a cornerstone of pest management, reducing reliance on insecticide sprays. In 2016, transgenic variants of major crops (soybean, maize, cotton), endowed with resistance to pathogens and herbicides, were planted on 185 million hectares worldwide.

## 10 Engineering Biomolecules

The large-scale preparation of enzymes and the increased stability conferred by immobilization techniques were key to industrial biotransformations [29]. Although the first immobilized enzyme, invertase, adsorbed on charcoal, was reported as early as 1916, efficient large-scale techniques for immobilization did not emerge before the 1960s. Another breakthrough for biochemical engineering was Alexander Klibanov's (1950-) observation, published in 1986 [30], that some enzymes could also function in organic solvents, enabling the conversion of hydrophobic substrates.

Recombinant DNA technology came with new options for in vitro mutagenesis and the creation of biomolecular diversity. In 1978, Clyde Hutchison III (1938-) presented a useful method for site-directed mutagenesis: A synthetic oligonucleotide, complementary to DNA around the targeted mutation site and containing a defined mutation, served for primer extension by DNA polymerase, copying it with the full transcript, thus producing a mutated strand that could be cloned. Several more efficient methods for site-specific mutation have been developed since then, and today, as the cost of synthetic DNA (see Sect. 14) has been steeply declining, the synthesis of complete genes is a viable, often superior method for introducing mutations.

Primer extension by DNA polymerase is the crucial step of PCR, invented in 1983 by Kary B. Mullis (1944-). He had the idea of in vitro amplifying DNA between two primers by performing repeated cycles of (1) melting of DNA to generate single strands, (2) hybridization of the primers to the single strands, and (3) subsequent polymerization, yielding new double strands that could be subjected to another round of amplification. Since the newly generated DNA served as a template for replication, the result was a chain reaction that exponentially amplified the DNA fragment. PCR permitted the isolation, analysis, and cloning of DNA from virtually any source, and has proven to be extremely useful for all fields of biomolecular research, medical diagnostics, genetics, and forensics.

The rational design of optimized biocatalysts by point mutations is complex and has rarely proven successful – even with the advantage of an available high-resolution crystal structure. A first proof-of-principle of enzyme engineering was the thermal stabilization of the T4 lysozyme, published in 1988 [31]. The team of Masazumi Matsumura (1952-) employed site-directed mutagenesis to deliberately exchange up to three pairs of amino acid residues of the natural protein with cysteine. The resulting disulfide bonds stabilized the enzyme, increasing the unfolding temperature by up to 23.4 K without affecting its activity. An alternative strategy to generate proteins and polynucleic acids with specific functions would be by mimicking the mechanisms of Darwinian evolution in vitro by repeated cycles of mutagenesis, recombination, and selection. The directed molecular evolution of proteins became feasible in 1985, when George P. Smith (1941-) employed engineered phages for linking the phenotype (a polypeptide displayed on the surface of the phage) and the genotype (its corresponding gene inserted into the phage genome). Phage display allowed for selecting individual molecules from 'phage libraries' that harbor variants of a DNA sequence, generated, e.g., by means of 'error-prone' PCR. In 1994, Willem P. C. Stemmer (1957–2013) published a method for the in vitro recombination of homologous genes, dubbed DNA shuffling, which proved useful for generating large libraries of variants of a gene. Recombination occurred when point mutated fragments from different parents annealed at regions of high sequence identity. After this assembly, PCR amplification served to generate full-length chimeric genes suitable for cloning into expression vectors. Directed evolution has been key to the engineering of novel therapeutic proteins such as fully humanized mAbs, fusion proteins, and vaccines, as well as enzymes tailored for specific industrial applications.

Numerous examples of increased enzyme activities and improved stability against temperature, pH, and solvents have been reported since Matsumura's pioneering work, often surpassing the effects of traditional enzyme immobilization. Since the 1990s, extremophiles have become a valuable source of enzymes already pre-selected by nature to function under harsh conditions. In the new century, major advances in bioinformatics, DNA, and screening technologies, and increased knowledge of protein structure and function have accelerated the development of novel biocatalysts [32, 33]. Enzymes engineered to catalyze novel reactions or accept non-natural substrates have opened new avenues to fine chemicals and pharmaceuticals [34, 35]. A prime example, reported by Codexis researchers in 2010, is an efficient biocatalytic process that replaced rhodium-catalyzed asymmetric enamine hydrogenation for the large-scale manufacture of the antidiabetic compound sitagliptin [36]. Starting from a transaminase lacking any activity toward the pro-sitagliptin ketone, the scientists applied a substrate walking, modeling, and mutation approach to create an enzyme active for the synthesis of the chiral amine. Further engineering by directed evolution allowed to synthesize chiral amines that were previously enantio-selectively inaccessible. The new biocatalytic process proved superior to the chemical process, since it reduced the total waste and eliminated all transition metals, while increasing the overall yield and the productivity by 53%.

Recent surveys have documented roughly 150 processes and over 500 industrial products made by enzymatic conversions [37, 38]. Since non-metabolizing cells often proved difficult for industrial biocatalysis, engineered enzymes dominate industrial biotechnology today. Microbial enzymes first reached the consumer market in 1956, when a detergent containing alcalase, a bacterial protease produced by *Bacillus licheniformis*, was introduced by Danish Novo Industry A/S (now Novozymes). In 1989, the company also introduced lipolase, the first recombinant washing enzyme. Since then recombinant proteases, lipases, amylases, oxidases, peroxidases, and cellulases, catalyzing the breakdown of fat and polymeric compounds, have become standard ingredients of household detergents.

## 11 Reading the Book of Life

In the late 1960s, Frederick Sanger, who had pioneered protein sequencing (see Sect. 7), applied two-dimensional chromatography to determine the sequences of viral RNA fragments that had been isolated from nuclease digests and labeled with radioactive $^{32}P$. Thus he could first show a nucleotide sequence (from R17 bacteriophage RNA) being correlated by the genetic code to a known amino acid sequence (of a coat protein of the phage). In 1972, Walter Fiers (1931-) first reported the full sequence of a gene coding for a 129-amino-acid-long coat polypeptide of bacteriophage MS2. By 1976, his team had sequenced the 3,569 ribonucleotides of the entire genome.

Between 1970 and 1973, Ray Jui Wu (1928–2008) demonstrated that DNA polymerization and specific nucleotide labeling could be employed to determine DNA sequences using synthetic location-specific primers. In 1977, Frederick Sanger and Walter Gilbert (1932-), the latter together with his graduate student Allan M. Maxam (1942-), independently presented techniques for rapid DNA sequencing. Gilbert's method involved careful fragmenting of DNA by chemical reagents that selectively cleaved double- or single-stranded DNA along its bases. Each reagent would break the DNA strands, radiolabeled with $^{32}P$-phosphate at the 5′-end, into fragments of varying length that could be separated by gel electrophoresis, thus revealing the position of each base. Sanger had developed Wu's method into a "chain-termination sequencing" process employing synthetic dideoxy versions of the four nucleotides (dideoxynucleotide-triphospates, ddNTPs). Starting from a labeled primer, each ddNTP specifically terminates the synthesis of the template copy being synthesized by DNA polymerase. As in the Gilbert-Maxam method, Sanger sequencing had to be performed separately for each of the four bases generating DNA chains to be separated by gel electrophoresis. The two methods allowed for reading the nucleotide sequences of entire genes consisting of up to 30,000 bp. In the same year as the presentation of techniques for rapid DNA sequencing (1977), Sanger published the sequence of the bacteriophage phi X174 genome, consisting of 5,386 nucleotides encoding 11 proteins [39]. It was the first DNA genome to be completely sequenced.

The concurrent development of recombinant DNA technology and PCR spurred the advancement of sequencing technologies, which made a major stride in the mid-1980s when capillary electrophoresis and labeling with fluorescent dyes enabled the automation of Sanger's method [40, 41]. Different fluorophores, either attached to the oligonucleotide primers or covalently bound to the four chain terminator ddNTPs, allowed the optical detection of DNA fragments corresponding with the base-specific termination reaction. The dye-terminator method permitted sequencing in a single reaction instead of four reactions. The reaction mixture could be co-electrophoresed through a narrow polyacrylamide gel tube (capillary); the read-out of separated fluorescent bands provided the nucleotide sequence.

These technologies were subsequently refined and commercialized as sequencing machines that would become indispensable for the Human Genome Project (HGP). Devised by prominent scientists in the late 1980s, the ambitious effort to sequence the 3,234.83 million bp that make up the human genome was officially started in October 1990. Collaborative genome sequencing projects also aimed at mapping and sequencing five model organisms, including the mouse. Mainly funded by the United States National Institutes of Health and the Department of Energy, the HGP soon received support from international partners and became a truly global co-operative effort, involving scientists from 40 countries. The HGP scientists used bacterial artificial chromosomes (BACs) containing fragmented human DNA for mapping the genome. Theoretically, the entire human genome can be covered by a 'BAC library' of 20,000 clones. Cut into smaller fragments of about 2,000 bp, the resulting subclones were subjected to sequencing and assembly, yielding contiguous stretches of sequence representing the human DNA. Milestones like the sequencing of large chromosomes were met ahead of schedule and a 'draft sequence' of the human genome, scheduled for 2005, was in fact available in June 2000. However, the HGP did not cross the finishing line alone. Celera Genomics (Rockville, MD, USA), a private company founded in 1998 to commercialize genome information, simultaneously published a complete assembly of the human genome. Led by J Craig Venter (1946-), the competing project succeeded by a 'shotgun' approach, which had been developed for sequencing *Haemophilus influenzae* Rd in 1995, providing the first complete genome sequence of a self-replicating autonomous organism. The sequence consisted of 1,830,137 bp and contained 1,749 genes. For 'shotgun sequencing', genomic DNA was randomly broken up into numerous segments and subjected to further rounds of fragmentation and sequencing, eventually resulting in multiple overlapping reads for the target DNA, which could be re-assembled into a continuous sequence by a computer. Both the HGP and the Celera human genome maps were presented at the White House and celebrated as seminal achievements. They have been completed and further refined since then. It came as a big surprise that only 1.5% of our genomic DNA codes for a total of 19,000–20,000 protein-coding genes, a number far lower than previously estimated. The lion's share of the genome contains information on non-coding RNA molecules, regulatory DNA sequences, inactive retrotransposons, introns, and sequences possibly involved in the controlling of chromosome architecture and epigenetic inheritance. The genome information proved extremely valuable for studying evolution, genetics, development, metabolism, and pathological processes. It is the basis of the individualized medicine that is currently being developed.

Besides advanced information technology and growing computing power, Sanger's chain-termination method was the core technology that enabled the HGP. It prevailed over two decades while it was constantly improved and adapted to automation. The race for sequencing genomes propelled the development of alternative sequencing technologies, while miniaturization, parallelization, and automation resulted in immense acceleration and reduction of costs [42]. Since the turn of the century several methods for high-throughput (HT) sequencing have matured into powerful technologies, of which Illumina's (San Diego, CA, USA) sequencing-by-synthesis, Applied Biosystems' (Foster City, CA, USA) sequencing-by-ligation (SOLiD), and Roche's (Basel, Switzerland) 454 pyrosequencing technology dominated the market. These massively parallel sequencing platforms relied on the production of libraries of clonally amplified DNA templates on microarrays combining DNA polymerization and generating read-outs in massively parallel runs. The first decade of the twenty-first century also saw the arrival of single-molecule sequencing platforms that could read the sequence of single strands without amplification [43]. Helicos BioSciences' (Cambridge, MA, USA) platform, featuring one-color reversible terminator sequencing of unamplified single-molecule templates, first hit the market place in 2009. Pacific Biosciences (Menlo Park, CA, USA) subsequently presented a 'real-time sequencing' platform, in which dye-labeled nucleotides were continuously incorporated into a growing DNA by a strand-displacing polymerase derived from φ29 phage. The enzyme was anchored within a zeptoliter waveguide detector, which allowed for continuous imaging of the labeled nucleotides as they entered the strand. Another 'real-time' sequencing concept employed membrane protein complexes, set into nanometer-sized holes (nanopores) in an electrically resistant synthetic membrane. Assisted by a polymerase or a helicase, a single DNA strand was ratcheted through the nanopore, nucleobase by nucleobase. Detection was achieved by measuring either ion flow, pH, or fluorescence. The first bench-top machines were marketed in 2014 by Oxford Nanopore Technologies (Oxford, UK). Not requiring DNA to be chopped into small pieces, nanopore sequencing generated longer "reads", which could be assembled more quickly and completely into a full sequence. However, nanopore sequencing required the development of de novo assembly algorithms and the use of long noisy data in conjunction with accurate short reads to produce high-quality reference genomes.

In April 2017 a team reported sequencing a record 882,000 bases of *E. coli* DNA – one-sixth of the bacterium's genome – in one pass with a nanopore device. More recently, assemblies of eukaryotic genomes, including yeasts, fungi, and *Caenorhabditis elegans*, have been reported, and the nanopore sequencing and assembly of a human reference genome was published in March 2018 [44].

Currently, performance is increasing while the cost of sequencing is declining (Fig. 2) [45]. The cost of sequencing a human genome, estimated at US$3 billion in

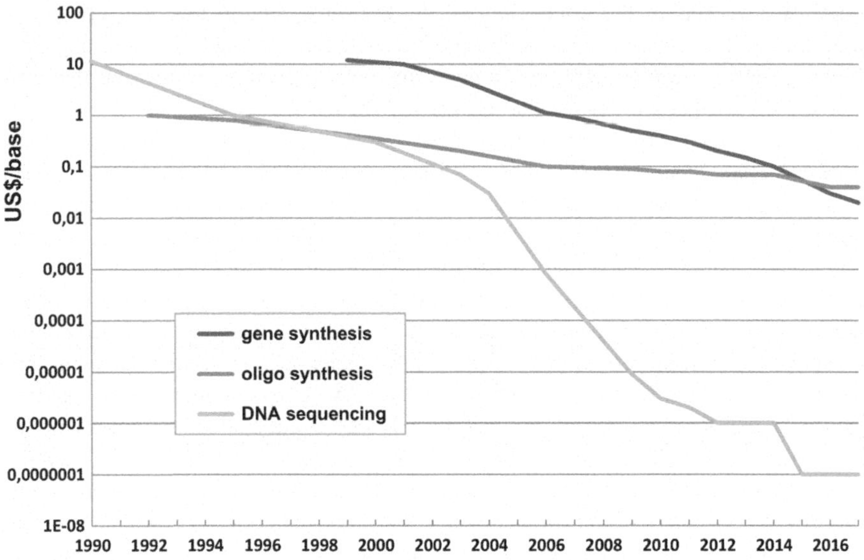


**Fig. 2** Cost of reading and writing DNA (see Sect. 14); adapted from [45], with permission from R. Carlson

2003, has dropped by six orders of magnitude to US$1,000 now and is expected to drop even further. Next-generation sequencing platforms have revolutionized biological research as they allow the sequencing of whole genomes within a few weeks. As of 2017, thousands of human genomes have been completely sequenced, and many more have been mapped at lower levels of resolution. Whole genome sequencing has transformed microbiology in particular. Within months after the publication of the *Haemophilus influenzae* genome, Venter's group published the genome of *Mycoplasma genitalium* (582,970 bp) [46] and genomes of other organisms soon followed. The optimizing of production strains based on microbial genomic information has boosted industrial biotechnology as it has become possible to identify novel genes and gene clusters and to thoroughly investigate the effects of gene loci and other regulatory elements.

Most importantly, the strides in sequencing technology and bioinformatics allowed for 'metagenome' analysis of the estimated 99% majority of microorganisms that cannot be cultivated in the laboratory [47]. As of April 2018, 9,892 complete sequences of prokaryotic genomes had been filed with the National Center for Biotechnology Information (NCBI) database.

## 12 Large-Scale, Quantitative Biology

Since the mid-1990s, miniaturization down to the nanometer-level, automation, ultra-sensitive detectors, and the ever-growing performance of computers have enabled the development of analytical platforms that can cope with the complexity of biological systems. High-throughput assays allowed for the performance of thousands of parallel experiments. Due to the availability of laser technology, fluorescent dyes, and green fluorescent protein (GFP), light microscopy saw impressive improvements, such as stimulated emission depletion (STED) microscopy, invented in 1994 by Stefan Hell (1962-), which has achieved resolutions below the classical Abbé limit and which is therefore often referred to as 'nanoscopy'.

The completion of major genome projects spurred the development of technologies for the genome-wide monitoring of gene activities (transcriptome analysis). Some of the basic methods, mainly in situ-hybridization, Northern blotting, and nucleotide synthesis, had been invented decades earlier. Combined with PCR, fluorescent dyes, and information technology, they could be integrated into powerful HT platforms. In the early 1990s, advanced production technologies (photolithography, programmable micromirror devices, ink-jet-based printing of nucleotides, semiconductor-based electrochemistry, etc.) enabled the production of DNA microarrays. These 'bio-chips' can carry fragments of thousands of genes, representing whole genomes, allowing for the analysis of gene activities in cells and tissues. Modern transcriptomics is based on HT DNA-sequencing technologies, often referred to as RNA-seq [48], and has become an indispensable standard method of biological and biomedical research. Since 2000, companies seeking to find cures from the identification of molecular drug targets in genome data have raised large sums from their investors. Incyte Genomics (Palo Alto, CA, USA), Celera Genomics, Human Genome Sciences (Rockville, MD, USA), Millenium Pharmaceuticals (Cambridge, MA, USA), and other newly founded companies have embarked on patenting putative 'disease genes'.

On the heels of genomics, the analysis of the protein inventory of cells and tissues (proteome) could also be developed into HT formats [49]. Proteomics comprises research on activities, modifications, and interactions of proteins, using powerful analytical technologies: For example, matrix-assisted laser desorption ionization (MALDI), a gentle method for the desorption and ionization of large molecules for mass spectrometry, as well as refined two-dimensional separation technologies, some of which had been invented in the 1970s, were developed into ultra-sensitive procedures for the massively parallel detection of peptides and proteins [50]. The 'yeast two-hybrid' (Y2H) system proved to be useful for studying protein-protein interactions (PPIs). The Y2H system employs a transcription factor split into a DNA-binding and an activating domain, each genetically fused with 'bait' or 'prey' proteins. If 'bait' and 'prey' associate forming a complex, the two domains cooperate like a transcription factor activating downstream reporter gene(s).

Coupling gas or high-performance liquid chromatography to mass spectrometry (GC-MS, HPLC-MS) resulted in ultra-sensitive analytical instruments which allowed for the separation and detection of thousands of small molecule metabolites from biological samples. During the 2000s this field of research, dubbed metabolomics, matured into a set of technologies that has even enabled imaging of the chemical composition of cells and the study of metabolic fluxes in cells. Information on fluxes has proven to be especially useful for the investigation of biosynthetic pathways and quantitative biology.

The Lotka-Volterra laws of ecological equilibrium (1931), the Jacob-Monod model of gene regulation (1959), the 1950s theoretical studies in nerve signal transduction by Andrew Fielding Huxley (1917-2012) and Alan Lloyd Hodgkin (1914-1998), or Denis Noble's (1936-) biophysical model of the heart (1960), are often referred to as early examples of quantitative biology. However, integrative quantitative studies of living systems had not been feasible before powerful 'omics' technologies emerged and the internet provided access to large databases, enabling real-time collaboration across oceans. At the threshold of the new millennium numerous systems biology projects were launched, collecting large amounts of biological data to be fed into in silico models. Incrementally refined through feedback cycles of hypotheses and experiments, they would eventually evolve into tools for predicting biological behavior. Multi-omic characterization has become key to the study of the effects of flux re-routing in central metabolism [51]. So far, systems biology approaches have helped to uncover the design principles of signaling networks of cells and identify bottlenecks in the production of metabolites. Terpenoid production in *E. coli*, 3-hydroxypropionic acid production in *Saccharomyces cerevisiae*, and L-lysine production in *Corynebacterium glutamicum* are the prime examples.

High-throughput "omics" assays, generating terabytes of data per run, required powerful analysis software and large computing capacities. Genome research, in particular, spurred the development of bioinformatics and the analytical software needed to retrieve information from the avalanches of data. With the cost of sequencing declining and the processes becoming more common, data is on the rise and both bioinformatics and computing capacity have become essential for biotechnology research [52, 53].

## 13 Engineering Life

It is not entirely false to regard cells as chemical factories that convert thousands of chemical compounds through thousands of chemical reactions. The main substrate is sugar, and the number of possible products is virtually infinite. Combining systems level information and information on metabolic fluxes with genetic engineering tools has enabled the rational design and engineering of biochemical pathways. Dubbed 'metabolic engineering' in 1990 publications, this field of research aims at enhancing the productivity of cells and endowing them with novel biosynthetic capacities by the re-routing of pathways and modification of enzymes [54–57]. In 1990, Gregory Stephanopoulos (1950-) used a complete method for pathway enumeration from a database of biochemical reactions to identify possible routes in the biosynthesis of lysine. The method provided information on possible maximum yields, critical bottlenecks, and key intermediates. The first genome-scale metabolic model was generated in 1995 for *Haemophilus influenzae*, and a multicellular organism, *Caenorhabditis elegans*, was first reconstructed in 1998.

Metabolic network reconstruction and simulation proved useful for pathway design and the removal of bottlenecks. The primary method for analyzing the networks is metabolic flux analysis (MFA): fluxes are determined under different conditions and their deviations from control conditions hint at kinetic limiting steps. The outcome is a metabolic flux map of the biochemical reactions. Metabolic control analysis (MCA) quantitatively correlates variables, such as fluxes and species concentrations, to network parameters. MCA has led to the development of purely computational methods to estimate fluxes under different conditions, termed flux balance analysis (FBA). The large-scale production of L-lysine is a remarkable success story of metabolic engineering employing these methods: Since its discovery in 1956 *Corynebacterium glutamicum* has been mutated to high-performance L-lysine production strains. Growing information on transcription regulation and carbon flux distribution studies performed in the 1990s indicated that the lysine yield was constrained by the flexibility of the phosphoenolpyruvate/pyruvate node, and that the rate of lysine biosynthesis never exceeded the rate of oxaloacetate synthesis via the anaplerotic pathways. The coordinated overexpression of two genes, encoding pyruvate carboxylase and aspartokinase, confirmed that pathway flux control is shared, and resulted in a significant improvement of lysine productivity [13].

## 14 Synthetic Biology

Complementing the top-down approach of systems biology, a bottom-up approach was envisioned, aiming at forward-engineering regulatory networks and the rational manipulation of biological systems, by either systematically tuning or re-arranging their modular molecular constituents. Synthetic biology serves both the study of the functional organization of natural systems and the creation of artificial regulatory networks, both of which have potential biotechnology and medical applications [58]. Genetic circuits engineered to carry out designed functions were first reported in 2000. One was a genetic switch, making cells toggle between two stable expression states in response to external signals. The other was an oscillatory circuit consisting of a triple negative-feedback loop of sequential repressor–promoter pairs, which resulted in a periodic oscillation of repressor protein expression. Circuit engineering has helped us to investigate the relationship between network design and the expression of prokaryotic and eukaryotic genes, providing valuable knowledge for the engineering of production organisms.

Editing genomes – either by specific and multiple genetic modifications or by writing genetic information via the synthesis of long polynucleotides – is at the heart of synthetic biology. Since the invention of genetic engineering in the 1970s new, versatile, and powerful tools have been developed, improved, and refined [59–61]. During the past two decades large-scale editing of genomes by targeted modifications of multiple sites has become a real option. The toolbox contains engineered proteins consisting of a DNA-binding and an effector unit, usually a nuclease. Site-specific recombinases, zinc finger nucleases (ZFN, introduced in 2003), transcription activator-like effector nucleases (TALENs, introduced in 2009), and clustered regularly interspaced short palindromic repeats (CRISPR)/Cas9 complexes (introduced in 2012) have become the most popular systems to knock out genes with high efficiency. Recombinases, ZNFs, and TALENs specifically recognize target DNA sequences via binding domains, whereas Cas9 is an RNA-guided endonuclease recognizing target DNA through base pairing. The CRISPR system, of which three distinct types are known, is an adaptive 'bacterial immune system' defending bacteria against invading viruses by targeting their DNA sequences and recording ('memorizing') them in the genome as 'spacers' between CRISPR. The development of CRISPR/Cas9 into a versatile technique for the in vivo editing of genomes from all kingdoms is another striking example of immediate transfer from basic research to application. When Jennifer Doudna (1964-) and Emmanuelle Charpentier (1968-) discovered that, in type II systems, a partially hybridized complex of two CRISPR transcripts precisely guides the nuclease to the target sequence, they concluded that a synthetic single-strand version of the RNA complex, single guide RNA (sgRNA), should function as well. In a 2012 paper they reported on a 'programmable dual-RNA-guided DNA endonuclease' allowing for the targeting of any sequence of genomic DNA [62]. With no need to laboriously re-engineer DNA-binding protein domains and with several convenient options of transfer into target cells available, CRISPR/Cas9 soon gained attraction as a versatile, precise tool [63]. Modifications to the Cas9 enzyme extended its applications beyond 'gene knock-out' to selectively activating or repressing target genes, purifying specific regions of DNA, or imaging DNA in live cells. CRISPR/Cas9 was also suited to exchange genetic information by utilizing homologous recombination, a cellular repair mechanism. The ease of generating sgRNAs has made it a widely scalable genome editing technology, that is, among other uses, useful for genome-wide screens.

Synthetic di-oligonucleotides were first reported in 1955 by A.M. Michelson and Todd (see Sect. 6). In the following decade, Har Gobind Khorana (see Sect. 7) improved phosphodiester, H-phosphonate, and phosphotriester approaches to synthesize longer oligomers. In the 1980s, Marvin Caruthers (1940-) developed solid-phase phosphoramidite chemistry, which well suited automated synthesis. Improvements in raw materials, automation, processing, and purification soon enabled the routine synthesis of long strands, of ~100-nucleotide length, at costs of US $0.05–0.15 per nucleotide and with error rates of 0.5% or better. DNA microarray technology allowed for large numbers of unique oligo sequences to be generated in parallel at costs 2–4 orders of magnitude lower than those of column-based oligomers (see Fig. 2).

Synthetic oligomers served as building blocks for gene synthesis. Between 1970 and 1972 Har Gobind Khorana synthesized the first synthetic gene, a 77-nucleotide

DNA coding for yeast alanine t-RNA, employing T4 DNA ligase to seal oligos together. By 1979 he had successfully applied this strategy to synthesize the functional 126-nucleotide-long gene for the precursor to tyrosine suppressor tRNA from *E. coli* that could be transcribed in vitro. Since then, efficient protocols for assembly by restriction and ligation, usually relying on PCR to isolate and amplify full-length from partially assembled fragments, have yielded polynucleotides exceeding the length of genes. Eckard Wimmer's (1936-) synthesis of polio virus genome cDNA (7.4 kb) in 2002 and Hamilton Smith's synthesis of the genome of bacteriophage phi X174 (5.4 kb) a year later were milestones. The synthetic replica of a 91-kb section of yeast chromosome IX, published in 2010 [64], pushed the technical limit of restriction-ligation-based methods. Constructing DNA sequences beyond this threshold called for assembly methods based on homologous recombination: in 2008, researchers at the JC Venter Institute successfully combined in vitro and yeast-based in vivo assembly to reconstruct the 583-kb genome of *Mycoplasma genitalium*, the smallest genome of an independently replicating cell [65]. The first 'synthetic genome' was made through a four-stage hierarchical scheme, beginning with sequence-verified synthetic DNA fragments. The final stages of the synthesis involved the production of large fragments, of approximately 144,000 bp, in artificial bacterial chromosomes that were subsequently cloned into *Saccharomyces cerevisiae* to synthesize the genome. Two years later, the researchers reported the assembly of the entire 1.079-Mb genome of *Mycoplasma mycoides* and its transfer and rebooting in a recipient cell [66]. The molecule had been formed in three stages from 1,079 overlapping 1-kb DNA cassettes produced from synthetic oligos. The cassettes were subsequently assembled into 10-kb strands in yeast. These were then stitched together to generate molecules 1/10 of the size of the genome which served for the final assembly of the complete genome in yeast. The synthetic chromosome could be transplanted into chromosome-free *Mycoplasma capricolum* recipient cells, thus creating *Mycoplasma mycoides* JCVI-syn1.0. Controlled only by the synthetic chromosome, the cell was capable of continuous self-replication and showed the expected phenotypic properties. The next step was building the smallest possible genome that could sustain a viable bacterial cell and would allow for defining the function of every gene (Fig. 3).

In 2016, after the deletion of genes deemed non-essential from JCVI-syn1.0 had failed, a refined mutagenesis strategy and retention of genes essential for growth resulted in a viable cell [67]. The minimal organism's 531-kb genome encoded 438 proteins and 35 RNAs. Surprisingly, 149 of the genes could not be assigned to a known function.

Since 2008, research teams across the globe have embarked on the collaborative "bottom-up" resynthesis of the 16 chromosomes of the 12 million-bp genome of baker's yeast. The Synthetic Yeast Genome Project (Sc2.0) is aimed at generating a synthetic eukaryotic genome free from known non-essential features (repetitive sequences, introns, and transposons) having all stop codons replaced by TAA triplets, and harboring unique open reading frames and recombination sites. In 2011, the scientists reported the creation of the first synthetic yeast chromosome arm, and by 2014, chromosome 3 was synthesized, being the first fully synthetic

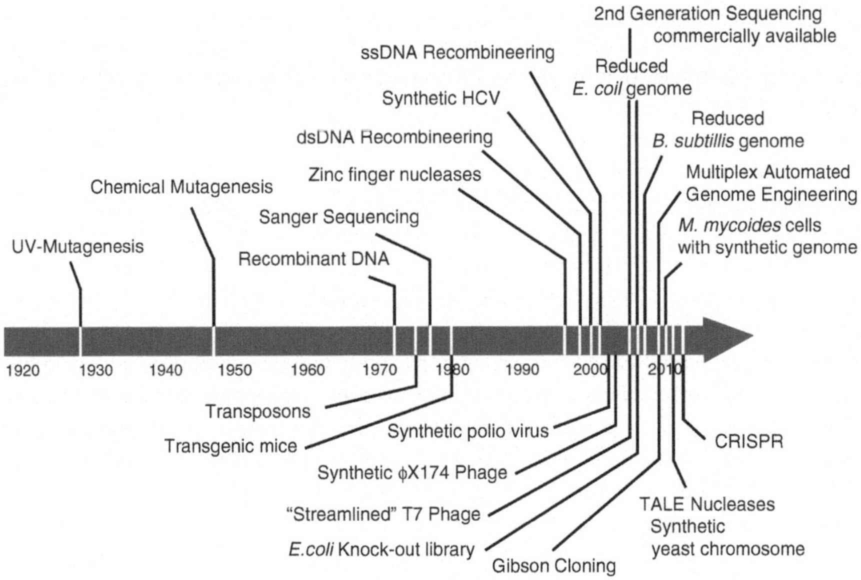


**Fig. 3** Timeline of advances toward genome-scale engineering (adapted from [59], with permission from Wiley-Blackwell). *CRISPR* clustered regularly interspaced short palindromic repeats, *HCV* hepatitis C virus, *TALE* transcription activator-like effector, *UV* ultraviolet

yeast chromosome (*synIII*). By May 2017, the Sc2.0 consortium had achieved the synthesis of six chromosomes.

It is apparent that DNA assembly is crucial for advancing synthetic biology, as seamless assembly cloning is increasingly replacing conventional cloning methods. Besides standardized restriction enzyme assembly protocols (BioBricks [68], BglBricks [69], Golden Gate [70]), new sequence-independent overlap techniques (in-fusion assembly, SLIC [71], Gibson isothermal assembly, ligase cycling reaction) and whole-cell methods (advanced quick assembly) have become popular for larger assemblies, while in vivo DNA assembly in yeast and *Bacillus* sp. has proven to be suitable for chromosome fabrication.

## 15 New Routes

Employing synthetic biology methods, metabolic engineering has facilitated the microbial production of chemicals – amino acids, (poly)alcohols, organic acids, nucleic acids, isoprenoids, flavonoids, polyketides, and complex precursors of pharmaceuticals have become available from engineered microorganisms [72–78]. Through the combined efforts of metabolic and bioprocess engineering, the list of commercial products has been growing: Today, D-lactic acid, the starting material for numerous chemicals and biodegradable plastics, is exclusively produced from starch or sugar by yeast strains tolerant to low pH [79]. Since 2008, it has also been available from cellulosic feedstock. Succinic acid [Bioamber, BASF (Ludwigshafen, Germany)], produced from *E. coli* or yeast, can be widely applied as a building block for the production of various chemicals and polymers, as well as for ingredients of food, drugs, and cosmetics [80]. 1,4 Butanediol (BDO) [81, 82], not known to be produced naturally, is an important precursor compound for the plastics and the textiles industry. In 2011, Genomatica (San Diego, CA, USA) reported the first biotechnological route to BDO from carbohydrate feedstocks. A systems-based metabolic engineering approach to strain design and development had yielded an *E. coli* strain capable of producing 18 g/L BDO. The process was soon licensed to chemical companies BASF, Lanxess (Leverkusen, Germany), and Novamont (Novara, Italy), scaling-up the production to the 650,000 t/a level (BASF). Since 2007 1,3-PDO (Dupont) has been industrially produced from renewable resources by a recombinant *E. coli* strain. 1,3-Propanediol (1,3-PDO) [81] is an important platform chemical with an annual demand of over one million tons. It is widely used for the manufacture of various polymers, for the production of drugs, cosmetics, lubricants, etc. Strain development took more than 7 years and 36 genes had to be altered. A total of 13 genes, including a P450 system, had to be engineered to produce the complex product hydrocortisone (Sanofi, Paris, France) by recombinant yeast. The flavors nalencene and nootkatone (Evolva, Basel, Switzerland) are commercially produced by engineered terpene-producing yeast cells. Vanillin (Evolva) made by fermentation has been commercialized since 2014. Engineered microbes were also employed for industrial-scale conversions of waste gases into fuels and chemicals [83]. LanzaTech (Skokie, IL, USA) developed a 'gas fermentation' process that maximizes the conversion of carbon monoxide to ethanol in *Clostridium autoethanogenum*. In 2012, the company started operating a 100,000 gal/year demonstration facility in Shanghai, China, using carbon monoxide from an adjacent steel mill. Since then, further plants have been constructed in China and Taiwan and more are to be built in India and South Africa. Another pilot-plant in New Zealand has been producing CO-based 2,3BDO at 15,000 gal/year. The chemical is a precursor of butadiene, which is a versatile starting compound, e.g., in the synthesis of adiponitrile, an intermediate in the manufacture of Nylon 6,6.

**A Case Study**

Synthetic biology methods have helped in the engineering of biosynthetic pathways to complex natural products such as hydrocortisone [11] or artemisinin. Biotechnologically produced artemisinic acid, the immediate precursor of artemisinin, is considered a prime example of such methods [84]. The sesquiterpene lactone endoperoxide, found in sweet wormwood plants (*Artemisia annua*) has been used as a potent anti-malaria drug in traditional Chinese medicine. Its discovery won Chinese scientist Youyou Tu a share of the 2015 Nobel Prize in Physiology and Medicine. In 2006, a

(continued)

team led by Jay Keasling (1964-) successfully engineered *Saccharomyces cerevisiae* to produce artemisinic acid. The yeast produced amorphadiene by the overexpression of nine genes of the mevalonate pathway and amorphadiene synthase of *Artemisia annua*. Subsequent enzymatic oxidation steps by a P450 oxidase and two dehydrogenases of plant origin yielded artemisinic acid, which could be photochemically converted to artemisinin. Fine tuning was necessary to prevent the oxidase from killing the yeast cells with reactive oxygen. Through synthetic biology efforts, as well as fermentation process optimization, titers of artemisinic acid have been improved by a factor of > one billion since the initial inception of the project in 1999: Plant genes were resynthesized to achieve better codon usage, improving the yield by a factor of 142. The yeast mevalonate pathway was up-regulated, resulting in a 90-fold improvement, while down-regulation of the native ergosterol pathway and further debugging yielded another 50-fold increase. Optimization of fermentation gave a 25-fold boost, and finally, tethering some key enzymes via a scaffold resulted in an additional 75-fold improvement. In 2013, Keasling's team reported fermentation titers of 25 g/L of artemisinic acid. Amyris (Emeryville, CA, USA), a biotech company, developed the bioproduction of artemisinic acid and the subsequent photochemical conversion to artemisinin into a viable industrial process. The Pharma giant Sanofi took charge of scale-up and commercialization. The production reached 35 metric tons of artemisinin in 2013, and 60 metric tons in 2014 – about a third of the global demand for the production of artemisinin-based combination therapies (ACTs).

Though rightfully regarded as a milestone of industrial biotechnology and chemical engineering, the hope that 'semi-synthetic' artemisinin (SSA) would offer a cheap and plentiful supply of drugs to tackle malaria – a disease that claims almost half a million lives every year – soon met with economic reality [85]. The agricultural supply of artemisinin has always been a roller-coaster of shortages and gluts associated with highly volatile market prices. In 2007, for example, when farmers in China and Vietnam expanded the acreage of Artemisia, the price of artemisinin plummeted from over US$ 1,100/kg to around US$ 200/kg, putting some 80 processing companies and a large number of farmers out of business. When Sanofi entered the market in 2013, natural artemisinin sold for less than US$ 250/kg – way below the company's 'no profit–no loss' margin. Moreover, traditional ACT manufacturers from South-East Asia were reluctant to buy drug ingredients from a direct competitor, and owing to efforts to more precisely diagnose malaria, the growth of the artemisinin market slowed down. In 2015, Sanofi produced no SSA at all and eventually sold the manufacturing site in Italy to a small company that had not been in the ACT business before.

Research efforts directed toward the conversion of biomass into liquid transportation fuels have their origins in the first energy crisis of October, 1973, owing to the OPEC oil embargo. At the beginning of the twenty-first century, global warming led to renewed efforts to provide an essentially $CO_2$-neutral fuel supply. Phototrophic microorganisms producing fuel-like products were an obvious option for mitigating the $CO_2$ problem. From the 1950s to the late 1970s, the large-scale cultivation of microalgae had been exhaustingly studied in Japan, Europe, and the United States, but without materializing as commercially viable production processes and plants. Since 2005, when the crude oil price peaked again, efforts to produce renewable biofuels from algae and from the microbial conversion of biomass have been resumed, making use of novel research tools and technologies. Start-ups from the synthetic biology field forged alliances with oil companies to explore the potential of modern biotechnology to produce biofuels other than bioethanol from corn and sugar cane, which dominated the United States and Brazilian markets. Despite high technical and economic challenges facing sustainable biofuel production, the large-scale biotechnological production of alcohols, fatty acid esters (biodiesel), and hydrocarbons was launched. Gevo (Englewood, CO, USA) retrofitted ethanol-producing plants to produce 38 million gallons of isobutanol per year from renewable feedstock. In 2011, Solazyme (San Francisco, CA, USA) started commercializing biodiesel from engineered microalgae. Through the fermentation of sugars by yeast, Amyris produced trans-β-farnesene, which could be hydrated to diesel fuel-like farnesane. Like squalene, which has been commercially produced from the farnesene platform strain since 2011, trans-β-farnesene is used as an ingredient of surfactants, lubricants, and personal care products. Today, alkanes, alkenes, and other fuel-like products from engineered fatty acid metabolism can be produced by fermentation, albeit mostly on laboratory- and pilot-scales [86, 87].

Recently, opened markets, increased shale gas production, and a slowing global economy have exerted pressure on the crude oil price, pushing biofuel production below the threshold of profitability in non-regulated markets. Pioneers like Amyris, Sapphire Energy (San Diego, CA, USA), and Solazyme had to abandon their biofuel businesses and diversified their portfolios toward high-value products for the food and health markets.

## 16 A Glimpse at the Future

Feeding a world population of ~10 billion people by 2050 calls for another 'Green Revolution', the marvelous success story of agricultural research in the 1950s, which is credited with having saved over a billion people from starvation, mainly owing to newly developed hybridized seeds and high-yielding varieties of cereals [88]. Today, huge R&D efforts are needed again: Besides improving factors such as irrigation infrastructure, logistics, transportation, storage and processing, the productivity of crops has to be significantly increased to keep up with a global food demand forecast to have doubled by 2050. Plant biotechnology is expected to provide novel crops that produce biomass and metabolites at maximum rates while being able to withstand stress by pests, flooding, drought, and saline soils. Modern plant breeding technologies, with genome editing allowing for multiplex modifications, in particular, may put forward these solutions [89]. Novel sustainable production processes, based on renewable raw materials, are needed to meet an equally growing demand for chemicals and fuels. The inevitable transition from fossil resources to biomass and $CO_2$ as raw materials means a shift toward a 'bio-economy' based on industrial biotechnology and engineered production organisms.

Emerging technologies, unimagined a few years ago, should provide the solutions: The first two decades of the twenty-first century saw big strides made toward quantitative biology and the rational engineering of cells [76, 90]. Modern bioanalytics technologies shed light on the inner life of individual cells and the synergy of systems and synthetic biology has the potential to eventually elucidate the detailed functioning of cells – their communication, division, maturation, and differentiation– providing valuable information for biological research and bioprocess engineering. The ever-increasing computing power is a major driving force. Data collected from large multi-omics assays can be processed to identify optimum variants of cells, large biomolecules, and metabolites. Growing information on biomolecular structures and protein–protein interactions is providing a strong underpinning to the computational design of novel protein folds [91–93]. Eventually, the de novo design of biocatalysts for any reaction, widely regarded as the 'Holy Grail' of biochemistry and enzymology, should become a reality. Even unnatural amino acids may be routinely incorporated into proteins via an expanded genetic code [94, 95].

A deeper understanding of cells is key to establishing complex biosynthetic pathways in cell factories. Genome engineering is considered a game changer, and knowledge derived from creating 'minimal organisms' will feed into the design of more complex genomes and 'chassis organisms'. Academic and industrial research teams around the globe are currently competing to create artificial pathways that, by using either sunlight, hydrogen, or electricity as an energy source, efficiently convert $CO_2$ or lignocellulose into value-added products or feedstocks [90, 96]. Transferring 'silent' biosynthetic gene clusters into suitable production hosts allows researchers to tap into the metabolic diversity of the 99% majority of microorganisms that are uncultivable in the laboratory. This approach yields access to novel antibiotically active compounds that are urgently needed to combat the surge of multi-resistant strains. Metabolic engineering is rapidly evolving at different levels, even including the construction of newly designed metabolisms consisting of *novel* pathways from *novel* reactions based on entirely *novel* enzymatic mechanisms, expanding the chemical diversity of fermentation beyond natural products.[1] Further studies on the implementation of non-natural, synthetic biochemical routes have been reported, as well as artificial pathways for biochemical $CO_2$-fixation that were up to 30% more energy efficient than natural routes [97].

[1]Tobias Erb defined metabolic engineering according to the level of modification: (1) optimizing existing pathways in natural hosts, (2) transfer and exchange of known (sub-) pathways in new hosts, (3) creating novel pathways from known reactions, (4) creating novel pathways from novel reactions based on known enzyme mechanisms, and (5) creating novel pathways from novel reactions based on entirely novel enzymatic mechanisms (synthetic metabolism) [97].

Scale-up from laboratory to commercial production is not trivial, and establishing viable large-scale production processes has always met with huge technical challenges: The optimum transfer of heat, gases, and nutrients in an Erlenmeyer shaker culture is quite different from that in a 1,000-L bioreactor vessel. Numerous specific problems need to be solved for developing a biotechnological process that usually also has to comply with strict regulatory guidelines. Sound knowledge and a wealth of experience have always been critical for successful bioprocess engineering, which requires the cooperation of engineers, biologists, (bio) chemists, materials scientists, and experts in systems control and other disciplines. The above-mentioned scientific and technological advancements are expected to have a huge impact on strain development, bioreactor design, process control, and downstream processing. For example, a better knowledge of intercellular communication, gained, for example, from single-cell experiments, can help us to understand the dynamics and heterogeneity within cell consortia that often determine the fate of a fermentation process. This 'upstream information' is essential for strain engineering, scale-up, and bioreactor design, taking into account the physiological constraints from large-scale production. Recent advances in systems and synthetic biology have also facilitated the population-level control of ecosystem dynamics, which is key to generating synthetic microbial consortia for novel bioprocessing applications that cannot be achieved by homogenous cultures [98].

Eventually, industrial biotechnology will be at the heart of the chemical industry. Numerous 'biorefinery' projects for converting biomass to renewable fuels, chemicals, and polymers are currently being developed. Coupling biotechnological and chemical processes will be key to producing biobased chemicals [99]. Recent advances in metabolic engineering have allowed for the development of biological catalysts that selectively de-functionalize biomass to yield platform molecules such as lactic acid, succinic acid, and alcohols, which can be upgraded using high-efficiency continuous processing by heterogeneous chemical catalysis. Linking microbial production with chemical upgrading overcomes the difficulties of both the activation of C–OH bonds by heterogeneous chemical catalysts and the production of petroleum analogues by biological catalysis. Carboxylic acids, pyrones, and alcohols are increasingly being used as highly flexible platform molecules bridging biological and chemical catalysis to produce biobased chemicals and 'drop-in' chemicals identical to traditional industrial chemicals. For example, ethanol can be chemically converted to ethene, 1-butanol, and butadiene, the latter being alternatively available from biologically produced 2,3-butanediol. A wide range of derivatives is produced from lactic acid: thermoplastic polylactic acid (PLA) and heteropolymers therof, 1,2-propanediol, a widely used commodity chemical, and acrylic acid, used in a wide range of polymer applications, just to name a few. The chemical upgrading of acetone-butanol-ethanol (ABE) mixtures (see Sect. 3) to fuels is achieved by catalyzed condensation, yielding diesel-range ketones. However, engineered *Clostridia* strains, which produce superior components such as 2-propanol in optimum mixtures, are about to substitute the chemical process [100]. Biotechnology has also facilitated the recovery of metals from ores [101]. At the largest copper mine in the world, in Escondida, Chile, copper has been produced by bioleaching since the 1980s. The process involves bacteria (e.g., *Acidithiobacillus* and

*Acidithiobacillus thiooxidans*) discovered in the 1940s, which oxidize the sulfidic ore and regenerate the chemical oxidant ferric iron, leaving soluble products that can be purified and refined to yield the desired metal. Currently, non-photosynthetic chemolithoautotrophic bacteria are being developed to produce biochemicals from $CO_2$ [102]. Metabolic engineering of these bacteria could be an alternative to classical heterotrophic bioproduction and may prove useful for the recycling of precious metals from waste, considered a cornerstone of a future circular economy.

## 17 Conclusion

A historical view of industrial biotechnology [12, 17, 29, 103–105], especially of its recent reports, is, essentially, an account of the advances in research and technology. The direct transfer of insights from basic research to applications and novel products has always been characteristic of biotechnology.

Novel technologies based on rapidly growing biological knowledge are equally transforming industry, society, and individuals. Through re-engineering complex replicating machines, which is what cells essentially are, we are tapping into the potential of evolution to solve mankind's major problems. A future sustainable 'bio-economy' is inconceivable without bioprocesses based on microbial factories that efficiently convert biomass and $CO_2$ into chemical products. The advancements of bioengineering in medicine are breathtaking as applications come into sight [106]. Patient-derived cells could soon be routinely engineered for individualized therapeutics that selectively attack pathogens and cancer cells [107], or cultivated for patient-specific tissues and 'spare organs'. Insights from brain research are likely to inspire the design of novel computers and artificial intelligence software that will revolutionize our life and question the uniqueness of the human mind. Even our germlines will not be exempt from engineering: Genome editing of human embryonic cells, first reported in August 2017 [108], has irreversibly ushered in the era of engineered humans – a vision that is both promising and disquieting.

## References

1. Eichholtz F (1960) Silage und ähnliche Gärerzeugnisse. Die Wissenschaft, vol 96. 2nd edn. Springer Fachmedien, Wiesbaden
2. The history of cheese - cheese history from an ancient nomad's horseback to today's luxury cheese cart. http://www.thenibble.com/REVIEWS/main/cheese/cheese2/history.asp. Accessed 8 Aug 2017
3. Shurtleff WAA (2017) A brief history of fermentation, East and West. http://www.soyinfocenter.com/HSS/fermentation.php. Accessed 12 June 2017
4. Max B, Salgado JM, Rodríguez N, Cortés S, Converti A, Domínguez JM (2010) Biotechnological production of citric acid. Braz J Microbiol 41(4):862–875. https://doi.org/10.1590/s1517-83822010000400005

5. Guleria S, Zhou J, Koffas MA (2017) Nutraceuticals (vitamin C, carotenoids, resveratrol). In: Wittmann C, Liao JC, Lee SY, Nielsen J, Stephanopoulos G (eds) Industrial biotechnology: products and processes, vol 4. Wiley-VCH, Weinheim, pp 309–336
6. Fry M (2016) Landmark experiments in molecular biology. Academic Press, Amsterdam
7. Walsh CT, Wencewicz TA (2014) Prospects for new antibiotics: a molecule-centered perspective. J Antibiot 67(1):7–22. https://doi.org/10.1038/ja.2013.49
8. Wright GD (2012) Antibiotics: a new hope. Chem Biol 19(1):3–10. https://doi.org/10.1016/j.chembiol.2011.10.019
9. Harvey AL, Edrada-Ebel R, Quinn RJ (2015) The re-emergence of natural products for drug discovery in the genomics era. Nat Rev Drug Discov 14(2):111–129. https://doi.org/10.1038/nrd4510
10. Renneberg R, Berkling V, Loroch V (2017) Chapter 4 - White biotechnology: cells as synthetic factories. Biotechnology for beginners2nd edn. Academic Press, Boston, pp 119–163. https://doi.org/10.1016/B978-0-12-801224-6.00004-7
11. Szczebara FM, Chandelier C, Villeret C, Masurel A, Bourot S, Duport C, Blanchard S, Groisillier A, Testet E, Costaglioli P, Cauet G, Degryse E, Balbuena D, Winter J, Achstetter T, Spagnoli R, Pompon D, Dumas B (2003) Total biosynthesis of hydrocortisone from a simple carbon source in yeast. Nat Biotechnol 21(2):143–149. https://doi.org/10.1038/nbt775
12. Demain AL (2010) History of industrial biotechnology. In: Soetaert W, Vandamme EJ (eds) Industrial biotechnology: sustainable growth and economic success. Wiley-VCH/John Wiley, Chichester/Weinheim, pp 17–77
13. Wendisch VF (2017) L-Lysine. In: Wittmann C, Liao JC, Lee SY, Nielsen J, Stephanopoulos G (eds) Industrial biotechnology: products and processes, vol 4. Wiley-VCH, Weinheim, pp 361–390
14. Hirasawa T, Shimizu H (2016) Recent advances in amino acid production by microbial cells. Curr Opin Biotechnol 42:133–146. https://doi.org/10.1016/j.copbio.2016.04.017
15. Adrio JL, Demain AL (2010) Recombinant organisms for production of industrial products. Bioeng Bugs 1(2):116–131. https://doi.org/10.4161/bbug.1.2.10484
16. Hansen ASL, Lennen RM, Sonnenschein N, Herrgard MJ (2017) Systems biology solutions for biochemical production challenges. Curr Opin Biotechnol 45:85–91. https://doi.org/10.1016/j.copbio.2016.11.018
17. Demain AL (2007) The business of biotechnology. Ind Biotechnol 3:269–283. https://doi.org/10.1089/ind.2007.3.269
18. Jones N (2014) Crystallography: atomic secrets. Nature 505(7485):602–603. https://doi.org/10.1038/505602a
19. Callaway E (2015) The revolution will not be crystallized: a new method sweeps through structural biology. Nature 525(7568):172–174. https://doi.org/10.1038/525172a
20. PDB Statistics: Overall growth of released structures per year (2018.) https://www.rcsb.org/stats/growth/overall. Accessed 20 Mar 2018
21. Beadle GW (1945) Biochemical genetics. Chem Rev 37(1):15–96. https://doi.org/10.1021/cr60116a002
22. Watson JD, Crick FH (1953) Molecular structure of nucleic acids; a structure for deoxyribose nucleic acid. Nature 171(4356):737–738
23. Watson JD (1968) The double helix: a personal account of the discovery of the structure of DNA. Weidenfeld & Nicolson, London
24. Thieffry D, Sarkar S (1998) Forty years under the central dogma. Trends Biochem Sci 23 (8):312–316
25. Roberts RJ (2005) How restriction enzymes became the workhorses of molecular biology. Proc Natl Acad Sci U S A 102(17):5905–5908. https://doi.org/10.1073/pnas.0500923102
26. Rasmussen N (2014) Gene jockeys: life science and the rise of biotech enterprise. Johns Hopkins University Press, Baltimore
27. Therapeutic antibodies account for two-thirds of total biologics sales in 2016 - 2016 was another year of record biologics sales (2017) La Merie. https://www.lamerie.com/. Accessed 20 Aug 2017

28. Monoclonal Antibody Therapeutics Market (2017.) https://www.transparencymarketresearch.com/monoclonal-antibody-therapeutics-market.html. Accessed 20 Aug 2017
29. Vasic-Racki D (2006) History of industrial biotransformations - dreams and realities. In: Liese A, Seelbach K, Wandrey C (eds) Industrial biotransformations. 2nd completely revised and extended edn. Wiley-VCH, Weinheim, pp 1–36
30. Zaks A, Klibanov AM (1985) Enzyme-catalyzed processes in organic solvents. PNAS 82:3192–3196. https://doi.org/10.1073/pnas.82.10.3192
31. Matsumura M, Becktel WJ, Matthews BW (1988) Hydrophobic stabilization in T4 lysozyme determined directly by multiple substitutions of Ile 3. Nature 334:406–410. https://doi.org/10.1038/334406a0
32. Bornscheuer UT, Huisman GW, Kazlauskas RJ, Lutz S, Moore JC, Robins K (2012) Engineering the third wave of biocatalysis. Nature 485(7397):185–194. https://doi.org/10.1038/nature11117
33. Reetz MT (2012) Directed evolution of enzymes. Enzyme catalysis in organic synthesis. Wiley-VCH Verlag GmbH & Co. KGaA, Weinheim, pp 119–190. https://doi.org/10.1002/9783527639861.ch5
34. Lalonde J (2016) Highly engineered biocatalysts for efficient small molecule pharmaceutical synthesis. Curr Opin Biotechnol 42:152–158. https://doi.org/10.1016/j.copbio.2016.04.023
35. Strohmeier GA, Pichler H, May O, Gruber-Khadjawi M (2011) Application of designed enzymes in organic synthesis. Chem Rev 111(7):4141–4164. https://doi.org/10.1021/cr100386u
36. Savile CK, Janey JM, Mundorff EC, Moore JC, Tam S, Jarvis WR, Colbeck JC, Krebber A, Fleitz FJ, Brands J, Devine PN, Huisman GW, Hughes GJ (2010) Biocatalytic asymmetric synthesis of chiral amines from ketones applied to sitagliptin manufacture. Science 329 (5989):305–309. https://doi.org/10.1126/science.1188934
37. Adrio JL, Demain AL (2014) Microbial enzymes: tools for biotechnological processes. Biomol Ther 4(1):117–139. https://doi.org/10.3390/biom4010117
38. Straathof AJ (2006) Quantitative analysis of industrial biotransformations. In: Liese A, Seelbach K, Wandrey C (eds) Industrial biotransformations. Wiley-VCH, Weinheim, pp 515–520
39. Sanger F, Air GM, Barrell BG, Brown NL, Coulson AR, Fiddes JC, Hutchison Iii CA, Slocombe PM, Smith M (1977) Nucleotide sequence of bacteriophage φX174 DNA. Nature 265:687–695. https://doi.org/10.1038/265687a0
40. Stranneheim H, Lundeberg J (2012) Stepping stones in DNA sequencing. Biotechnol J 7 (9):1063–1073. https://doi.org/10.1002/biot.201200153
41. Shendure J, Balasubramanian S, Church GM, Gilbert W, Rogers J, Schloss JA, Waterston RH (2017) DNA sequencing at 40: past, present and future. Nature 550(7676):345–353. https://doi.org/10.1038/nature24286
42. Loman NJ, Pallen MJ (2015) Twenty years of bacterial genome sequencing. Nat Rev Microbiol 13(12):787–794. https://doi.org/10.1038/nrmicro3565
43. Deamer D, Akeson M, Branton D (2016) Three decades of nanopore sequencing. Nat Biotechnol 34(5):518–524. https://doi.org/10.1038/nbt.3423
44. Jain M, Koren S, Miga KH, Quick J, Rand AC, Sasani TA, Tyson JR, Beggs AD, Dilthey AT, Fiddes IT, Malla S, Marriott H, Nieto T, O'Grady J, Olsen HE, Pedersen BS, Rhie A, Richardson H, Quinlan AR, Snutch TP, Tee L, Paten B, Phillippy AM, Simpson JT, Loman NJ, Loose M (2018) Nanopore sequencing and assembly of a human genome with ultra-long reads. Nat Biotechnol 36(4):338–345. https://doi.org/10.1038/nbt.4060
45. Carlson R (2016) On DNA and transistors. Bioeconomy capital, bioeconomy dashboard. http://www.synthesis.cc/synthesis/2016/03/on_dna_and_transistors. Accessed 14 Oct 2017
46. Fraser CM, Gocayne JD, White O, Adams MD, Clayton RA, Fleischmann RD, Bult CJ, Kerlavage AR, Sutton G, Kelley JM, Fritchman JL, Weidman JF, Small KV, Sandusky M, Fuhrmann J, Nguyen D, Utterback TR, Saudek DM, Phillips CA, Merrick JM, Tomb J-F, Dougherty BA, Bott KF, Hu P-C, Lucier TS, Peterson SN, Smith HO, Hutchison CA, Venter JC (1995) The minimal gene complement of mycoplasma genitalium. Science 270:397–404. https://doi.org/10.1126/science.270.5235.397

47. Ziemert N, Alanjary M, Weber T (2016) The evolution of genome mining in microbes - a review. Nat Prod Rep 33(8):988–1005. https://doi.org/10.1039/c6np00025h
48. van Vliet AH (2010) Next generation sequencing of microbial transcriptomes: challenges and opportunities. FEMS Microbiol Lett 302(1):1–7. https://doi.org/10.1111/j.1574-6968.2009.01767.x
49. Anderson L (2014) Six decades searching for meaning in the proteome. J Proteome 107:24–30. https://doi.org/10.1016/j.jprot.2014.03.005
50. Aebersold R, Mann M (2016) Mass-spectrometric exploration of proteome structure and function. Nature 537(7620):347–355. https://doi.org/10.1038/nature19949
51. O'Brien EJ, Monk JM, Palsson BO (2015) Using genome-scale models to predict biological capabilities. Cell 161(5):971–987. https://doi.org/10.1016/j.cell.2015.05.019
52. Pennisi E (2011) Human genome 10th anniversary. Will computers crash genomics? Science 331(6018):666–668. https://doi.org/10.1126/science.331.6018.666. 331/6018/666 [pii]
53. Green ED, Rubin EM, Olson MV (2017) The future of DNA sequencing. Nature 550 (7675):179–181. https://doi.org/10.1038/550179a
54. Woolston BM, Edgar S, Stephanopoulos G (2013) Metabolic engineering: past and future. Annu Rev Chem Biomol Eng 4:259–288. https://doi.org/10.1146/annurev-chembioeng-061312-103312
55. Stephanopoulos G (2012) Synthetic biology and metabolic engineering. ACS Synth Biol 1 (11):514–525. https://doi.org/10.1021/sb300094q
56. Zhang J, Babtie A, Stephanopoulos G (2012) Metabolic engineering: enabling technology of a bio-based economy. Curr Opin Chem Eng 1(4):355–362. https://doi.org/10.1016/j.coche.2012.09.003
57. Chen PW, Theisen MK, Liao JC (2017) Metabolic systems modeling for cell factories improvement. Curr Opin Biotechnol 46:114–119. https://doi.org/10.1016/j.copbio.2017.02.005
58. Cameron DE, Bashor CJ, Collins JJ (2014) A brief history of synthetic biology. Nat Rev Microbiol 12(5):381–390. https://doi.org/10.1038/nrmicro3239
59. Esvelt KM, Wang HH (2013) Genome-scale engineering for systems and synthetic biology. Mol Syst Biol 9:641. https://doi.org/10.1038/msb.2012.66
60. Kosuri S, Church GM (2014) Large-scale de novo DNA synthesis: technologies and applications. Nat Methods 11(5):499–507. https://doi.org/10.1038/nmeth.2918
61. Haimovich AD, Muir P, Isaacs FJ (2015) Genomes by design. Nat Rev Genet 16(9):501–516. https://doi.org/10.1038/nrg3956
62. Jinek M, Chylinski K, Fonfara I, Hauer M, Doudna JA, Charpentier E (2012) A programmable dual-RNA-guided DNA endonuclease in adaptive bacterial immunity. Science 337 (6096):816–821. https://doi.org/10.1126/science.1225829
63. Wright AV, Nunez JK, Doudna JA (2016) Biology and applications of CRISPR systems: harnessing nature's toolbox for genome engineering. Cell 164(1–2):29–44. https://doi.org/10.1016/j.cell.2015.12.035
64. Blake WJ, Chapman BA, Zindal A, Lee ME, Lippow SM, Baynes BM (2010) Pairwise selection assembly for sequence-independent construction of long-length DNA. Nucleic Acids Res 38(8):2594–2602. https://doi.org/10.1093/nar/gkq123
65. Gibson DG, Benders GA, Andrews-Pfannkoch C, Denisova EA, Baden-Tillson H, Zaveri J, Stockwell TB, Brownley A, Thomas DW, Algire MA, Merryman C, Young L, Noskov VN, Glass JI, Venter JC, Hutchison 3rd CA, Smith HO (2008) Complete chemical synthesis, assembly, and cloning of a *Mycoplasma genitalium* genome. Science 319(5867):1215–1220. https://doi.org/10.1126/science.1151721
66. Gibson DG, Glass JI, Lartigue C, Noskov VN, Chuang RY, Algire MA, Benders GA, Montague MG, Ma L, Moodie MM, Merryman C, Vashee S, Krishnakumar R, Assad-Garcia-N, Andrews-Pfannkoch C, Denisova EA, Young L, Qi ZQ, Segall-Shapiro TH, Calvey CH, Parmar PP, Hutchison 3rd CA, Smith HO, Venter JC (2010) Creation of a bacterial cell controlled by a chemically synthesized genome. Science 329(5987):52–56. https://doi.org/10.1126/science.1190719
67. Hutchison 3rd CA, Chuang RY, Noskov VN, Assad-Garcia N, Deerinck TJ, Ellisman MH, Gill J, Kannan K, Karas BJ, Ma L, Pelletier JF, Qi ZQ, Richter RA, Strychalski EA, Sun L,

Suzuki Y, Tsvetanova B, Wise KS, Smith HO, Glass JI, Merryman C, Gibson DG, Venter JC (2016) Design and synthesis of a minimal bacterial genome. Science 351(6280):aad6253. https://doi.org/10.1126/science.aad6253
68. Sleight SC, Bartley BA, Lieviant JA, Sauro HM (2010) In-Fusion BioBrick assembly and re-engineering. Nucleic Acids Res 38(8):2624–2636. https://doi.org/10.1093/nar/gkq179
69. Shetty RP, Endy D, Knight TF (2008) Engineering BioBrick vectors from BioBrick parts. J Biol Eng 2(1):5. https://doi.org/10.1186/1754-1611-2-5
70. Engler C, Kandzia R, Marillonnet S (2008) A one pot, one step, precision cloning method with high throughput capability. PLoS One 3(11):e3647. https://doi.org/10.1371/journal.pone.0003647
71. Li MZ, Elledge SJ (2007) Harnessing homologous recombination in vitro to generate recombinant DNA via SLIC. Nat Methods 4(3):251–256. https://doi.org/10.1038/nmeth1010
72. Keasling JD (2010) Manufacturing molecules through metabolic engineering. Science 330 (6009):1355–1358. https://doi.org/10.1126/science.1193990. 330/6009/1355 [pii]
73. Lee JW, Kim TY, Jang YS, Choi S, Lee SY (2011) Systems metabolic engineering for chemicals and materials. Trends Biotechnol 29(8):370–378. https://doi.org/10.1016/j.tibtech.2011.04.001
74. Lee JW, Na D, Park JM, Lee J, Choi S, Lee SY (2012) Systems metabolic engineering of microorganisms for natural and non-natural chemicals. Nat Chem Biol 8(6):536–546. https://doi.org/10.1038/nchembio.970
75. Sun X, Shen X, Jain R, Lin Y, Wang J, Sun J, Yan Y, Yuan Q (2015) Synthesis of chemicals by metabolic engineering of microbes. Chem Soc Rev 44(11):3760–3785. https://doi.org/10.1039/c5cs00159e
76. Nielsen J, Keasling JD (2016) Engineering cellular metabolism. Cell 164(6):1185–1197. https://doi.org/10.1016/j.cell.2016.02.004
77. Wang Z, Cirino PC (2016) New and improved tools and methods for enhanced biosynthesis of natural products in microorganisms. Curr Opin Biotechnol 42:159–168. https://doi.org/10.1016/j.copbio.2016.05.003
78. Chen Y, Nielsen J (2016) Biobased organic acids production by metabolically engineered microorganisms. Curr Opin Biotechnol 37:165–172. https://doi.org/10.1016/j.copbio.2015.11.004
79. Benninga H (1990) A history of lactic acid making: a chapter in the history of biotechnology. Kluwer Academic Publishers, Dordrecht
80. Ahn JH, Jang YS, Lee SY (2016) Production of succinic acid by metabolically engineered microorganisms. Curr Opin Biotechnol 42:54–66. https://doi.org/10.1016/j.copbio.2016.02.034
81. Sabra W, Groeger C, Zeng AP (2016) Microbial cell factories for diol production. Adv Biochem Eng Biotechnol 155:165–197. https://doi.org/10.1007/10_2015_330
82. Burgard A, Burk MJ, Osterhout R, Van Dien S, Yim H (2016) Development of a commercial scale process for production of 1,4-butanediol from sugar. Curr Opin Biotechnol 42:118–125. https://doi.org/10.1016/j.copbio.2016.04.016
83. Heijstra BD, Leang C, Juminaga A (2017) Gas fermentation: cellular engineering possibilities and scale up. Microb Cell Factories 16(1):60. https://doi.org/10.1186/s12934-017-0676-y
84. Paddon CJ, Keasling JD (2014) Semi-synthetic artemisinin: a model for the use of synthetic biology in pharmaceutical development. Nat Rev Microbiol 12(5):355–367. https://doi.org/10.1038/nrmicro3240
85. Peplow M (2016) Synthetic biology's first malaria drug meets market resistance. Nature 530 (7591):389–390. https://doi.org/10.1038/530390a
86. Cheon S, Kim HM, Gustavsson M, Lee SY (2016) Recent trends in metabolic engineering of microorganisms for the production of advanced biofuels. Curr Opin Chem Biol 35:10–21. https://doi.org/10.1016/j.cbpa.2016.08.003
87. Zargar A, Bailey CB, Haushalter RW, Eiben CB, Katz L, Keasling JD (2017) Leveraging microbial biosynthetic pathways for the generation of 'drop-in' biofuels. Curr Opin Biotechnol 45:156–163. https://doi.org/10.1016/j.copbio.2017.03.004

88. Borlaug N (2008) The green revolution revisited. Special 30th anniversary lecture. The Norwegian Nobel Institute, Oslo
89. Hilscher J, Burstmayr H, Stoger E (2017) Targeted modification of plant genomes for precision crop breeding. Biotechnol J 12(1):1600173. https://doi.org/10.1002/biot.201600173
90. Jullesson D, David F, Pfleger BF, Nielsen J (2015) Impact of synthetic biology and metabolic engineering on industrial production of fine chemicals. Biotechnol Adv 33(7):1395–1402. https://doi.org/10.1016/j.biotechadv.2015.02.011
91. Davids T, Schmidt M, Bottcher D, Bornscheuer UT (2013) Strategies for the discovery and engineering of enzymes for biocatalysis. Curr Opin Chem Biol 17(2):215–220. https://doi.org/10.1016/j.cbpa.2013.02.022
92. Currin A, Swainston N, Day PJ, Kell DB (2015) Synthetic biology for the directed evolution of protein biocatalysts: navigating sequence space intelligently. Chem Soc Rev 44(5):1172–1239. https://doi.org/10.1039/c4cs00351a
93. Arnold FH (2015) The nature of chemical innovation: new enzymes by evolution. Q Rev Biophys 48(4):404–410. https://doi.org/10.1017/S003358351500013X
94. Liu CC, Schultz PG (2010) Adding new chemistries to the genetic code. Annu Rev Biochem 79:413–444. https://doi.org/10.1146/annurev.biochem.052308.105824
95. Chin JW (2017) Expanding and reprogramming the genetic code. Nature 550(7674):53–60. https://doi.org/10.1038/nature24031
96. Woo HM (2017) Solar-to-chemical and solar-to-fuel production from CO2 by metabolically engineered microorganisms. Curr Opin Biotechnol 45:1–7. https://doi.org/10.1016/j.copbio.2016.11.017
97. Erb TJ, Jones PR, Bar-Even A (2017) Synthetic metabolism: metabolic engineering meets enzyme design. Curr Opin Chem Biol 37:56–62. https://doi.org/10.1016/j.cbpa.2016.12.023
98. Hays SG, Patrick WG, Ziesack M, Oxman N, Silver PA (2015) Better together: engineering and application of microbial symbioses. Curr Opin Biotechnol 36:40–49. https://doi.org/10.1016/j.copbio.2015.08.008
99. Wheeldon I, Christopher P, Blanch H (2017) Integration of heterogeneous and biochemical catalysis for production of fuels and chemicals from biomass. Curr Opin Biotechnol 45:127–135. https://doi.org/10.1016/j.copbio.2017.02.019
100. Lutke-Eversloh T, Bahl H (2011) Metabolic engineering of Clostridium acetobutylicum: recent advances to improve butanol production. Curr Opin Biotechnol 22(5):634–647. https://doi.org/10.1016/j.copbio.2011.01.011
101. See chapter by Glombitza, Kermer, and Reichel
102. Banerjee I, Burrell B, Reed C, West AC, Banta S (2017) Metals and minerals as a biotechnology feedstock: engineering biomining microbiology for bioenergy applications. Curr Opin Biotechnol 45:144–155. https://doi.org/10.1016/j.copbio.2017.03.009
103. Demain AL, Vandamme EJ, Collins J, Buchholz K (2017) History of industrial biotechnology. In: Wittmann C, Liao JC, Lee SY, Nielsen J, Stephanopoulos G (eds) Industrial biotechnology: products and processes, vol 3a. Wiley-VCH, Weinheim, pp 3–84
104. Fiechter A (2000) History of modern biotechnology. 1. Springer, London
105. Buchholz K, Collins J (2013) The roots—a short history of industrial microbiology and biotechnology. Appl Microbiol Biotechnol 97(9):3747–3762. https://doi.org/10.1007/s00253-013-4768-2
106. Xie M, Haellman V, Fussenegger M (2016) Synthetic biology-application-oriented cell engineering. Curr Opin Biotechnol 40:139–148. https://doi.org/10.1016/j.copbio.2016.04.005
107. Fischbach MA, Bluestone JA, Lim WA (2013) Cell-based therapeutics: the next pillar of medicine. Science Transl Med 5(179):179ps177. https://doi.org/10.1126/scitranslmed.3005568
108. Ledford H (2017) CRISPR fixes disease gene in viable human embryos. Nature 548(7665):13–14. https://doi.org/10.1038/nature.2017.22382

Adv Biochem Eng Biotechnol (2020) 173: 53–74
DOI: 10.1007/10_2018_70

Published online: 26 July 2018

# Economic Aspects of Industrial Biotechnology

**Gunter Festel**

## Contents



**Abstract** Industrial biotechnology is a key tool in the switch from a petro-based to a bio-based economy. For biotechnological processes to compete competitively in cost with chemical synthesis, the best available state-of-the-art technologies are necessary. In the last few years, industrial biotechnology has undergone fast technological development, resulting in a high number of basic technologies emanating from research efforts at universities and research institutions. Academic spin-offs have great importance in technological development because of their innovation from academic backgrounds. Technology transfer through spin-offs can help significantly in translating research at European universities and research institutions into commercial applications. More business oriented and experienced people, similar to founding or business angels, should join such new ventures to achieve successful realization of technology transfer.

G. Festel (✉)
Festel Capital, Fuerigen, Switzerland

Technische Universität Berlin, Berlin, Germany

University of Basel, Basel, Switzerland
e-mail: gunter.festel@festel.com

**Graphical Abstract**

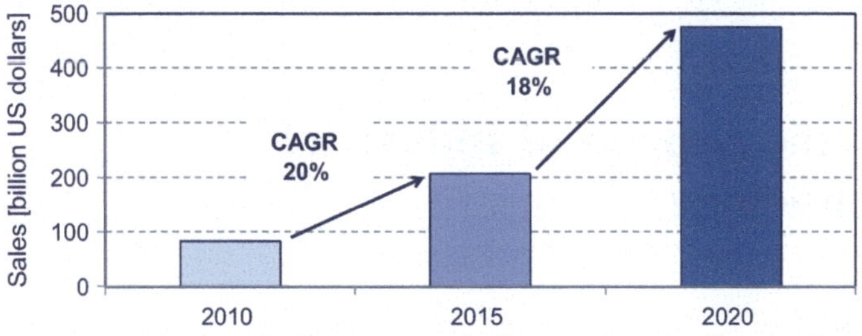


**Keywords** Biorenewables, Industrial biotech, Spin-off, Start-up, Technology transfer, Venture capital

## 1 Introduction

Industrial biotechnology is seen as a key production technology of the future [1, 2]. Instead of using chemical synthesis, industrial biotechnology works with biological systems, such as microorganisms or enzymes, using renewable resources [3]. Industrial biotechnology also enables the development of new products that cannot be made using traditional synthetic methods and processes. An increasing number of chemicals and materials such as base chemicals and polymers, as well as high-value products such as consumer chemicals and specialty chemicals, can therefore be produced using biotechnology in one or more process steps [4–8].

Industrial biotechnology is a key tool in the switch from a petro-based to a bio-based economy and therefore in the future development of the global economy, offering dynamic growth opportunities for different industries [9–11]. Whereas the agro and forestry industries mainly provide raw materials such as agricultural products and wood, the chemical and other related process industries have the research and development (R&D) and global marketing and sales (M&S) capabilities. Governments recognize the importance of industrial biotechnology and are increasing their support in order to take advantage of the potential and to remove barriers for growth in Europe [12], the United States [13], and other regions.

Products produced by biotechnological processes usually do not achieve higher prices than their chemically produced counterparts. Only in niche markets within some industrial segments, such as the food industry, can higher prices sometimes be achieved. Existing production facilities for chemical syntheses cannot be changed to biotechnological production without substantial new investments, so companies have to manage the capital requirements to build up new production facilities [14, 15]. To illustrate the diversity of molecules made by biotechnology processes, some examples are described in Box 1. As a rule, biotechnology production in the starting phase is more expensive than traditional chemical production routes, as the synthesis of existing products by chemical procedures with already amortized production facilities is frequently well established. Furthermore, established chemical production processes profit from economies of scale and learning curve effects. For biotechnological processes to compete competitively in cost with chemical synthesis, the best available state-of-the-art technologies are necessary [16].

**Box 1 Examples of Biotechnology Products**

Riboflavin, also known as vitamin B2, is a micronutrient with a key role in maintaining health in humans and animals. Various biotechnological processes have been developed for industrial-scale riboflavin biosynthesis using different microorganisms, including filamentous fungi and bacteria, which are genetically modified to increase the bacteria's production of riboflavin [17]. BASF, for example, produces riboflavin using filamentous fungi.

Lysine is an amino acid essential for humans and an important additive to animal feed because it is a limiting amino acid when optimizing the growth of certain animals, such as pigs and chickens, for the production of meat. Production exceeds 600,000 tons/year and the main producers are Archer Daniels Midland, BASF, and Evonik. Lysine is usually manufactured through a microbial fermentation process using bacteria from a base mainly composed of sugar. Genetic engineering research is actively pursuing bacterial strains to improve the efficiency of production and to allow lysine to be made from other substrates [18].

Succinic acid or 1,4-butanedioic acid is a dicarboxylic acid used as a precursor for polyesters and a component of alkyd resins. It is also used in the food and beverage industry, primarily as an acidity regulator. Global production is estimated at 30,000 tons/year. This molecule is produced by several chemical methods, such as the hydrogenation of maleic acid, oxidation of 1,4-butanediol, or carbonylation of ethylene glycol, but more and more succinic acid is being produced through the fermentation of glucose from renewable feedstock and purification of raw bio-based succinic acid [19, 20]. Companies such as BioAmber, Reverdia, Myriant, BASF, and Purac are progressing from demonstration to commercial scale.

1,3-Propanediol is a three-carbon diol mainly used as a building block in the production of polymers such as polytrimethylene terephthalate. It can also be formulated into a variety of industrial products including adhesives, coatings, and paints, as well as composites, laminates, and moldings. In the cosmetics industry it is used as a solvent, humectant, emollient, or hand-feel modifier. 1,3-Propanediol has a production volume of more than 100,000 ton per year. It can be chemically synthesized by the hydration of acrolein or by the hydroformylation of ethylene oxide and subsequent hydrogenation. A newer biotechnology process enables the conversion from corn syrup effected

(continued)

**Box 1** (continued)
by a genetically modified strain of bacteria by DuPont Tate and Lyle BioProducts [21].

## 2 Market Overview

The following market data derive from a database of Festel Capital (Box 2). The database has been built up since 2003 by Festel Capital, based on desk research using publicly available sources such as databases, articles, and company disclosures, as well as interviews with experts from the industry, universities and research institutions, the investment sector, and representatives from governmental institutions [9, 22].

**Box 2 Source of the Market Data**
Biorenewables are products made from renewable resources through biotechnological processes. Sales figures for biorenewables are shown as part of chemical sales because the most important market players are established chemical companies or companies in related industries. The database uses the segments and sub-segments used by the Conseil Européen de l'Industrie Chimique (CEFIC) for global chemical sales, for example, for their yearly publication "Facts & Figures" describing the chemical industry in Europe and its global position. The segments are as follows, with the sub-segments in brackets: base chemicals (inorganics, fertilizers and gases, organic chemicals, polymers and fibers), specialty chemicals (agrochemicals, adhesives and sealants, paints and coatings, food additives, other specialty chemicals), and consumer chemicals (detergents, cosmetics). The sub-segment polymers and fibers, including building blocks and intermediates to produce these products, is also shown on a segment level because of its size and importance.

Within the database, the sales of biorenewables are estimated on a sub-segment level (in some cases also on a more detailed product level) as a rolling forecast. Subsequently, the sub-segment or product level data are aggregated to obtain the figures on a segment level. This is done separately for the regions Europe (EU-27 countries and Switzerland), North America (Canada, Mexico, and the United States), Asia Pacific, and the rest of the world. Biomass-derived energy (including biofuels) and pharmaceuticals (pharmaceutical end products as well as active pharmaceutical ingredients and intermediates) are not included in the sales figures.

The sales of all products made from renewable resources using biotechnological conversion processes are considered. If single chemical process steps are involved, these products are also considered, but renewable raw materials

(continued)

**Box 2** (continued)
converted only through chemical processes and non-renewable raw materials converted through biotechnological processes are not included in these market data. Not considered are captive production and inter-company sales based on transfer prices. The sales figures in this database generally only show product sales between two independent companies based on market prices.

Sales of industrial biorenewables in 2010 were 82.94 billion US dollars, representing 5.3% of total chemical sales (Fig. 1). The four categories – base chemicals, polymers and fibers, specialty chemicals, and consumer chemicals – contributed nearly equally, with specialty chemicals being the largest segment, followed by polymers and fibers. Showing a more detailed market analysis, chemical sales in 2010 totaled 1,574.10 billion US dollars with a 75.4 billion US dollar share for biorenewables (Fig. 2). Although basic chemicals made up around 34.2% of total chemical sales in 2010 (539 billion US dollars), only 3.3% of those (17.71 billion US dollars) were biorenewables. Biorenewables had a share of 21.12 billion US dollars of sales, equaling 4.9% of chemical sales in the polymers and fibers segment, which totaled 431.20 billion US dollars. Within specialty chemicals, which accounted for 366.52 billion US dollars, the share of biorenewables was 6.6% of chemical sales (24.09 billion US dollars). The highest share of biorenewables within chemical sales with 237.16 billion US dollars was consumer chemicals with 8.5% (20.02 billion US

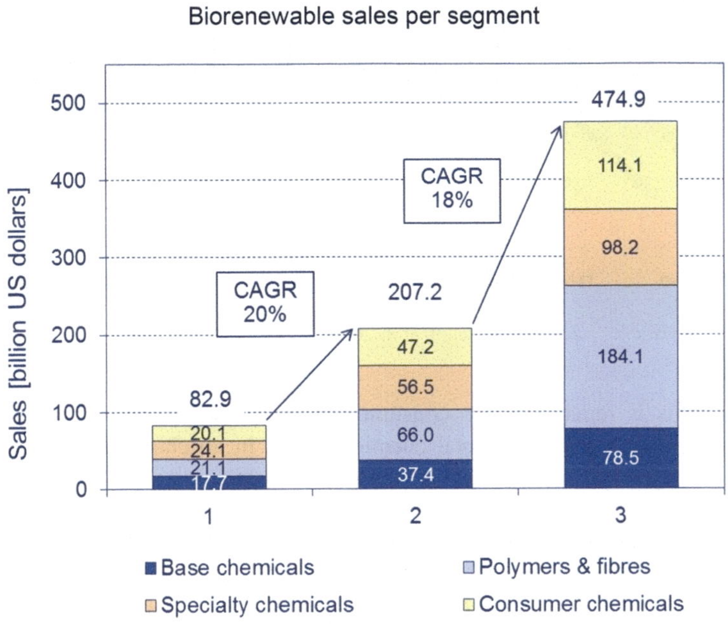


**Fig. 1** Market size of industrial renewables in 2010, 2015, and 2020

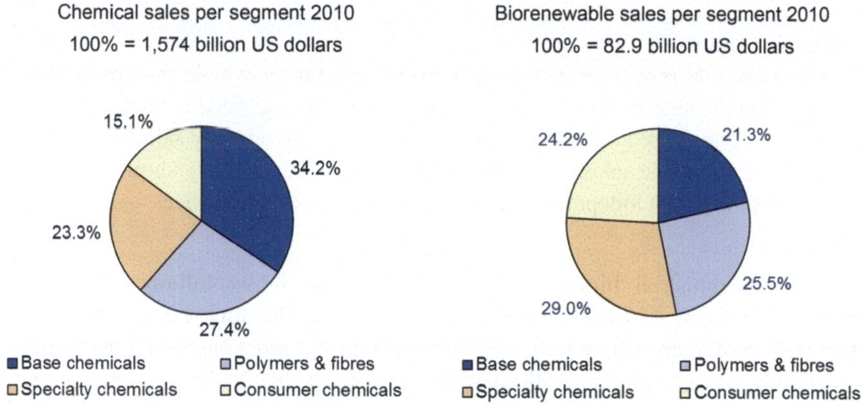


**Fig. 2** Chemical and biorenewables sales per segment in 2010

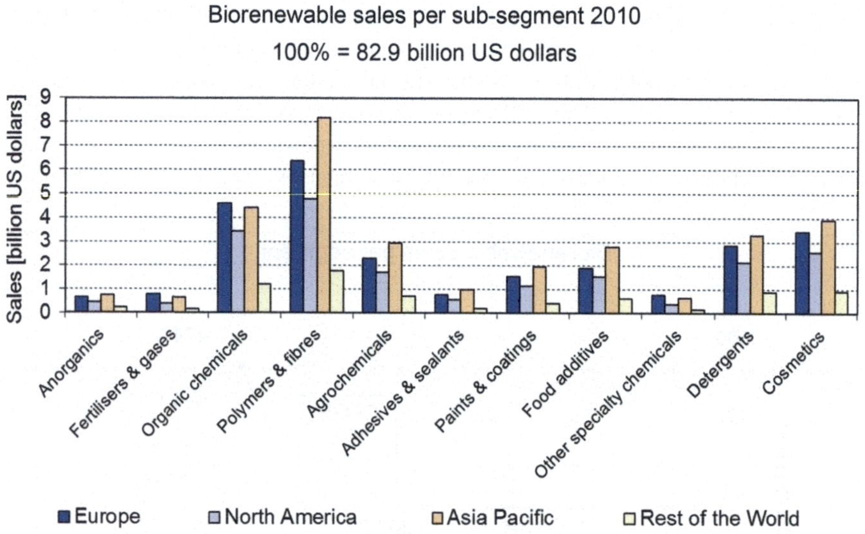


**Fig. 3** Biorenewables sales per sub-segment in 2010

dollars). This is remarkable, as consumer chemicals was the smallest chemical segment with only 15.1% of total chemical sales. In all chemical segments it is expected that the percentage of biorenewables will increase significantly over the next few years.

The most important sub-segment in 2010 was bio-based polymers and fibers, followed by the sub-segments organic chemicals, cosmetics, and detergents (Fig. 3). Bio-based polymers and fibers had almost 6.6 billion US dollars sales in Europe, around 4.95 billion US dollars sales in North America, and around 8.25 billion US dollars sales in Asia Pacific. Organic chemicals had more than 4.4 billion US dollars sales in Europe as well as Asia Pacific, and around 3.3 billion US dollars sales in North America. Cosmetics and detergents accounted for around 3.3 billion US dollars sales in Europe and Asia Pacific. The sales figures in North America were around 2.2 billion US dollars. Important sub-segments with around 2.2 billion US dollars sales in each major region were agrochemicals and food additives. Biorenewables sales per region show that Asia Pacific is the strongest region with 37.9% (31.46 billion US dollars), followed by Europe with 29.5% (24.42 billion US dollars) and North America with 22.2% (18.48 billion US dollars) (Fig. 4). The rest of the world accounts for 10.4% (8.58 billion US dollars) and this figure is mainly driven by Brazil.

In 2015, chemical sales reached 1,964 billion US dollars (Fig. 5). Sales for industrial biorenewables were around 207.13 billion US dollars representing 10.5% of total chemical sales, resulting in a compound annual growth rate (CAGR) for 2010–2015 of 20.1%. Whereas basic chemicals contributed 30.5% of total chemical sales (599.61 billion US dollars), only 6.2% (37.4 billion US dollars) were biorenewables. Biorenewable polymers and fibers strongly increased to 66 billion US dollars, which was almost a third of total biorenewables sales and 12.1% of 545.49 billion US dollars of chemical sales in that segment. Specialty chemicals showed 490.93 billion US dollars including 56.54 billion US dollars for biorenewables which was 27.3% of total biorenewables sales and 11.5% of chemical sales. Consumer chemicals with 327.25 billion US dollars included biorenewables sales of 47.19 billion US dollars. At 14.4%, this was the largest share of biorenewables sales in the chemicals segment. The smallest segment was base chemicals with 37.4 billion US dollars. The most important sub-segments in 2015 were the same as in 2010 (Fig. 6). Polymers and fibers accounted for 22 billion US dollars in both Europe and Asia Pacific. In North America, with 14.3 billion US

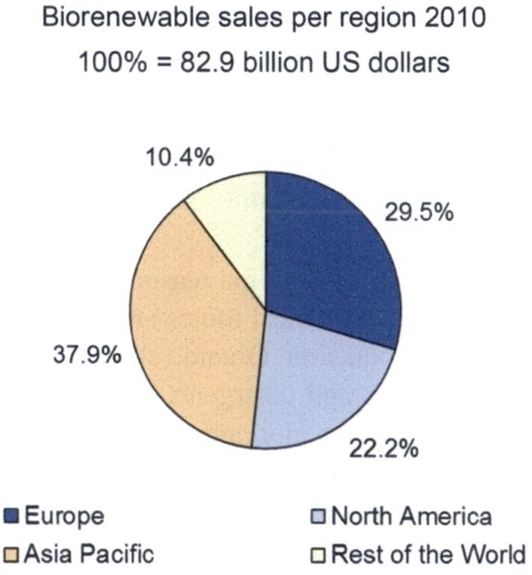


**Fig. 4** Biorenewables sales per region in 2010

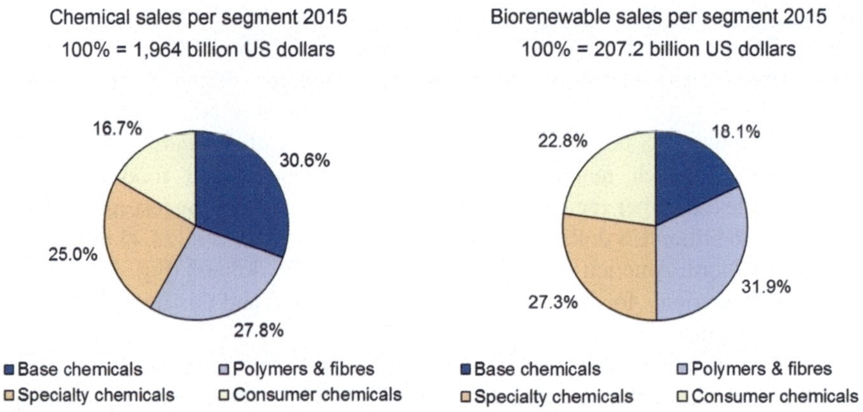


**Fig. 5** Chemical sales and biorenewables sales per segment in 2015

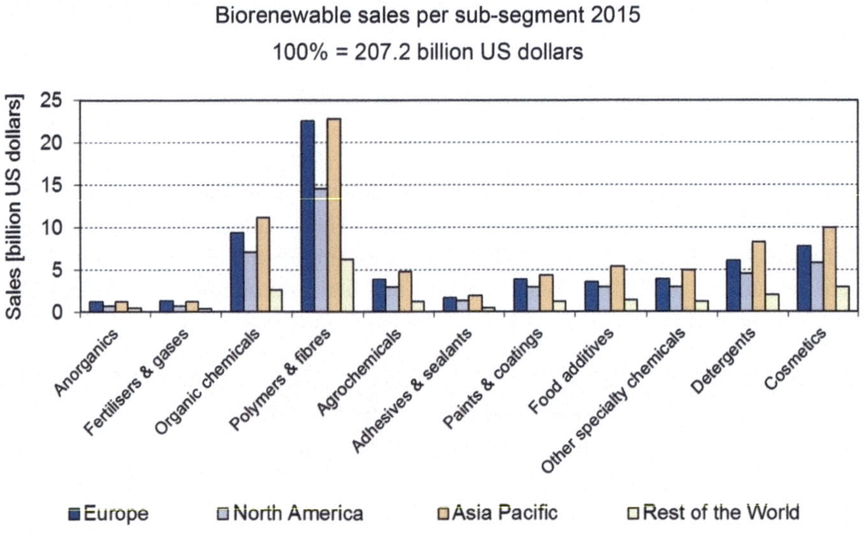


**Fig. 6** Biorenewables sales per sub-segment in 2015

dollars, this value was significantly lower, the reason being that producers in North America stuck more to traditional chemical routes and a major part of the available biomass was used for biofuel production. Organic chemicals had 7.7–11 billion US dollars in the three major regions and detergents as well as cosmetics of 4.4–9.9 billion US dollars. The sub-segments inorganics and fertilizers and gases saw the lowest sales figures of 1.1–2.2 billion US dollars.

It was predicted in 2010 that chemical sales would increase to 2,447.5 billion US dollars by 2020, whereby 474.98 billion US dollars, representing 19.5% of total chemical sales, would belong to industrial biorenewables (Fig. 7). The CAGR from 2015 to 2020 reaches, with 18.1%, almost the same level as the CAGR from 2010 to

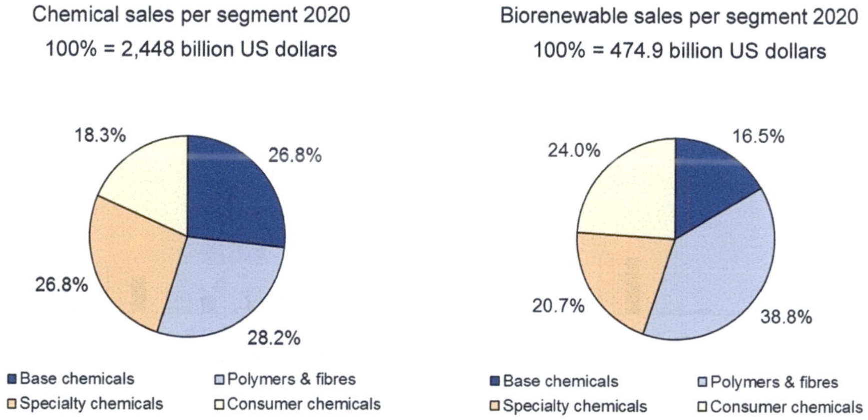


**Fig. 7** Chemical sales and biorenewables sales per segment in 2020

2015. The importance of base chemicals will further decrease with 26.8% of total chemical sales in 2020 compared to 34.2% in 2010. Predicted sales of 654.72 billion US dollars include 78.54 billion US dollars for biorenewables, representing a share of 12%. Polymers and fibers will achieve the highest biorenewables sales with 184.14 billion US dollars. This will be almost 40% of total biorenewables sales and 26.7% of chemical sales, which will be 689.48 billion US dollars. Specialty chemicals, with 89.3 billion US dollars, will account for 15% of the 655.05 billion US dollars of total chemical sales in that segment. Consumer chemicals will be the second largest biorenewables segment with 114.07 billion US dollars. This is 24% of total biorenewables sales and 25.5% of the 448.14 billion US dollars total sales for consumer chemicals. Looking at the sub-segment level, polymers and fibers will be strongest with 55 billion US dollars in Europe, nearly 44 billion US dollars in North America, and more than 66 billion US dollars sales in Asia Pacific (Fig. 8). The second largest sub-segment, organic chemicals, will account for around 22 billion US dollars in all major regions. Detergents and cosmetics will show between 11 and 24.2 billion US dollars. As in 2015, inorganics and fertilizers and gases will have the lowest sales figures of 1.1–2.2 billion US dollars.

## 3 Company Types

Companies active in the area of industrial biotechnology range from small and medium enterprises (SMEs) to multinational enterprises (MNEs). Based on the definition of the European Union, SMEs have less than 250 employees and less than 55 million US dollars annual turnover. Companies with more employees or higher annual turnover are seen as MNEs because they normally have operations in more than one country. The differentiation into specific company types based on

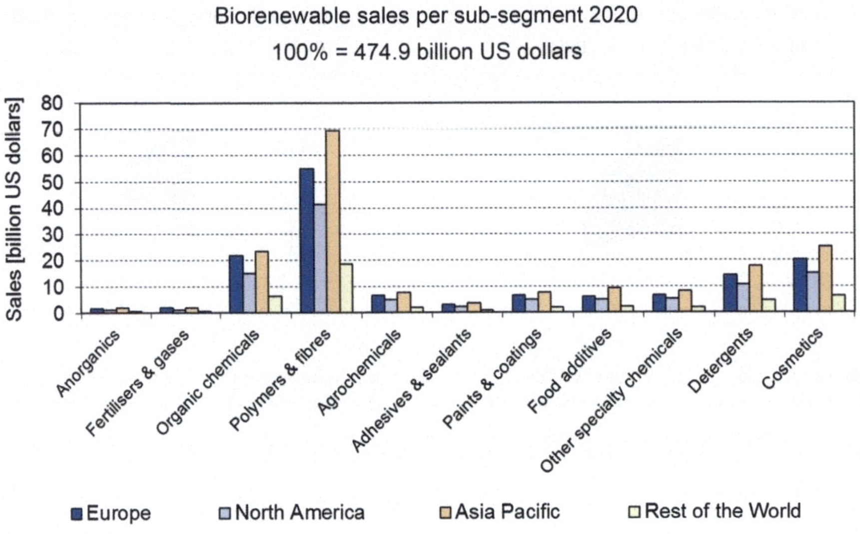


**Fig. 8** Biorenewables sales per sub-segment in 2020

size, that is MNEs vs SMEs, and areas of activity, that is dedicated to industrial biotechnology vs diversified over a broader range of areas, is necessary to understand the industrial biotechnology sector, as industrial biotechnology is of different importance for these company types (Fig. 9) [9, 23].

Academic spin-offs are very important for technological development because of their high innovation based on their academic backgrounds, but have only limited impact on commercial development. The reasons are limited resources and know-how for product development and missing M&S resources during the early years, which also have an impact on the business model. It is characteristic for spin-offs to start with a technology that is immature and requires further development. The proof-of-concept is normally done on a laboratory scale. Before larger investments in production and M&S are made, it is necessary to reach the technical proof-of-concept. A typical example within industrial biotechnology is upscaling: the development of a cost-effective world-scale production process through process and plant engineering.

The need for further development of the technology is directly linked to additional financial requirements and other resources to facilitate R&D work. Because of restricted resources in their early years, academic spin-offs focus mainly on a service-oriented business approach, offering their particular know-how in support of other companies. The intellectual property (IP) from these cooperations normally belongs to the customer, resulting in limited growth as well as value creation potential. However, the business risk is also limited as there are only low capital requirements to realize this business model. The spin-offs avoid the time- and cost-consuming development of own products, and their customers can transfer the spin-offs' technologies into new products.

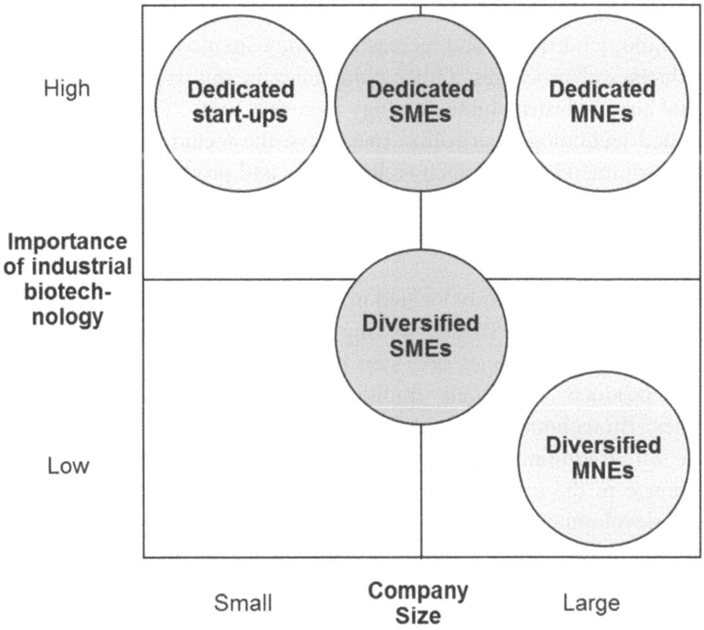


**Fig. 9** Company types within the field of industrial biotechnology

Nevertheless, the development of own IP and products is necessary for the further growth of new ventures. It can be observed that, over time, service-oriented spin-offs take on a more IP/product-oriented business approach. This is possible as it is accepted that a significant part of the developed IP within research cooperations belongs to the technology provider. For example, companies are developing biocatalysts for established companies within R&D cooperations whereby a special biocatalyst, including all related IP, belongs to the customer, and new IP regarding further developments of the technology belongs to the spin-off. As a result, with growing maturity, spin-offs are increasingly able to develop and commercialize own technologies and products.

Many SMEs have been founded during the last 10–30 years, mainly as academic spin-offs. After performing intensive R&D work, especially as service providers for larger companies during the early years, they are now focused on the development and market introduction of their own products. Many industrial biotech SMEs have financed their business through their operational income and often risk capital, but this limits growth because of restricted financial resources. As they are focused on R&D, their importance for technological development is high. In contrast, their importance for commercial development of industrial biotechnology is only medium, as they are still building up their product portfolios and M&S capabilities.

Dedicated MNEs are dominated by companies that have been active in the area of natural products for decades, often coming from the agro industry. They have normally used optimized biotechnological processes for many years for traditional markets such as starch or yeast products. Industrial biotechnology is one cornerstone in their technology portfolios and increasingly they are moving toward new biotech-based products and processes. Other companies in this segment are more R&D oriented and have industrial biotechnology as core activity. Their strength is a broad and integrated technology portfolio. They have the technical as well as financial resources to commercialize biotech technologies and products worldwide. Because of this and their M&S capabilities/resources, they have medium to high importance for the technological and high importance for the commercial development of industrial biotechnology.

Diversified MNEs are mainly located in established industrial sectors, such as the chemical or food industry. Already serving developed markets with highly specialized products, these companies are, step by step, introducing biotechnology processes and products into their traditional markets to realize future growth opportunities. Biotechnology is only one technological approach and always has to compete with traditional chemical synthesis routes. Diversified MNEs have only low importance in the technological development and medium importance in the commercial development of industrial biotechnology because of their strong focus on already established products and processes.

# 4 Cooperation Models

As described in the previous section, the different company types have very different roles regarding the technological and commercial development of the industrial biotechnology sector. Commercial development is mainly driven by MNEs, whereas SMEs contribute primarily to technological development (Fig. 10). This has an impact on the different cooperation models between industrial biotechnology companies [24–26].

Dedicated and diversified MNEs normally have enough in-house resources to realize most of the technology developments in-house. Additional R&D capacity and cost reduction (reducing fixed costs or people on the payroll) is not relevant for working together with external partners. However, these companies have a strong interest in additional, external know-how which is unavailable in-house, too expensive, or too risky if it was to be built up internally. Expanding in-house capabilities through external expertise is seen as the most important advantage of using external technologies by way of cooperation with service providers. An important task for established companies is to integrate internal and external knowledge optimally within the innovation process so as to be able to benefit from synergy effects. This strategy has often been used in the past and almost all industrial biotech companies take part in such cooperation.

The situation for SMEs is different compared to MNEs, as they are more dependent on technology transfer from academic research to develop new products internally or together with partners because of limited financial and management resources. They see cooperation with academics as an effective method to capture

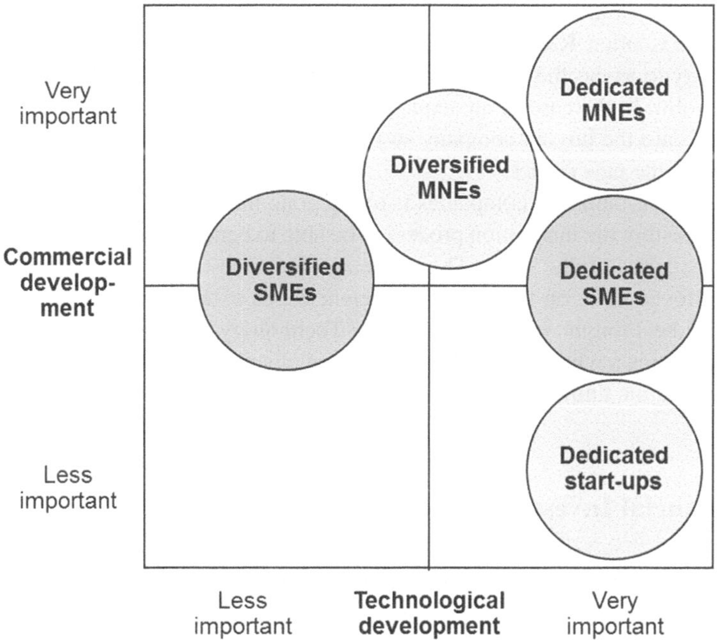


**Fig. 10** Importance of the different company types for the further development of industrial biotechnology

capacity and expertise without investing much money in in-house resources. The preferred option to access new technologies involves R&D cooperation with universities and R&D institutions but also specialized spin-offs.

The technology transfer gap between R&D at universities or research institutions and the commercialization by established companies can be closed, especially by academic spin-offs [27]. After the spin-off process, the technology is further developed within the new venture, normally using additional resources from external investors. As soon as the technology reaches a certain grade of maturity, the spin-offs can cooperate with an established company and work for them as a service provider, or be acquired. Spin-offs make state-of-the-art technological expertise from academic research available to established companies who can use this to leverage their product development and global M&S capabilities.

Different technology transfer models have been used between established companies and spin-offs, and the chosen technology transfer approach depends on the type of company. Whereas MNEs are very active in making new technologies available, both by acquiring spin-offs and by engaging them as service providers, SMEs, because of limited financial resources, are more focused on research cooperation with spin-offs, especially by engaging as service providers. After building up an attractive technology or product portfolio with correlating IP protection or, if the technological and market proof-of-concept is shown, technology transfer through the acquisition of these spin-offs by MNEs or SMEs is an option. The first step of an acquisition is often R&D cooperation, which gives the established company the opportunity to assess the technology of the spin-off and the fit into its own technology portfolio. In the case of an acquisition, the spin-offs are normally more or less integrated into the buying company so that the complete know-how and IP is fully available to the new owner.

A task for established companies is to integrate internal and external knowledge optimally within the innovation process to be able to benefit from the positive effects each activity has on the other. The advantage for the established companies is that they can focus more on their core competencies, as external technological competence can be brought into the company. Technology transfer from academia to industry creates a win-win situation for all participants, leading to faster dissemination of academic knowledge into practice.

## 5 Financial Investors

In the last few years, industrial biotechnology has gone through a fast technological development, resulting in a high number of basic technologies based on research efforts at universities and research institutions. Technological development in the industrial biotech sector has been supported by substantial investment of venture capital (VC) over the years. VC is risk capital provided by specialized investment companies with technical expertise in the investment field. Numerous new ventures have been founded thanks to these investments [28, 29].

Industrial biotech was recognized as an attractive investment target for VC funds in the early 2000s, and since that time financial engagement has increased substantially. In the period 2002–2006, however, VC activity remained rather limited. Net VC stock (the balance of investments vs divestments or exits) grew slightly from 766 million US dollars in 2002 to 844 million US dollars in 2006, which corresponds to a compounded annual growth rate of ~2%. In 2007, both investment and divestment activity expanded substantially.

The industrial biotech sector suddenly attracted the attention of VC investors, resulting in large investments in the biofuel and biochemical area in 2007. Two factors were critical in stimulating the flow of investment into the sector. First, there was a rise in oil prices from 30 US dollars per barrel in 2003 to 80 US dollars in 2006, suggesting that bio-based solutions could approach competitiveness with fossil fuels; second, the renewable fuel standard was introduced in the United States as a federal program that requires transportation fuel sold in the United States to contain a minimum volume of renewable fuels. The standard originated with the Energy Policy Act of 2005 and was expanded and extended by the Energy Independence and Security Act of 2007.

As new VC investment has exceeded divestment every year since, the net stock balance of VC has steadily increased. From 2007 to 2013 the compounded annual growth rate was as high as 25%, resulting in a net worth of stock of VC in industrial

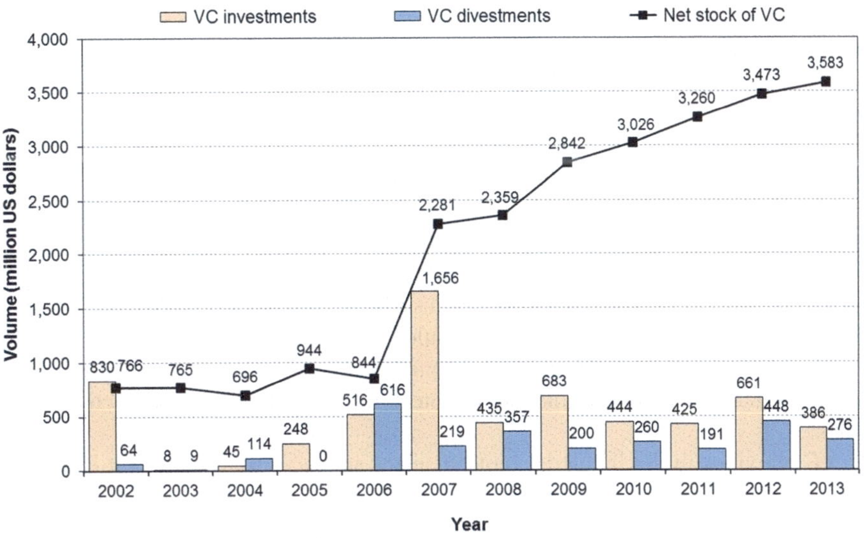


**Fig. 11** Volume of VC investments/divestments and net stock of VC

biotech companies of 3.6 billion US dollars in 88 companies by the end of 2013. The main driver for this development was an exceptional windfall in 2007, when there was a large net capital injection of VC of nearly 1.5 billion US dollars (Fig. 11). VC companies invested a total of 6.3 billion US dollars in 107 industrial biotech firms, which was only partly offset by 2.8 billion US dollars in divestments or exits. The rather low divestment volume points to the premature nature of the industrial biotech industry.

At a more disaggregated level, in 2013 the gross amount of VC flowing to the companies in our analysis was 386 million US dollars, distributed among 20 recipient firms. This corresponds to an average amount of 19.3 million US dollars to each target company. Despite the trend of increasing VC investments to the industrial biotech segment, the huge difference in the magnitude of total VC investment between industrial biotech and biopharma reveals that this sector still plays a subordinate role in the biotech investment universe as a whole.

VC is not the only external capital source for industrial biotech companies. Other investors, such as multinational chemical and food companies, also discovered industrial biotech to be an attractive asset class and substantially increased their investment volumes. These multinationals can offer not only finance but also other important support, such as access to their distribution networks and R&D skills. The accumulated net investment from non-VC sources increased between 2002 and 2013 from a few million dollars to three billion US dollars. The main sources of this investment were stock market transactions (initial public offerings (IPOs) and subsequent placement of shares) of ~ 1.8 billion US dollars and multinational corporate investors, with almost 0.9 billion US dollars.

This general increase in financing from non-VC sources was reinforced by the global financial crisis toward the end of the past decade. Non-VC investments reached a peak of 1.1 billion US dollars of gross money inflow to industrial biotech entities in 2009. One major consequence of the global financial crisis was the general reduction in money supply to VC funds themselves. Institutional capital managers turned away from VC funds because of the meagre returns, high risk, long lockup periods, and substantial management fees. Many VC fund managers, who, in turn, distributed the money collected from institutional capital providers to biotech companies, were no longer able to convince their capital suppliers of their future return prospects. Thus, they were unable to attract sufficient funding for the VC industry. In particular, they failed to prove that they could deliver higher earnings than the public small caps market, which has the added benefit of being more liquid. Given this, the rising capital contribution in industrial biotech indicates that it is an attractive asset class.

VC activities in the industrial biotech sector are focusing on research-intensive firms, defined as those owning at least one active patent (Fig. 12). Sixty-seven percent of the VC investment volume on average was allocated to companies that had received a patent before the capital was distributed. In turn, 62% of the VC volume that was transferred by the end of 2012 was targeted to industrial biotech firms that had obtained at least one additional patent in subsequent years. However, a company's patent status is not fixed and could change, as the companies could file a patent in the future. Thus, the 62% average is provisional and potentially an underestimate. Nonetheless, slightly more VC volume was distributed to companies that already owned an active patent in the year in which they received the funding.

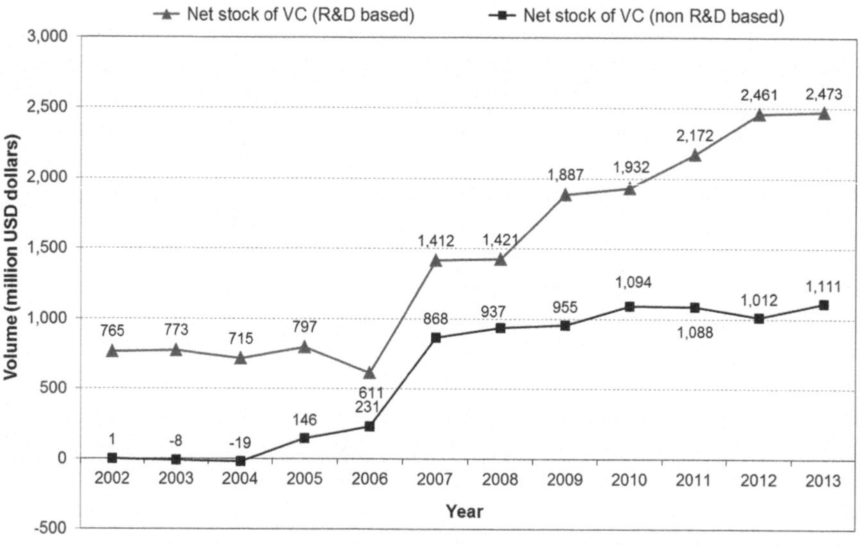


**Fig. 12** VC investments in R&D-based and non-R&D-based companies

In 2013, the net stock of VC in research-intensive firms was 2.5 billion US dollars compared to 1.1 billion US dollars in non-research-intensive firms (Fig. 12). VC investors prefer research-intensive firms over firms that do not have patents. From the perspective of the supported companies, VC is a motor of growth in this emerging industry because of the increasing volume of VC invested in the industrial biotech sector as a whole, and in particular in research-intensive companies.

Substantial VC funding volumes were assigned to firms less than 4 years old (Fig. 13). In 2002, for example, 816 million US dollars (98% of the total investments in this year) was invested in this segment, in 2006, 593 million US dollars (83%), and in 2007, 634 million US dollars (37%). This suggests that VC is a significant financial pillar for companies in the early stages. As young companies in industrial biotech are often technology driven, this result supports the observation that VC mainly targets research-intensive companies and is thus an engine for innovation. Since 2007, >50% of annual VC investments were allocated to relatively older companies (>4 years). Of note is the substantial drop in VC investment to new ventures in 2009, which may be explained by the financial market turmoil in late 2008 and 2009.

The growing interest of VC investors in industrial biotech over the past decade is also evident from the increasing number of companies receiving VC investment over time. In 2002 only six industrial biotech companies obtained VC investment; in 2005 that number rose to 14. Subsequently, the number of industrial biotechnology companies with VC investment rose sharply to 88 by the end of 2013 (Fig. 14).

The strategy of VC companies regarding the future development of their portfolio companies strongly depends on the business models of the portfolio companies. Production technology-oriented companies (companies developing new processes to

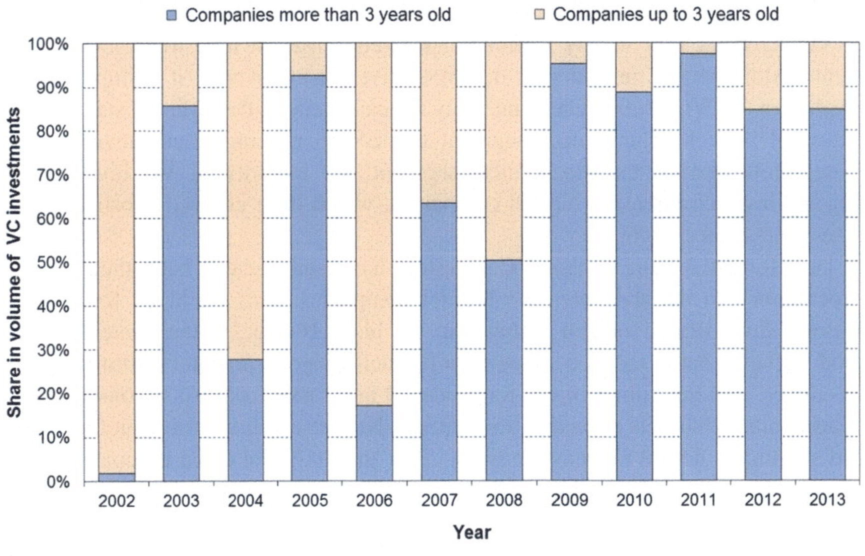


**Fig. 13** VC investments by age of the company at time of investment

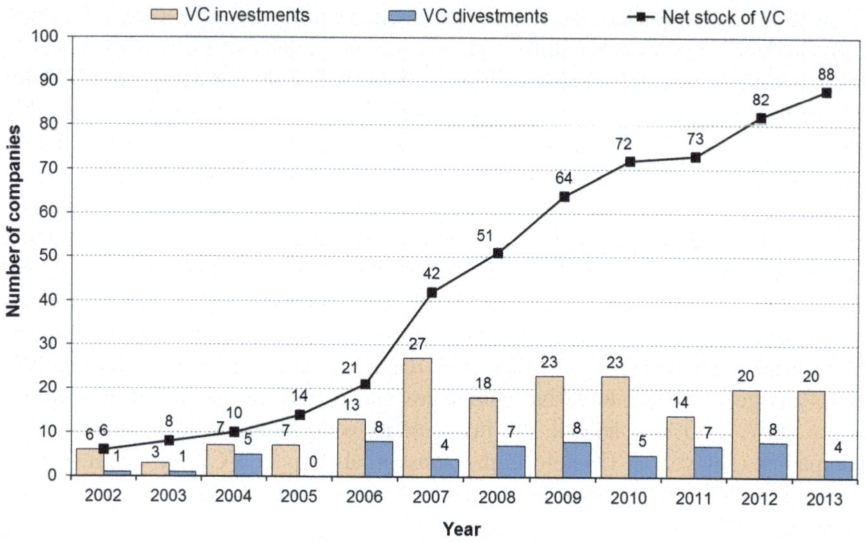


**Fig. 14** Number of VC investments/divestments and net stock of VC

produce already existing products), which are mainly found in the biofuels and bulk biochemicals area, mostly follow the strategy to license out their technology after successfully running a demonstration plant. These companies are meant to act in the future as stand-alone companies with revenues coming from licensing fees as well as product sales. Product-oriented companies (companies developing and producing new products), mainly in the specialty biochemicals and bioactives areas, are often positioned by VC companies as acquisition targets for established companies.

VC activities vary widely within and between different industrial biotech segments. Although VC investments dominate divestments overall, it is important to emphasize that VC investment is unevenly allocated among the different subfields of industrial biotech. The biofuel segment topped the ranking of net investments, closely followed by the biochemical segment; the majority of VC investments targeted research-intensive biofuel companies, which have enjoyed substantial net investments since 2006.

Data for 2013 indicate that VC investors have since scaled back their capital expenditures to research-intensive biofuel companies. This could be for several reasons: diminishing societal enthusiasm for biofuels (e.g., in the context of the food vs fuel debate), reduced support for biofuels by governmental institutions and incentives, and the falling prices for crude oil and fossil fuels. As a consequence, divestments in biofuels exceeded investments. Non-research-intensive biofuel firms had significant net investments between 2006 and 2009, followed by considerable divestments in 2012. In the bioactive segment, annual investments and divestments were highly volatile throughout the time period, driven by a few big deals.

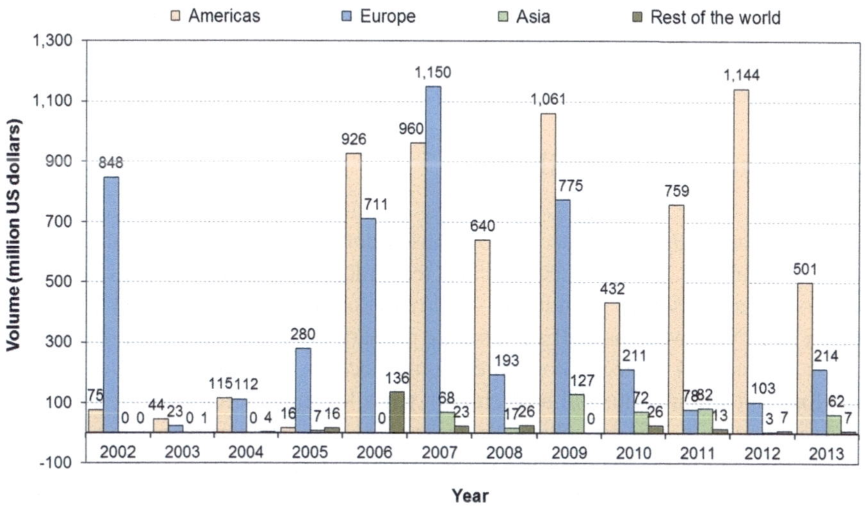


**Fig. 15** Volume of VC investments by region

The regional breakdown of M&A (mergers and acquisitions) activities in the sector reveals that the United States is the leading region with a total transaction amount of ~ 7.0 billion US dollars, followed by Europe with about 4.8 billion US dollars (Fig. 15). This mirrors regional patterns of VC investment in the biopharma sector. Notably, since 2006 the annual M&A transaction volume in the United States remained high, with a low of about 0.4 billion US dollars in 2010 and a high of 1.1 billion US dollars in 2012. Conversely, in Europe there has been an overall downward trend since the peak of 1.2 billion US dollars in 2007. In Asia and the rest of the world the volume of M&A transactions has been modest, ~ 0.7 billion US dollars for both regions. Despite the emergence of industrial biotech markets of substantial size in Asia, the data suggest that VC-backed companies are rarely based in emerging Asian economies – again a factor that mirrors the situation for VC-based biotech as whole. Given the stated intentions of many local governments to promote industrial biotech, Asia is still woefully underrepresented in VC allocation strategies.

The relatively low number of VC exits in the industrial biotech area has different causes. For one thing, markets for bio-based products develop slowly because of the challenges regarding raw material supply and low oil prices. In addition, technological challenges were underestimated, especially with regard to scaling-up processes from laboratory and especially pilot and demonstration scales to an industrial scale. Especially in the biofuels area, the hype in 2006 and 2007 has been replaced by a sober assessment of the situation. As a result, many mineral oil companies who could act as strategic investors or potential acquirers of biofuel companies stopped or reduced dramatically their engagements in the biofuel industry. The chemical industry, the second large partner industry for industrial biotech companies, was not very active either as strategic investor or acquirer of biotech companies. Profitability of chemical operations is still low compared to that of mineral oil and pharmaceutical companies, leading the industry to be much more conservative regarding acquisitions.

## 6 Conclusions and Outlook

In recent years, industrial biotechnology has undergone fast technological development, resulting in a high number of basic technologies based on research efforts at universities and research institutions. As a rule, commercial development is mainly driven by MNEs, whereas SMEs contribute primarily to technological development. As in the medical biotechnology area some decades ago, the fast technological development within industrial biotechnology has led to numerous new ventures.

Academic spin-offs can close the technology transfer gap between academic research and industrial application in the area of industrial biotechnology. Spin-offs make state-of-the-art technological expertise from academic research available for established companies, which can use these to leverage their product development and global sales capabilities. After the spin-off process, the technology is further developed within the new venture, normally using additional resources from external investors. As soon as the technology reaches a certain grade of maturity, the spin-offs can cooperate with an established company and work for them as a service provider or be acquired. Whereas MNEs are very active in making new technologies available both by acquiring spin-offs or engaging them as service providers, SMEs are focused on partnering with spin-offs because of limited financial and management resources.

Technology transfer from academia to industry creates a win-win situation for all participants, leading to a faster dissemination of academic knowledge into practice and resulting in an economic advantage [30]. Companies should use the advantages of new ventures, such as more target-oriented R&D work or faster time-to-market, to improve the innovation capabilities within their companies. R&D managers in established companies should be more open to use new ventures actively for technology transfer and understand that entrepreneurial behavior can support technology transfer to improve innovation processes.

However, creating spin-offs is not yet systematically used for technology transfer from universities and research institutions into industrial applications. Despite some elements of "entrepreneurial thinking" within the Horizon 2020 program and some national initiatives within governmental funding programs, there is still no general awareness about the value of entrepreneurial thinking. As high quality research at universities and research institutions in Europe has not been sufficiently translated into commercial applications, policy makers should foster technology transfer through spin-offs. One way to do this is to provide incentives for business-oriented and experienced people, similar to founding angels or business angels, to join new ventures and to realize technology transfer. These incentives could be tax incentives for new ventures (e.g., preferred depreciation models for R&D expenses), entrepreneurs, and investors (e.g., reduced tax rates on exit profits).

In addition, EU programs and European initiatives should be designed in such a way that eases access for young high-tech firms. In the Horizon 2020 program, eligibility criteria and administrative requirements for participating in the program need to consider the specific situation of young firms and their limited capacities to fulfil bureaucratic demands. The same applies to the multilateral, EU co-funded program Eurostars which provides R&D funding for SMEs.

## References

1. EuropaBio (2003) White biotechnology: gateway to a more sustainable future. EuropaBio, Brussels
2. SusChem (2005) Industrial or white biotechnology. SusChem, Brussels
3. Hermann BG, Blok K, Patel MK (2007) Producing bio-based bulk chemicals using industrial biotechnology saves energy and combats climate change. Environ Sci Technol 41 (22):7915–7921
4. Coons R (2010) Industrial biotechnology. Chem Week 172(27):22–26
5. Fröhling M, Hiete M (2018) Sustainability and life cycle assessments in industrial biotechnology: a review of current approaches and future needs. In: Fröhling M, Hiete M (eds) Sustainability and life cycle assessment in industrial biotechnology. Springer, Berlin
6. MacLean H, Saville B (2018) Environmental aspects of industrial biotechnology. In: Fröhling M, Hiete M (eds) Sustainability and life cycle assessment in industrial biotechnology. Springer, Berlin
7. Osseweijer P, Posada J, Asveld L (2018) Societal and ethical aspects of industrial biotechnology. In: Fröhling M, Hiete M (eds) Sustainability and life cycle assessment in industrial biotechnology. Springer, Berlin
8. Schürrle K (2018) History, current state and emerging applications of industrial biotechnology. In: Fröhling M, Hiete M (eds) Sustainability and life cycle assessment in industrial biotechnology. Springer, Berlin
9. Festel G (2010) Industrial biotechnology: market size, company types, business models, and growth strategies. Ind Biotechnol 6(2):88–94
10. Kircher M (2011) Industrial biotechnology becomes a key competitive factor. J Bus Chem 8 (1):3–4
11. Nieuwenhuizen P, Lyon D (2010) Anticipating opportunities in industrial biotechnology: sizing the market and growth scenarios. J Commer Biotechnol 17(2):159–164
12. Joint Research Centre (JRC) (2007) Consequences, opportunities and challenges of modern biotechnology for Europe. Joint Research Centre, Sevilla
13. United States International Trade Commission (USITC) (2008) Industrial Biotechnology - Development and Adoption by the U.S. Chemical and Biofuel Industries, Investigation No. 332-481, USITC Publication 4020, Washington
14. Festel G, Knöll J, Götz H (2004) Industrial biotech - influencing production. Chem Ind 7 (5):21–22
15. Festel G, Knöll J, Götz H, Zinke H (2004) Der Einfluss der Biotechnologie auf Pro-duktionsverfahren in der Chemieindustrie. Chemie Ingenieur Technik 76(3):307–312
16. Festel G (2011) Drivers and barriers for industrial biotechnology. Int Sugar J 113(1345):19–23
17. Stahmann KP, Revuelta JL, Seulberger H (2000) Three biotechnical processes using Ashbya gossypii, Candida famata, or Bacillus subtilis compete with chemical riboflavin production. Appl Microbiol Biotechnol 53(5):509–516
18. Pfefferle W, Möckel B, Bathe B, Marx A (2003) Biotechnological manufacture of lysine. Adv Biochem Eng Biotechnol 79:59–112

19. Otero JM, Cimini D, Patil KR, Poulsen SG, Olsson L, Nielsen J (2013) Industrial systems biology of Saccharomyces cerevisiae enables novel succinic acid cell factory. PLoS One 8(1): e54144
20. Thakker C, Martínez I, San KY, Bennett GN (2017) Succinate production in Escherichia coli. Biotechnol J 7(2):213–224
21. Biebl H, Menzel K, Zeng AP, Deckwer WD (1999) Microbial production of 1,3-propanediol. Appl Microbiol Biotechnol 52(3):289–297
22. Festel G (2016) Market structure and growth rates of industrial biorenewables. In: de Maria PD (ed) Industrial biorenewables: a practical viewpoint. Wiley, Hoboken, pp 245–254
23. OECD (2011) Future prospects for industrial biotechnology. OECD Publishing, Paris
24. Festel G (2013) Catalyzing innovation – technology transfer models between industrial biotechnology companies and academic spin-offs. Ind Biotechnol 9(5):252–257
25. Festel G (2013) Technology transfer by new ventures within the chemical and pharmaceutical industry. J Bus Chem 10(3):115–130
26. Festel G, Rittershaus P (2014) Fostering technology transfer in industrial biotechnology by academic spin-offs in Europe. J Commer Biotechnol 20(2):5–10
27. Festel G (2013) Academic spin-offs, corporate spin-outs and company internal start-ups as technology transfer approach. J Technol Transfer 38(4):454–470
28. Festel G, Rammer C (2015) Importance of venture capital investors for the industrial biotechnology industry. J Commer Biotechnol 21(2):31–42
29. Festel G, Rammer C (2015) Fostering innovation in industrial biotechnology through venture capital investments. Ind Biotechnol 11(3):146–150
30. Czarnitzki D, Rammer C, Toole A (2014) University spinoffs and the "performance premium". Small Bus Econ 43(2):309–332

# Part III
# Industrial Biotechnology from an Assessment Perspective

Adv Biochem Eng Biotechnol (2020) 173: 77–120
DOI: 10.1007/10_2019_98

Published online: 9 August 2019

# Environmental Aspects of Biotechnology

**Aranya Venkatesh, I. Daniel Posen, Heather L. MacLean, Pei Lin Chu, W. Michael Griffin, and Bradley A. Saville**

## Contents



A. Venkatesh
KeyLogic Systems, Inc., Morgantown, WV, USA

I. D. Posen
Department of Civil and Mineral Engineering, University of Toronto, Toronto, ON, Canada

H. L. MacLean (✉) and P. L. Chu
Department of Civil and Mineral Engineering, University of Toronto, Toronto, ON, Canada

Department of Chemical Engineering and Applied Chemistry, University of Toronto, Toronto, ON, Canada
e-mail: heatherL.maclean@utoronto.ca

W. M. Griffin
Department of Engineering and Public Policy, Carnegie Mellon University, Pittsburgh, PA, USA

B. A. Saville
Department of Chemical Engineering and Applied Chemistry, University of Toronto, Toronto, ON, Canada

**Abstract** A key motivation behind the development and adoption of industrial biotechnology is the reduction of negative environmental impacts. However, accurately assessing these impacts remains a formidable task. Environmental impacts of industrial biotechnology may be significant across a number of categories that include, but may not be limited to, nonrenewable resource depletion, water withdrawals and consumption, climate change, and natural land transformation/occupation. In this chapter, we highlight some key environmental issues across two broad areas: (a) processes that use biobased feedstocks and (b) industrial activity that is supported by biological processes. We also address further issues in accounting for related environmental impacts such as geographic and temporal scope, co-product management, and uncertainty and variability in impacts. Case studies relating to (a) lignocellulosic ethanol, (b) biobased plastics, and (c) enzyme use in the detergent industry are then presented, which illustrate more specific applications. Finally, emerging trends in the area of environmental impacts of biotechnology are discussed.

**Graphical Abstract**

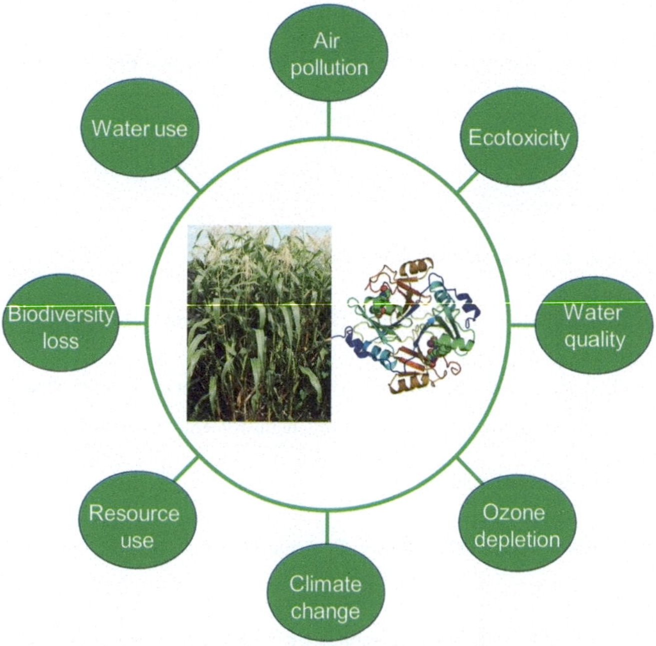


**Keywords** Biofuels, Bioproducts, Environmental impact, Industrial biotechnology

# 1 Key Environmental Aspects of Industrial Biotechnology

It is rare to encounter a discussion of biotechnology's potential that does not feature prominent claims about environmental benefits. BIO, the world's largest biotechnology trade association, begins its definition of industrial biotechnology by pointing out that "Industrial biotechnology is one of the most promising new approaches to *pollution prevention, resource conservation*, and cost reduction" [1]. The US National Bioeconomy Blueprint, developed during President Obama's era, aims to "allow Americans to live longer, healthier lives, reduce our dependence on oil, *address key environmental challenges*, transform manufacturing processes and increase the productivity and scope of the agricultural sector [...]" [2]. The Australian government highlights how industrial biotechnology can be used to "replace petroleum-based feedstocks" and "reduce the environmental impact of manufacturing" [3]. The European Union argues that "biotechnology helps the EU economy grow and provide new jobs, while also supporting sustainable development, public health and *environmental protection*" [4]. Though reduction of negative environmental impacts is a key motivation behind the development and adoption of industrial biotechnology, accurately assessing these impacts remains a formidable task.

In the broadest sense, nearly all environmental concerns stem from different types of resource use, the results of emissions (to air, water, or land), or modifications to natural environments and habitats. These activities can lead to a wide range of measurable impact categories (e.g., ozone depletion, acidification), which in turn cause damage to key categories of interest such as human health and ecosystem quality. A recent guidance document by the United Nations Environment Programme/Society of Environmental Toxicology and Chemistry [5] groups these final damages under three umbrella categories: intrinsic (human health and ecosystem quality), instrumental (socioeconomic assets, natural resources, ecosystem services), and cultural (cultural and natural heritage). This chapter will focus primarily on the more granular environmental impact categories. Numerous attempts have been made to classify environmental concerns into specific impact categories (e.g., Goedkoop et al. [6], Guinée [7], Bare et al. [8], Jolliet et al. [9], Federal Office for the Environment [10], Huijbregts et al. [11]), resulting in a staggering number of potential environmental metrics, including contribution to:

- *Nonrenewable resource depletion*: often separated into subcategories, such as the consumption of fossil fuels and dispersion of metals or other minerals, this category encompasses a suite of concerns regarding the use of resources which cannot be regenerated on a human timescale.
- *Water withdrawals and consumption*: water use is often differentiated based on whether the water is withdrawn and then returned to its source (as is common for cooling water) or consumed (e.g., evaporative losses, used in agricultural irrigation, or if there are spatiotemporal differences between withdrawal and return). Water consumption is especially problematic when it reduces the availability of freshwater, especially in water-stressed regions. Water withdrawals can also

create environmental issues, for example, if the water is returned with contaminants or at a higher temperature, potentially creating inhospitable environments for aquatic life.

- *Climate change*: the release of greenhouse gases (GHG) like $CO_2$, $N_2O$, and $CH_4$ can trap outgoing infrared radiation, upsetting the global heat balance and leading to changes in the climate that can impact sea-level rise, extreme weather, wildfires, and more.
- *Acidification*: the process that increases the acidity of water and soil, changing local ecosystems that can result in the death of plants and animals. Emissions of gases like nitrogen oxides ($NO_x$) and sulfur dioxide ($SO_2$) are common contributors to acidification.
- *Eutrophication*: the process whereby nutrients such as nitrogen and phosphorus compounds run off into water, leading to overgrowth of aquatic plant life. Common consequences include unpleasant odor and taste, production of chemicals that are toxic to humans, and death of marine life due to depletion of aquatic oxygen.
- *Ozone depletion*: release of gases such as chlorofluorocarbons (CFCs), which react to remove ozone in the upper atmosphere, reducing the protection the ozone layer would otherwise provide against UV radiation which can damage crops, human skin and eyes, plankton, etc.
- *Photochemical oxidation*: release of non-methane volatile organic compounds and $NO_x$ can induce photochemical oxidation reactions that lead to the creation of ground-level ozone. Ozone is a key contributor to some types of smog (i.e., summer smog), which is associated with negative impacts on plant life as well as increased human mortality and respiratory conditions.
- *Criteria air pollutants*: in addition to smog, other air contaminants, such as particulate matter, $NO_x$, and $SO_2$, are also associated with a range of human health complications including respiratory disease and premature mortality.
- *Other negative human health impacts*: a wide range of other pollutants can negatively impact human health, including cancer-causing agents like benzene, neurotoxins like lead and methanol ("wood alcohol") [12], and others. In many cases, the full range of human health impacts from different substances is not fully understood.
- *Ecotoxicity*: often delineated by whether the affected species are aquatic or terrestrial, this category captures all substances that produce toxic effects in nonhuman species. Ecotoxicity is a broad category, which often suffers from incomplete characterization of how substances impact different species.
- *Natural land transformation and/or occupation*: treated, respectively, as a problem of resource use (i.e., land availability) or for its secondary impacts (e.g., habitat destruction, loss of ecosystem services like carbon storage, water purification, and flood control), land transformations can create numerous environmental concerns. Further discussions are presented by Lindner et al. [13].
- *Biodiversity loss*: represents increases in the rate of local or global species extinctions, potentially driven by other impact categories like natural land

transformation and ecotoxicity. Often, special attention is paid to local reductions in the population of at-risk species.

The challenge of evaluating the "environmental impact" of any process is complicated both by the numerous potential impact categories to consider, as well as the numerous potential contributors within a given impact category. As one of the preeminent tools for quantitative environmental sustainability analysis, life cycle assessment (LCA) aims to quantify these impacts by (a) capturing relevant environmental flows across a product's life cycle (i.e., from raw material extraction and manufacturing through distribution, use, and disposal), (b) assigning these flows to relevant impact categories, and (c) converting flows within an impact category into common units using clearly defined characterization factors. While some impact categories have natural units (e.g., liters of water withdrawal), others use well-established though imperfect characterization factors (e.g., global warming potential, measured in units of kg $CO_2$ equivalents/kg gas), and still others – like biodiversity [14] – continue to pose measurement difficulties for the LCA community. Although it is conceptually impossible to capture all the potential environmental impacts of any technology, much less an entire class of technologies like industrial biotechnology, there exists a wide literature that has brought a number of key issues to the forefront. In the text that follows, we highlight prominent ways in which industrial biotechnology is known to influence many of the environmental concerns noted above. For this analysis, we differentiate between industrial biotechnology processes by whether biobased products/materials are (a) used as feedstock or (b) support process reactions (e.g., enzymatic processes). Processes in each of these categories will often likely have different considerations when evaluating environmental impacts.

## *1.1 Processes That Use Biobased Feedstocks*

Biobased feedstocks are used in a number of different processes to produce renewable energy, chemicals, and other materials/products. While in some cases they may be the primary feedstock of choice (e.g., producing sugar from sugarcane), in many cases they may be considered as substitutes for conventional feedstocks such as petroleum. These substitutions are often motivated by environmental considerations. For example, cellulosic ethanol is expected to have lower GHG emissions over its life cycle, compared to petroleum-based gasoline when used in vehicles.

Examples of biobased feedstocks include corn, wheat, sugarcane, dedicated energy crops such as switchgrass and *Miscanthus*, algae, wood and wood residues, as well as organic wastes such as municipal solid waste (MSW). While these feedstocks are typically considered renewable, the environmental impacts associated with feedstock production and processing may be considerable and should be considered appropriately. We describe these potential environmental impacts based on key stages that are typically part of bioproduct life cycles.

### 1.1.1 Feedstock Production

Feedstock production is often a substantial contributor to the environmental impact of biobased products [15]. The environmental impacts of producing biobased feedstocks may include but are not limited to:

- Nitrate pollution of rivers, lakes, streams, and other water bodies, caused by nitrogen fertilizers used to improve soil quality. Excess nutrient runoff into water bodies may cause eutrophication. Nitrous oxide produced by the nitrification/denitrification processes relating to fertilizer use is a GHG, contributing to global warming [16], and may also contribute to the decline in stratospheric ozone [17].
- Impact on water quality and food web structures due to phosphorus fertilizer use, when it is introduced in aquatic systems [18].
- Land use modifications for the purpose of biobased feedstock production, which may also result in biodiversity loss through change in habitat (e.g., fragmentation) [19].
- Consumptive freshwater use for irrigated crops, which can contribute to water stress. To put this in perspective, agriculture contributes to 80% of all consumptive water use in the USA [20].
- Changes in soil carbon depending on farming practices, which could result in increased or decreased $CO_2$ emissions to the atmosphere [21].
- GHG and other air pollutant emissions from energy (e.g., diesel) used to operate machinery for feedstock production (e.g., tilling, harvesting, or threshing biobased feedstocks) [22]
- Direct or indirect land use change impacts due to expanded biofeedstock production [23].
- Reduction in atmospheric GHG emissions due to sequestration of $CO_2$ during the growth phase of the plant or other organism (i.e., during photosynthesis) [24].

In all cases, the type of biobased feedstock used, farming/forestry and harvesting practices, their role in the regional or global markets, and the use of potential by-products are a few of the factors that determine the magnitude of environmental impacts associated with biobased feedstock production.

### 1.1.2 Feedstock Processing/Refining

After feedstock production, conversion processes are often the most energy-intensive stage in the life cycle of biobased products [15]. The Biorefineries Roadmap published by the German government [25] distinguishes between pretreatment and preparation of biomass, primary refining (typically, the separation of biomass into intermediates such as cellulose or starch) and secondary refining (subsequent conversion or processing). The environmental impacts of converting biobased feedstock to fuels and products are determined by the steps involved in processing or refining and therefore may differ based on the processing pathways used. As an example, converting cellulosic material to ethanol using a biochemical process typically requires chemicals and energy (e.g., electricity, natural gas, and/or lignin). To account comprehensively for environmental impacts across the full life cycle of fuel/product production, the environmental impacts of the chemicals and energy production would also be considered – these may include energy, materials, and water use, resulting in GHG and other air emissions that impact air quality, emissions to water and soil, as well as resource depletion for constrained resources. If any co-products are generated during processing, environmental burdens associated with feedstock production and processing are typically apportioned to the co-products. The environmental impacts of cellulosic ethanol production are elaborated in Case Study 1. Further discussions on industrial activities are presented in Saling [26].

### 1.1.3 Feedstock and Product Transportation

Biobased feedstocks and products can be transported using a number of different modes – by road (truck), rail, or pipeline. Typically, the energy and GHG impacts of the transportation phase for petroleum- and coal-based fuels and products are relatively low compared with other life cycle phases (e.g., Franklin Associates [27], Venkatesh et al. [28, 29], Choquette-Levy et al. [30]); transportation GHG emissions for biofuels and bioproducts usually constitute a greater share of the life cycle emissions. This generally results from the often low bulk density and high moisture content of feedstock (e.g., wheat straw and wood are examples of feedstock with these respective properties) and due in part to overall life cycle emissions typically being much lower for biofuels/bioproducts. For example, in a case study [31] of wood pellet production for use in an electricity-generating station in Ontario, Canada, forest operations (for biomass supply), transportation (to mill and generating station), and pelletization each accounted for approximately one-third of the total GHG emissions, whereas in the coal reference case, transportation was a very minor contributor, representing approximately 2% of life cycle emissions. The values for transportation-related emissions were similar in magnitude for both pathways. Life cycle emissions for the pellet and coal pathways were on the order of 80 and 1,000 g of carbon dioxide equivalent ($gCO_2e$)/kWh, respectively. Transportation also contributes to non-GHG environmental impacts. For example, using diesel trucks to transport feedstock or finished products may result in considerable particulate matter, $NO_x$, $SO_2$, and other pollutants that impact local air quality – especially important if this transportation (e.g., of the final product to the point of sale) occurs within highly populated urban environments. The installation and use of pipelines (e.g., as proposed by the Asia-Pacific Economic Cooperation Energy Working Group for dedicated biofuel transportation [32]) over large areas may result in habitat fragmentation, affecting ecosystem services and local species' populations (e.g., Abrahams et al. [33] analyzed similar issues relating to pipeline development associated with Marcellus shale gas).

### 1.1.4 Product Use

Substituting more natural products like ethanol, wood, and lactic acid derivatives for products like gasoline, coal, and other petroleum derivatives, respectively, may reduce use-phase exposure to carcinogenic or otherwise harmful compounds (e.g., aromatics) typically found in coal and petroleum. Nevertheless, being biobased is no guarantee of reduced harm, with the toxic properties of products like methanol being well known [12]. The impact of biobased energy products on GHG emissions has been much debated in recent years. Like fossil fuels, the energy in biomass is stored primarily as carbon-carbon and carbon-hydrogen chemical bonds. As a result, combustion of biomass releases similar quantities of $CO_2$ per unit energy to combustion of fossil fuels [34]. Thus, claims of reduced GHG emissions from bioenergy relative to fossil fuels rely on arguments about changes to the carbon cycle elsewhere in the system. For energy crops that are part of short-term carbon cycles, life cycle studies typically assign a credit for the $CO_2$ that is removed from the atmosphere during plant growth [35, 36]. Thus, $CO_2$ emissions from biomass combustion are treated as carbon neutral. Although this method has been challenged in recent years [37, 38], it remains standard and – in our view and that of others [24, 39] – reasonable. Non-GHG impacts associated with bioproduct/fuel use may be more substantial and include air pollutants such as $NO_x$ and $SO_2$ that may be generated as a by-product when combusting biofuels, affecting local air quality. The impact on air quality depends on both the nature of the fuel itself, as well as the characteristics of the boiler/engine in which it is combusted. With appropriately calibrated engines, the existing literature suggests that increased use of biofuels like ethanol and biodiesel has little adverse impact (and possibly some benefits) for air quality ($NO_x$, non-methane hydrocarbons, particulate matter, and mobile source air toxics) relative to fossil fuels [40]. Additionally, ecotoxicological and human health risks may be associated with biofuel spills [41] that may be similar to petroleum fuels.

### 1.1.5 Product End-of-Life

The biofuel life cycle typically ends during product use, with combustion for energy, as described in the previous subsection. On the other hand, depending on how biobased products are treated and/or degraded after use, the associated environmental impacts could be significantly different. Options for disposal generally include landfill, recycling, composting (for biodegradable products), incineration (with or without energy recovery), or other forms of energy recovery such as anaerobic digestion and gasification. Each of these strategies will have a different profile of environmental impacts for a given product, facing similar issues to conventional products. Key considerations for biobased products include their longevity compared with fossil-based products (e.g., highly biodegradable products like thermoplastic starch may have limited reuse value, potentially increasing the rate at which new products must be manufactured), recyclability (possibly hampered by product biodegradation and low production volumes), and the ultimate fate of the carbon within biodegradable products (e.g., released aerobically as $CO_2$, anaerobically as $CH_4$ – a more potent GHG – or stored in a long-lived product or landfill).

### Bioproducts' Displacement of Conventional Products

Underlying most claims of environmental benefit of products sourced from biobased feedstocks is the notion that they can replace conventional products that have greater negative environmental impact. In rare cases, production of biobased products can itself have environmental benefits, for example, by sequestering carbon dioxide into inert products [42–44]. Otherwise, assumptions surrounding the reference (conventional) product are key to understanding the impact of these biobased products. For example, it is common for biorefineries to produce a surplus of electricity, which may be exported to the grid. Generating this electricity may greatly reduce greenhouse gas emissions and air pollutants if it replaces a dirty fuel like coal (e.g., Whitaker et al. [45]) but have little benefit if it replaces a relatively clean power source like hydropower (e.g., Kumar et al. [46]). Similarly, renewable natural gas produced by anaerobic digestion (and biogas upgrading) of organic waste can most naturally replace conventional (fossil-based) natural gas, but it may also be used to substitute a range of other fuels like coal or diesel throughout the industrial sector. In all cases, the reference system can substantially influence conclusions about whether or to what extent the biobased product has a positive impact.

Even within a product category, results may differ depending on the exact source of the reference product. For example, deployment of transportation biofuels like ethanol is generally assumed to displace petroleum fuels. The net environmental impact of this change depends on the source of the displaced petroleum, with estimates of greenhouse gas emissions, say, varying by up to 60% when comparing the life cycle of crude oil sourced from different fields [47]. Assessing which petroleum source is likely to be displaced by biofuels requires analysis to determine which oil fields are at the economic margin [48]. In a first-order assessment of fuels produced in California refineries by Wallington et al. [48], they show that gasoline produced from a marginal oil could have 10% higher GHG emissions than gasoline produced from a conventional oil. Similar considerations of "marginal producers" are required when considering the displacement of any reference product with substantial heterogeneity in its environmental impact.

Additional challenges exist when the biobased product is not identical to the reference product. For example, ethanol is typically compared with gasoline on an energy-equivalent basis (e.g., per megajoule). While accurate in some cases, using ethanol may also enable increases in vehicle engine efficiency due to increases in fuel octane ratings [49]. It is therefore necessary to estimate the actual service provided by the different products (e.g., km of vehicle travel) when assessing the net impact of the biobased product. In some cases, it may not be possible to draw clear functional equivalence. For example, in the USA and Canada, ethanol is typically blended with gasoline at levels of up to 10–15%. Higher-level blends (e.g., 85% ethanol) require the installation of new fueling infrastructure, the use of specially designed "flex-fuel" vehicles, and potentially more frequent trips to the fueling station, due to the lower-energy density of ethanol, about 65–70% that of gasoline [50], complicating the assessment of ethanol's net environmental impact. A further complication is that ethanol, when used in higher compression engines, delivers higher combustion efficiency, which can at least partially offset the lower-energy density.

### Indirect and Market-Mediated Effects

In addition to the issue of identifying marginal producers, discussed above, the adoption of products sourced from biobased feedstocks has been the source of intense scrutiny due to the potential for an array of economically and behaviorally driven indirect effects. Here, we discuss a small subset of the most prominent such effects.

#### *Induced Land Use Change*

It is well established that agriculture is substantially more land intensive (e.g., per unit energy) than other sources of energy or carbon feedstock [51–53]. As a result, increased use of agriculture-based products can spur changes in global land use patterns, resulting in an array of negative environmental impacts. In addition to the tautological impact on natural land transformations, this induced land use change has been the subject of much scrutiny due to its potential for increasing greenhouse gas emissions. For example, if forests or grasslands are converted to cropland in response to an increased demand for agricultural products, then these land conversions can increase GHG emissions, most notably through the release of stored carbon (e.g., due to decomposition of removed biomass or changes in soil carbon stocks). Some studies have historically differentiated between *direct* and *indirect* land use change, depending on whether the new emissions occur on the land that is directly used to grow biofuels or elsewhere in the global agricultural system in response to diverting land or crops to industrial products. For the purpose of assessing the consequences of using biobased feedstocks, this distinction is often not useful as the net impact is the same. Increasingly, experts rely on blanket terms, such as induced land use change (ILUC), to capture the emissions from all land transformations resulting from increased use of biofuels.

ILUC is beyond the scope of traditional LCA but can be a deciding factor as to whether biofuels and bioproducts are likely to achieve net reductions in GHG emissions relative to their fossil counterparts. Various studies have projected ILUC emissions due to biofuel production, with estimates ranging from below 0 (i.e., removing carbon dioxide from the atmosphere) to over 200 g $CO_2e$/MJ [23], more than double the emissions of gasoline production and use (approx. 90 g $CO_2e$/MJ), depending on the biofuel source and ILUC model employed. Typically, ILUC models estimate the amount, type, and location of land transformations based on regional or global economic models that respond to "shocks" in demand for biobased products. The resulting (estimated) greenhouse gas emissions can then be used to formulate a measure of "carbon debt" [54, 55] or may be amortized over a period of time (e.g., 30 years of future production in the case of biofuels), to obtain a normalized value per unit product (like in the numbers reported above) [23].

Due to the complexity in their estimation, ILUC emissions are highly uncertain [23] and the subject of much controversy [56]. They are highly context specific, depending on economic conditions along with the scale and scope of the policy (if any) that induce the use of biobased products. Because ILUC stems from economy-wide interactions, the resulting emissions are attributed to different product types (e.g., corn ethanol used in the USA) and do not depend on the specific production process from any given company. For further discussion of land use impacts of industrial biotechnology, see Lindner et al. [13].

*Indirect Output Use Change*

Another prominent indirect effect associated with the use of biobased feedstocks extends the discussion of conventional product displacement to consider market-wide changes in product demand. Although biobased products can often substitute for fossil products on a 1:1 basis in physical terms, the resulting changes in market prices (e.g., a reduction in the price of the conventional product, spurring additional demand elsewhere) make 1:1 replacement unlikely across the entire market. This effect is rarely addressed in existing LCA studies. A nascent literature examines this effect, with authors using different terminology, but all essentially referring to the increase in fossil product consumption as a result of price changes when biobased products are introduced: the indirect fuel use effect (IFUE) [57], market leakage [58], indirect output use change (IOUC) [58], indirect demand change (IDC) [59], or the rebound effect in fuel markets [60]. In reviewing existing modeling efforts to quantify IOUC resulting from US and European biofuel policies, Smeets et al. [60] report IOUC values range from −20 to 119% (i.e., 1 unit of biofuel may displace up to 1.2 units of fossil fuel or may cause a net increase in fossil fuel consumption of 0.19 units), depending on the policy context and modeling framework. Most estimates suggest that each unit of biofuel displaces less than half a unit of gasoline [61], leading some authors to conclude that policies inducing greater use of biobased products may result in a net increase in (greenhouse gas) emissions, even if these biobased products have lower life cycle emissions than their fossil counterparts [61–64].

*Additional Indirect Effects*

Although indirect land use change and indirect output use change have generally received the most attention in recent years, there are additional economic and behavioral considerations that can affect the environmental impact of biobased products. For example, the US biofuel policy creates incentives for producers to import Brazilian sugarcane-based ethanol, which is typically found to have lower GHG emissions than US corn-based ethanol. This creates a situation in which the USA imports sugarcane ethanol from Brazil, while Brazil imports corn ethanol from the USA, resulting in additional emissions (e.g., due to duplicate transportation) than would otherwise be expected due to US use of sugarcane ethanol [65]. Similarly, the use of high percent ethanol blends may require consumers to drive further to find appropriate refueling stations, depending on availability of infrastructure (e.g., Seki et al. [66]). More firmly on the behavioral side, some research has suggested that consumers may be subject to a "mental rebound" effect – compensating for environmentally friendly behavior, with increased consumption or "license" to engage in other environmentally damaging behavior [67]. Applying the concept to the present study, there is the potential for consumers to respond to the use of renewable (biobased) products or fuels in ways that may negate the putative benefits of these products. Yet another behavioral effect may result if the use of biodegradable biobased plastic packaging facilitates diversion of food waste from landfills to compost facilities [68]. The above discussion provides just a small cross section of examples that illustrate the complexity of evaluating the net environmental impact of biobased fuels and products in the real world. Readers interested more broadly in understanding or evaluating indirect effects may consult the discussions in Miller and Keoleian [69] along with sources included above and the growing body of literature on "consequential" life cycle assessment [70–78].

## *1.2 Industrial Activity That Is Supported by Biological Processes*

The second broad category of applications of industrial biotechnology is processes that are supported by biological processes. Examples of such processes vary across different industries, from those producing fine chemicals and fermented foods to treating and managing waste. In general, the environmental impacts of these processes may be most relevant when compared to incumbent processes that use more conventional thermochemical methods. In our experience, the literature on environmental impacts relating to this category of applications is quite limited, compared to the extensive set of studies on processes that use biobased feedstocks. We therefore highlight some key insights from reviewing this literature, in brief.

Processes that use biocatalysts may be able to operate at lower temperatures and produce less toxic waste and by-products and emissions relative to conventional chemical processes. Biocatalysts with improved selectivity may also reduce the need for by-product separations, thus lowering energy demand and emissions [79]. For further discussion, see Saling [26].

Biological processes may contribute to reducing the negative environmental impact of waste management. For example, solid wastes and wastewater can be managed or processed using biocatalysts and enzymes to produce bio-methane. Anaerobic digestion offers the possibility of avoiding the release of $CH_4$ that occurs when organics are landfilled while simultaneously generating renewable energy in the form of biogas [80, 81]. Reducing the quantity of waste sent to landfills may also help avoid environmental issues such as leaching toxic materials into the local water streams. Bio-methane produced via fermentation processes can be used as an energy source and substitute for other conventional sources of energy. Ongoing work may also lead to the development of specialized microbes that can be used as a bioremediation strategy to break down otherwise persistent/inert plastics and reduce their accumulation in the environment [82].

Enzymes are used as biocatalysts in a range of different industrial reactions, such as in the pharmaceutical and food industries, and enzymatic processing methods have been considered as one of most promising environment-friendly alternatives to conventional production [83]. Although energy and chemicals are used to produce these enzymes, there may be trade-offs where the environmental impacts are reduced, relative to the conventional process it "displaces." In a recent review of studies published since the early 2000s, Jegannathan and Nielsen [83] show that using enzyme-based processes generally reduced environmental impacts across a number of categories including global warming, acidification, eutrophication, photochemical ozone, and energy use. These studies considered cases from a range of industries including pulp and paper, leather, textiles, detergents, food and beverages, animal feed, fine chemicals, pharmaceuticals, and cosmetics. From reviewing these studies, some of which used LCA approaches, they suggest that using enzymatic production methods may be a step toward cleaner industrial production. Some examples, largely gleaned from Jegannathan and Nielsen [83], are presented as follows. A further discussion is presented in Saling [26].

- *Pulp and paper industry*: Cellulase is used for a number of processing operations such as deinking of recycled paper, where cellulase softens paper so that ink is released easily, avoiding the use of chemicals such as NaOH, as well as reducing processing time (e.g., Skals et al. [84]).
- *Textiles*: Yarn and fabric are made by treating raw cotton in processes such as scouring, bleaching, and polishing, all of which are typically energy-, water-, and chemical-intensive. For example, scouring (removal of impurities from noncellulosic pectin components in raw cotton) usually requires high-temperature operating conditions with the use of chemicals such as NaOH, $H_2O_2$, and $Na_2CO_3$. Using the enzyme pectate lyase instead to break down pectin can reduce overall environmental life cycle impacts relating to energy, water, and materials, which offset the impacts of producing the enzyme, as shown by Nielsen et al. [85]. They also show similar impacts when a catalase enzyme is used for bleaching of knitted yarn and fabrics prior to dyeing.
- *Detergents*: Surfactants in detergents that are primarily responsible for removing stains from clothes are most active at high temperatures. Considerable energy is used to heat water during laundry processes for this purpose. After washing, the surfactants may be toxic to aquatic life if not treated appropriately before releasing into the environment. Enzymes, specifically lipases, are a good alternative due to their lower toxicity impacts and the ability to impact stains at lower temperatures and more than offset any environmental impacts associated with their production.

- *Food and beverages*: Enzymes have been used extensively in this industry to improve yield and produce key ingredients. As an example, the use of the enzyme phospholipase, to remove phosphatide gums from crude vegetable oil, which increases yield and reduces the use of chemicals such as caustic soda, has been shown to improve overall environmental life cycle impacts [86]. Another example is of enzyme use in increasing the shelf life of bread, which in turn prevents and/or reduces emissions from food waste and transport. A review by Oxenbøll and Ernst [87] showed that the addition of enzymes such as amylase reduces starch breakdown and moisture loss, which improve taste. The GHG emissions from amylase production were reported to be small compared to the avoided emissions from sending stale bread to landfills. They also showed that since some stale bread is converted to animal feed, there could be some additional environmental impact due to "displaced" animal feed production and therefore highlighted the need for detailed systems-level analysis that includes such secondary effects.
- *Pharmaceuticals*: Synthesis of pharmaceutical products through conventional chemical processing typically uses large amounts of energy and chemicals while producing hazardous waste. Enzyme use in pharmaceutical ingredient synthesis can lower overall process impacts. An example is in the production of 7-aminocephalosporic acid (7-ACA), which is used to make antibiotics. Enzymes used to produce 7-ACA reduce the total number of processing steps from seven in the conventional chemical process to three [88]. Environmental impacts across a number of categories including freshwater use, GHG emissions, photochemical ozone creation potential, and acidification are primarily lowered by reducing raw material production.

While using biomaterials to enhance processing that would have otherwise required more stringent operating conditions, higher materials, chemicals, or energy use, it is important to understand the environmental impacts of the incumbent, conventional processes, as well as to determine how the new process (including biomaterials consumption) would alter these impacts. One of the key gaps that emerged from our review of this literature is the lack of comprehensive systems-level analysis. Most studies had relatively limited boundaries of analysis (e.g., no indirect effects) and focused on a few environmental categories.

## 2 Further Issues in Accounting for Environmental Impacts

To evaluate and consistently compare the environmental impacts of industrial biotechnology processes with others, a number of factors should be considered. A few key issues are highlighted in the subsections that follow.

## 2.1 *Geographic and Temporal Scope of Processes and Environmental Impacts*

Supply chains linked to industrial biotechnology process may include stages across geographically dispersed regions, a reflection of global markets. In addition to influencing the actual pollutants emitted (e.g., through differences in local technology, land productivity, the makeup of electric grid), geography can also influence the importance or impact of a given environmental concern. While $CO_2$ is well-mixed globally in the atmosphere, other environmental pollutants or phenomena may be significant in a local or regional context. For example, feedstock produced in Iowa may be transported to Texas for biomaterials production. While the local impact of freshwater used for irrigation in some regions in Iowa may not be significant given sufficient local water supply, an equivalent amount of water used in Texas for feedstock processing may result in local water stress. Similarly, $NO_x$ emissions from biomass combustion may have a more significant impact in non-attainment areas (in the USA) where the air quality is lower than average. Therefore, the geographical scope of the processes being evaluated, location of environmental impacts, as well as its regional/global implications need to be considered carefully. This is especially critical when using a life cycle approach, where environmental impacts across life cycle stages are often aggregated, without any consideration to where those impacts occurred.

Similarly, the temporal aspects of environmental impacts relating to the biotechnology process life cycles should be carefully considered. The "carbon neutral" biogenic carbon accounting convention may not be appropriate when large differences exist in timing between carbon release and reuptake – either due to one-time land use change or regular crop rotations. As an example, forest biomass accumulates atmospheric carbon over decadal timescales, while the $CO_2$ emissions released during its combustion occur as a spike or step function, over much shorter timescales. As shown by McKechnie and MacLean [89], forest biomass combustion immediately increases greenhouse gas emissions, thus affecting the ability to reduce emissions in the near term. This may increase the overall cost of emissions reductions in the longer term due to the increased burden in the near term. As an example, forest thinning is practiced to harvest biomass in German forests, instead of clear cutting [90]. Therefore, the timing of carbon sequestration and release of biogenic $CO_2$ via processing and/or combustion need to be considered in the context of biotechnology processes that use biobased feedstocks. Nevertheless, with the exception of slow-growing biomass (e.g., trees) [91] or temporary carbon storage in long-lived products [92], timing of GHG emissions is generally not an important factor in the assessment of bioenergy systems [93, 94].

### 2.1.1 How Co-products Affect the Attribution of Environmental Impacts

The output of some industrial biotechnology processes may include a range of co-products, all used for similar or different purposes. The challenge is then in "distributing" the environmental impacts of the process (e.g., freshwater use) over these co-products in an appropriate fashion. Allocation approaches are commonly used in LCAs, mass, energy, or economic allocation methods, where the environmental impacts associated with the process are distributed across the co-products using their mass, energy content, or economic value, respectively. In some cases, if the co-product is not part of a mature market, all the environmental impacts may be allocated to the primary product. As an example, corn stover (leaves and stock left over after harvest of the corn grain) does not have a large-scale market currently, and therefore all the environmental impacts associated with corn and stover production may be attributed to corn only. System expansion is typically a more comprehensive approach, where the boundaries of the process systems being compared are expanded such that the outputs of both systems are equivalent in terms of co-products. For example, if biomass is used in a combined heat and power (CHP) system, the incumbent system selected to compare against it could include a coal power plant and a natural gas boiler that produce an equivalent amount of electricity and heat, respectively. McKechnie et al. [95] show that lignocellulosic ethanol produced from short rotation forestry feedstock for a specific case study has emissions that are 109 to 174% relative to petroleum-based gasoline – this range of emissions reductions is determined by different co-product and process co-location configurations. The negative emissions of the ethanol pathways imply that more GHGs are being sequestered than emitted (essentially net carbon sequestration), during the production and use of the lignocellulosic ethanol. Specifically in these cases, the co-product credit associated with the co-product of the ethanol production is larger than the sum of the GHG emissions from all other life cycle activities associated with the ethanol production.

### 2.1.2 Uncertainty and Variability in Characterizing Environmental Impacts

Uncertainty is inherent in environmental assessments of processes (including industrial biotechnology processes), for a number of reasons which may include, but are not limited to, modeling assumptions, the nature of available data, and temporal and spatial differences [28]. Variability may be introduced in these assessments, for example, due to natural variation in process feedstock, alternate transportation modes, or differences in process configurations between companies or operators due to technology licensing. Comparative analyses of processes must therefore appropriately address these effects in order to make more robust decisions [96]. Previous studies have estimated probability distributions representing environmental impacts of biotechnology processes over their life cycle [22, 97–102]. These studies typically use Monte Carlo simulation approaches where feasible. Analytical uncertainty propagation, fuzzy data sets, and scenario modeling have also been considered [103]. One of the key challenges is in collecting sufficient data to represent the uncertainty or variability in model inputs. When LCAs are conducted at a small scale, such as within organizations or institutions, the availability and control of relevant data may reduce some uncertainty in the results. However, in cases such as designing policy at the regional and national levels, large uncertainties, for example, due to insufficient data or improper accounting of spatial effects, may propagate significantly [28]. Other uncertainties may be introduced due to insufficient knowledge of environmental impacts of certain products or processes given the current state of science. An example of uncertainty in emissions from a range of biofuels, as estimated by Mullins et al. [22], is presented in Fig. 1. The figure shows substantial spread in the potential GHG emissions from the modeled biofuels, with most having at least some probability of surpassing the emissions from gasoline. Further discussion on evaluating uncertainty in environmental impacts is explored in Case Study 1.

### 2.1.3 Multidimensionality of Environmental Impacts

While this is no way unique to industrial biotechnology processes, it is worth noting that environmental impacts relating to a single process can occur along multiple

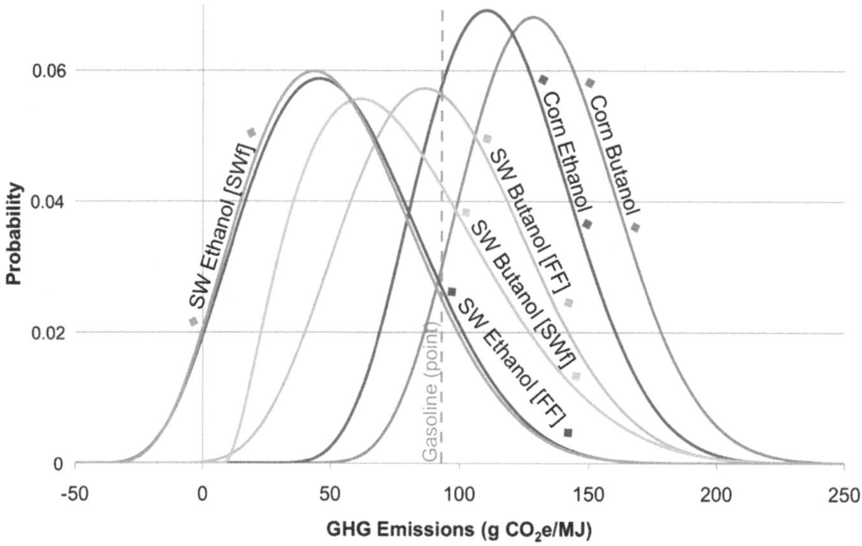


**Fig. 1** Probability distributions representing uncertainty in life cycle GHG emissions from a range of biofuels as presented in Mullins et al. [22]. *SW* switchgrass, *FF* fossil fuel energy for processing, *SWf* energy from switchgrass combustion for processing. Reprinted with permission from Mullins et al. [22]. Copyright 2011 American Chemical Society

dimensions – emissions to air, water, and soil leading to climate change, ozone depletion, resource depletion, acidification, biodiversity loss, and other impacts. In some cases, improvements in one impact category come with trade-offs in others. For example, a recent study [104] analyzing the life cycle environmental impacts of an E85 blend from sorghum processed in California showed that while GHG intensity of the cellulosic ethanol may be one-third to one-half that of gasoline, emissions contributing to local air and water quality may be higher. Though not required or standardized in LCA, methods exist within life cycle impact assessment that offer ways to potentially aggregate the environmental impacts of processes across different categories into single scores using characterization and normalization schemes [105, 106]. These methods are subject to some uncertainty due to normative decisions around appropriate characterization and weighting factors.

# 3 Case Studies: Comprehensive Analysis of Environmental Impacts Associated with Key Industrial Biotechnology Processes

A key challenge in assessing the environmental impact of any system is that details matter: exactly what technology is being employed? Where/when? How? At what scale? Under what regulatory regime? With what processes feeding into the supply chain? Industrial biotechnology is no exception to this rule, and so the challenges associated with understanding its environmental impact are difficult to discuss in generalities. Thus, in this section, we present three illustrative case studies of industrial biotechnology processes that have gained attention in the scientific literature and industry, namely, (a) lignocellulosic ethanol, (b) biobased plastics, and (c) enzyme use in the detergent industry. Throughout this chapter we have emphasized that it is important not to generalize the environmental impacts of biofuels and bioproducts (e.g., all lignocellulosic ethanol results in reduced environmental impacts compared to gasoline) and that the environmental performance of specific products whether biobased or others depend on the specifics of the entire life cycle. Through the case studies, we attempt to further illustrate some specifics, keeping in mind that the case studies themselves cover fairly broad categories of products and are not focused on a particular feedstock/process/product pathway with a particular spatial or temporal boundary.

## *3.1 Case Study 1: Lignocellulosic Ethanol*

There has been much controversy surrounding the net GHG balance of corn ethanol, which is the dominant feedstock for ethanol production to date in the USA. A number of studies have even suggested that corn ethanol could result in increased emissions relative to gasoline (e.g., Searchinger et al. [107], Mullins et al. [22], Cassidy [108]). The use of food crops for fuel production has also given rise to concerns about exacerbating global hunger [109, 110]. In contrast, cellulosic biofuels are generally reported to have a more favorable GHG balance [22, 111–113] and less impact on global food prices [110]. Even so, the entire biofuel endeavor remains controversial [109, 114].

The main drivers of lignocellulosic ethanol research, development, and deployment have been biofuel mandates and low carbon fuel standards [115, 116]. Attributes of the fuel include its potential to have a low life cycle GHG intensity, that it can be produced from diverse biomass feedstock that need not compete for cropland, and that it can displace a portion of gasoline used in the light-duty vehicle fleet. Ethanol can be blended with gasoline at low levels (E10/15) and used in most modern light-duty vehicles, or higher-level blends (up to E85) can be used in flexible fuel vehicles. While starch (e.g., corn) and sugar (e.g., sugarcane) ethanol technologies are considered mature, lignocellulosic technologies are still emerging. Few commercial scale lignocellulosic ethanol plants are in operation, and high-quality, reliable data needed for a detailed evaluation of environmental impacts is not publicly available.

In spite of the emerging nature of the lignocellulosic ethanol industry and many of the feedstocks that it would rely upon, and associated data uncertainty, numerous studies have examined environmental impacts of its production and use, often from a life cycle perspective [117]. These studies examined a wide range of feedstock, geographic and temporal settings, and to a lesser extent, conversion processes [118]. The majority of studies quantified GHG emissions [117], a considerable number also examined energy use, and a much smaller proportion of studies examined other environmental impacts such as air pollutant emissions (e.g., Spatari and MacLean [101]), photochemical oxidant formation, acidification, and eutrophication (e.g., González-García et al. [119]). Most studies assumed the use of the ethanol in a light-duty vehicle, generally in a high-level blend with gasoline in a flexible fuel vehicle or E100 in a dedicated vehicle. The conventional fuel that is almost always assumed to be displaced is petroleum gasoline used in a light-duty vehicle. A small number of studies quantitatively examined uncertainty in environmental impacts (GHG emissions, energy use, and air pollutants) [22, 101, 120], while other considered different scenarios (e.g., McKechnie et al. [95]) and others still have reported point estimates. Studies have generally not considered indirect effects.

### 3.1.1 Feedstock Production

Lignocellulosic ethanol feedstocks typically examined include agricultural residues (e.g., stover, straw, bagasse), forestry products/residues (e.g., poplar), or dedicated energy crops (e.g., switchgrass, *Miscanthus*, energy cane). The production method and class of land affect overall GHG emissions [121] as well as other environmental impacts (e.g., water use, eutrophication potential). Some feedstocks are used directly (e.g., energy crops), while others are residues of a primary crop (e.g., corn, wheat), which has implications for assigning environmental burdens. In the case of energy crops, environmental impacts associated with all of the feedstock production stages (e.g., production and use of fertilizers, planting, and harvesting) would be attributed to the ethanol production. In the case of corn stover, because the corn itself is the main product, often the only environmental impacts that are assigned to the stover are those associated with its harvest, as well as the production and use of any fertilizers that are needed to replace nutrients removed in the stover [122]. As mentioned, the majority of studies have focused on GHG emissions; however, studies considering residue feedstock have taken into consideration amounts of residue that can be "sustainably" removed considering maintenance of soil organic matter and soil fertility and mitigation of soil erosion. McKechnie et al. [122] examined the production of corn stover in various counties in the US Midwest building on the work of Kim et al. [123] and found considerable variation in GHG emissions for corn stover production from 6 g $CO_2$e/MJ ethanol (Macon County, Missouri) to 13 g $CO_2$e/MJ ethanol (Hardin County, Iowa). Major sources of the variability were location-specific soil carbon and $N_2O$ emissions responses to stover removal. The overall life cycle GHG emissions associated with the stover ethanol ranged from 1.5 to 22 g $CO_2$e/MJ ethanol depending on location and modeling assumptions. Co-product credits and the assumption of biomass carbon neutrality contributed to these low emissions. The results show that the feedstock production is a major contributor to life cycle emissions and the variation in stover production emissions and life cycle emissions emphasize the importance of examining uncertainty and variability associated with biofuel/product systems. A review of more than 100 life cycle-based studies of lignocellulosic ethanol concluded that feedstock production/source of biomass was a key driver of environmental performance, including air emissions but additionally impact categories such as ozone depletion, acidification, and eutrophication potential [117].

### 3.1.2 Conversion Process

While cellulosic ethanol can be produced through biochemical and thermochemical processes, the discussion here focuses on the biochemical processes as these have received more attention in industry and literature. Most studies examining environmental impacts of lignocellulosic ethanol have assumed the US Department of Energy's National Renewable Energy Laboratory's (NREL) process that uses dilute acid pretreatment, with electricity as the sole co-product [118]. There are several other technology options that have been proposed and adopted for commercial plants, with additional technologies considered at the pilot scale [124]. Distinctions in the pretreatment processes of these technologies are the most obvious, but there are implications for downstream processes, including enzymatic hydrolysis and fermentation and, ultimately, ethanol and co-product yields, all of which have implications for environmental impacts. Key factors affecting environmental impacts of the conversion process of cellulosic ethanol include the type of pretreatment process, energy inputs (whether biomass based or fossil fuels) and chemical inputs including enzymes, co-products of the process, and how these are assigned environmental impacts. Below we elaborate on these aspects.

### 3.1.3 Process Chemical Inputs

In the corn stover ethanol study of McKechnie et al. [122], biorefinery GHG emissions, which were based on the 2011 biorefinery model developed by the US National Renewable Energy Laboratory (NREL), were the single greatest emissions source (18 g $CO_2e$/MJ ethanol). These emissions were found to be approximately double than those assessed for an earlier version of the NREL biorefinery model (from 2002), due primarily to the inclusion of GHG-intensive inputs (e.g., ammonia, glucose) in the updated 2011 model. Earlier versions of key software models that estimate life cycle GHG and air pollutant emissions associated with cellulosic ethanol production did not include emissions associated with the production and use of process chemical inputs. Initial work by MacLean and Spatari [125] and later work by Hong et al. [126] brought attention to the importance of including these in estimating life cycle energy use and GHG emissions impacts of cellulosic ethanol production and provided insights into the distinctions in environmental impact of on-site and off-site enzyme production. MacLean and Spatari [125] concluded that even when assuming considerable improvement compared to then-current enzyme performance, the inputs for the near-term lignocellulosic technologies studied were found to be responsible for 30–40% of fossil energy use and 30–35% of GHG emissions, not insignificant fractions given that the models represented technology developers' nth plant performance. Hong et al. [126] estimated enzyme GHG emissions to be 258 g $CO_2e\ L^{-1}$ of ethanol for on-site production versus 403 g $CO_2e\ L^{-1}$ for off-site production, based on a 150 million L/year ethanol plant (assuming 11.5 mg enzyme $g^{-1}$ substrate and a cellulase fermentation yield of 90%).

### 3.1.4 Co-products

As noted above, the majority of lignocellulosic ethanol studies assume a single co-product, electricity and that excess electricity is sold to a local grid. Generally, the ethanol is assumed to receive a "credit" (i.e., negative energy use or emissions) associated with the electricity that would have had to be produced if the co-product electricity were not produced/sold to the grid. This lowers the energy use/emissions that are associated with the ethanol production. Most often the average grid mix is assumed in calculating the credit, but in some cases analysts have considered the marginal electricity source (e.g., Scown et al. [127]). The benefit of the electricity credit in lowering energy use and emissions associated with lignocellulosic ethanol production can be considerable. For example, some studies report negative overall life cycle GHG emissions for lignocellulosic ethanol as the electricity credit (reported as negative emissions) is larger in magnitude than the sum of emissions associated with all other aspects of the life cycle (e.g., Spatari and MacLean [101], Posen et al. [59]). A key consideration in efforts to increase ethanol yield is that when more of the biomass feedstock is converted to ethanol, less is available to be incorporated into co-products and so increased ethanol yield reduces the co-product credit. A low GHG electricity grid or marginal source also results in a lower credit than an electricity source with a large amount of coal. In essence, increasing ethanol yield can lead to higher emissions for ethanol production [95].

Co-products other than electricity can be produced in lignocellulosic ethanol production processes. Pourbafrani et al. [128] examined life cycle energy use and greenhouse gas (GHG) emissions implications of alternative pretreatment technologies (dilute acid hydrolysis, ammonia fiber expansion, and autohydrolysis) and co-products (electricity, pellet, protein, and xylitol) for ethanol production from corn stover. Results showed that the choices of pretreatment technology and co-product(s) impacted ethanol yield, life cycle energy use, and GHG emissions. Compared to producing only electricity as a co-product, the co-production of pellets and xylitol decreased life cycle GHG emissions associated with the ethanol, while protein production increased emissions. The life cycle GHG emissions of blended ethanol fuel (85% denatured ethanol by volume) ranged from −38.5 to 37.2 g $CO_2$e/MJ of fuel produced, reducing emissions by 61–141% relative to gasoline. As noted above for electricity co-product, for the broader set of co-products, lower ethanol yields led to lower GHG emissions. In a lignocellulosic biorefinery process model analyzed by Laure et al. [129] and developed at the pilot scale, lignin was assumed for use either as a thermoplastic or as a substitute for phenol in the production of polyurethane foams. The LCA, which allocated emissions to the co-products based on their mass, focused on all the products from the biorefinery including lignin (reference product: phenol), glucose (reference product: sugar solution from sugar beet), and hydrolysis lignin (reference products: C-5 sugars and heat from wood). Using the mass allocation approach, the authors found that environmental impacts could be reduced by 50–80% from the reference case across a number of impact categories.

### 3.1.5 Feedstock and Ethanol Transportation

The transportation of feedstock and final ethanol product results in GHG and air pollutant emissions if the transportation modes (e.g., truck) combust fuels during their operation. Upstream impacts associated with fuel production also must be considered. Due to the low density and bulky nature of much biomass feedstock, steps such as densification, drying, and locating fuel production facilities near to feedstock production have been taken to lower transportation costs, and these can lessen negative environmental impacts. Environmental impacts of transportation such as habitat disruption, fragmentation of landscapes, etc., however, have not generally been considered in assessments of lignocellulosic ethanol.

### 3.1.6 End Use: Ethanol Use in a Light-Duty Vehicle

Most studies assume that lignocellulosic ethanol would be combusted in a low- or high-level blend with gasoline in a light-duty vehicle and that one megajoule of ethanol would replace one megajoule of gasoline. While a common assumption, there can be octane and vehicle efficiency benefits of the use of ethanol if vehicles are optimized for the fuel. As well, if indirect effects are considered, the IOUC effect discussed earlier in the chapter brings into question the validity of the 1:1 replacement assumption.

While the combustion of ethanol results in GHG emissions, as discussed earlier, the most common assumption is that due to the biogenic nature of the carbon in the ethanol, the $CO_2$ emissions associated with its combustion do not contribute to atmospheric $CO_2$. This results in GHG emissions associated with E85 use in a vehicle being small – essentially the GHG emissions associated with the 15% gasoline and very small amounts of non-$CO_2$ GHGs associated with the ethanol. Air pollutants, both regulated (e.g., NOx, particulate matter, formaldehyde) and unregulated (e.g., acetaldehyde), also result from the combustion of ethanol/gasoline blends in vehicles. Due to vehicle exhaust and evaporative emissions being regulated, vehicles utilizing any ethanol blends must meet the same gram per kilometer levels of emissions as gasoline vehicles. The situation is more complicated if one really wants to understand the real-world emissions of vehicles running on ethanol blends (or a gasoline reference vehicle) as these depend on the ethanol blend that is used, model year of the vehicle, aftertreatment system and its condition, driving cycle, etc. One study that examined emissions of model year 1984–2007 vehicles using ethanol blends using a US certification driving cycle (Federal Test Procedure) (which does not reflect *real-world* driving) found decreases in some emissions (e.g., $NO_x$ emissions in newer vehicles, total hydrocarbon emissions, non-methane hydrocarbons, and CO except E85 blend) and increases in others (e.g., formaldehyde and acetaldehyde generally and NOx emissions in older vehicles) [130]. A more recent meta-analysis of literature studies focused on midlevel ethanol blends in production engines (the engines were not optimized for the ethanol blend) reported average results for E20/25 exhaust emissions (termed "end-of-pipe" in the study) compared to E0 (i.e., gasoline) [131]. Results show on average a 3% increase in specific fuel consumption on an energy basis; an increase in thermodynamic efficiency (5%); reductions in $CO_2$ (2%), which include biogenic (from biobased ethanol in the blend) and thermogenic (from fossil fuel in the blend) emissions; CO (20%); and hydrocarbons (5%) and equivalent emissions for NOx. The authors noted that additional work would be needed to evaluate particulate matter emissions, but in general there was the opportunity for these to be lower with E20/25 than E0. Overall, as noted earlier in Sect. 1.1.4, with appropriate engine and emission control systems, the use of ethanol is not expected to negatively impact emissions and in most cases is expected to result in emissions benefits.

### 3.1.7 Indirect Effects

Land use change emissions are expected to be less of an issue for lignocellulosic ethanol than for corn and sugarcane ethanol as lignocellulosic feedstock does not *necessarily* compete with other demands. Lignocellulosic ethanol can be produced from "waste" (residue) feedstock for which there is little to no current demand. Lignocellulosic energy crops can potentially be grown on marginal or degraded lands that may not be suitable for production of food/feed crops. This has the potential to mitigate impacts on grain price that lead to induced land use change. Only energy crops grown on fertile cropland are likely to directly compete with food/feed crops and induce significant amounts of land conversion. However, growing energy crops on marginal or degraded lands may result in increased inputs (fertilizer, energy) and lower crop yields, which have negative implications for the environment.

Other potential indirect effects of increased ethanol production (these effects are not specific to *lignocellulosic* ethanol) include those on gasoline consumption and price [57]. Regarding fuel market effects, studies comparing GHG emissions of ethanol to those of gasoline typically assume displacement on an energy-equivalent basis. However, if an increase in ethanol consumption causes gasoline prices to decrease (the market response), it may cause an increase in consumption of fuel, including gasoline [57]. These effects have received less attention in the ethanol literature than iLUC. For further discussion of land use impacts, refer to Lindner et al. [13].

### 3.1.8 Trends in Life Cycle Results

Overall, lignocellulosic ethanol production is expected to offer GHG emissions reductions and other benefits compared to first-generation ethanol, but choices of feedstock and specifics of the production process matter, and it is possible that there are pathways that do not result in improvements. Important factors driving environmental impacts and key trends in results from prior studies and meta-analyses include (specific references are included in the list below):

- Majority of studies on environmental impacts have focused on fossil fuel energy use and GHG emissions with almost all studies reporting lower fossil energy use and GHG emissions for lignocellulosic ethanol compared to gasoline. Results vary greatly depending on feedstock and conversion process assumed and even across studies that examine the same feedstock/conversion process [113, 117, 118, 122, 132].
- Far fewer studies have examined other key environmental impacts such as water use, acidification potential, eutrophication potential, and human health implications [133, 134].
- Borrion et al. [111] provide an excellent review of 52 life cycle studies of lignocellulosic ethanol and reported the following overall trends:

  - All studies reviewed reported a reduction in fossil energy use for the ethanol pathways.
  - Fifty of the studies concluded that there is a reduction in GHG emissions for cellulosic ethanol compared to a fossil fuel reference system (the two studies that reported an increase in GHG emissions did so when using economic value for co-product allocation but reported reductions when using other bases for allocation), but different studies report different main contributors to GHG emissions (biomass cultivation stage vs ethanol conversion process). There is a large variation in GHG emissions reported in the studies. Compared to the gasoline baseline of 0.26 kg $CO_2$ eq/km driven, the E85 results ranged from $-1.15$ to 0.79 g $CO_2$ eq/km driven.
  - The studies provide conflicting results as to whether ethanol results in more acidification, eutrophication, and ecotoxicity potential and human health impacts than gasoline [111].
  - Water consumption is reported to vary significantly depending on feedstock, irrigation practices, and whether feedstock or ethanol conversion is the primary contributor [111].

- A number of studies considered variability and uncertainty associated with the lignocellulosic ethanol life cycle and implications for environmental impacts and found these aspects to be important. Mullins et al. [22] reported a 10% probability that lignocellulosic (switchgrass) ethanol could have higher GHG emissions than a gasoline reference (after accounting for ILUC).
- Overall, the choice of feedstock and production methods, process technology, and co-products can materially impact life cycle environmental performance.

## 3.2 Case Study 2: Biobased Plastics

In addition to their value as an energy/fuel source, biobased products have attracted attention for their ability to displace fossil fuels in a wide range of chemical and material applications, ranging from solvents and catalysts to final products like packaging and consumer products. Of these, the polymer market is the largest by volume, with biobased plastics being a potential linchpin category in the operation of biorefineries [135]. Although biobased plastics represent a broad category of products, there has been particular emphasis on a small subset of biobased thermoplastic families, including bioethylene-based plastics, polylactic acid (PLA), polyhydroxyalkanoates (PHAs), and thermoplastic starch (TPS) [99, 136]. These plastics are all based on either starch- or sugar-containing feedstocks, but the products require different processing steps and exhibit an array of different properties and applications, complicating any assessment of their environmental impacts.

Bioethylene can be used as an ingredient into conventional ethylene-containing plastics like polyethylene (PE), polyethylene terephthalate (PET), polyvinyl chloride (PVC), and polystyrene (PS), replacing the more typical natural gas- or petroleum-based ethylene sources. In these cases, the resulting plastics are physically and chemically identical to their fossil counterparts. In contrast, PLA and PHA are often proposed as substitutes for certain applications of conventional plastics – typically PET/PS and PE/PS/polypropylene (PP), respectively (Posen et al. [99] and citations therein) – though with some change in the resulting properties. Going a step further, TPS has relatively poor mechanical properties and sensitivity to moisture [137] and so can only replace conventional plastics in specialized applications (e.g., as one-time use packaging) or by acting as a filler material [136, 138, 139]. As the name suggests, TPS is essentially a processed form of starch. Bioethylene, PLA, and PHA can be produced from any cellulose-, starch-, or sugar-containing source of biomass with sample production steps illustrated in Fig. 2 for production from corn grain. Each of the illustrated stages requires inputs of energy and often process chemicals, all of which contribute to the life cycle environmental impact of bioplastic production. Rather than walk through these stages in detail, this case study highlights several key issues related to the assessment of the environmental impact of bioplastics.

As a key driver in the development of biobased plastics [42], the GHG emissions from biobased plastics have been assessed in more detail than other impact categories [140]. Thus, as instructive examples, much of the analysis below focuses on how several factors influence GHG emissions results.

### 3.2.1 Feedstock Choice

Characteristic of many bioproduct assessments, feedstock choice is a key driver of GHG emissions. Studies by Posen et al. [59, 99] have shown dramatic differences in the life cycle GHG emissions of plastics produced from US corn – often observing a net increase in emissions relative to fossil products – versus more favorable pathways such as Brazilian sugarcane or cellulosic crops like US-based switchgrass. Key emission drivers in feedstock production tend to be associated with land use change as well as fertilizer production and the resulting $N_2O$ emissions from fertilizer application. Similar results were reported by Bos et al. [141], who showed the highest life cycle GHG emissions for the production of PLA and bioethylene-based PE from corn and wheat, compared to the potential for negative life cycle emissions (i.e., net $CO_2$ removal from the atmosphere) when using crops like sugarcane or *Miscanthus*. Negative GHG emissions can arise as a result of increases in soil carbon (i.e., carbon stored in the ground via root systems) or assumptions about carbon being stored in the final product (particularly relevant if short rotation crops are transformed to long-lived plastics). Waste-based pathways substantially reduce or eliminate the impact of feedstock production (though possibly at the expense of additional processing requirements) and may result in even lower emissions [142]. Studies have also shown the land management practices, such as the choice between till or no-till farming, can be a major driver of GHG emissions for biobased plastics [143]. The role of cropping systems, land quality, and land management practices has likewise been recognized as a relevant GHG emissions factor within any system relying on biobased feedstocks [144, 145].

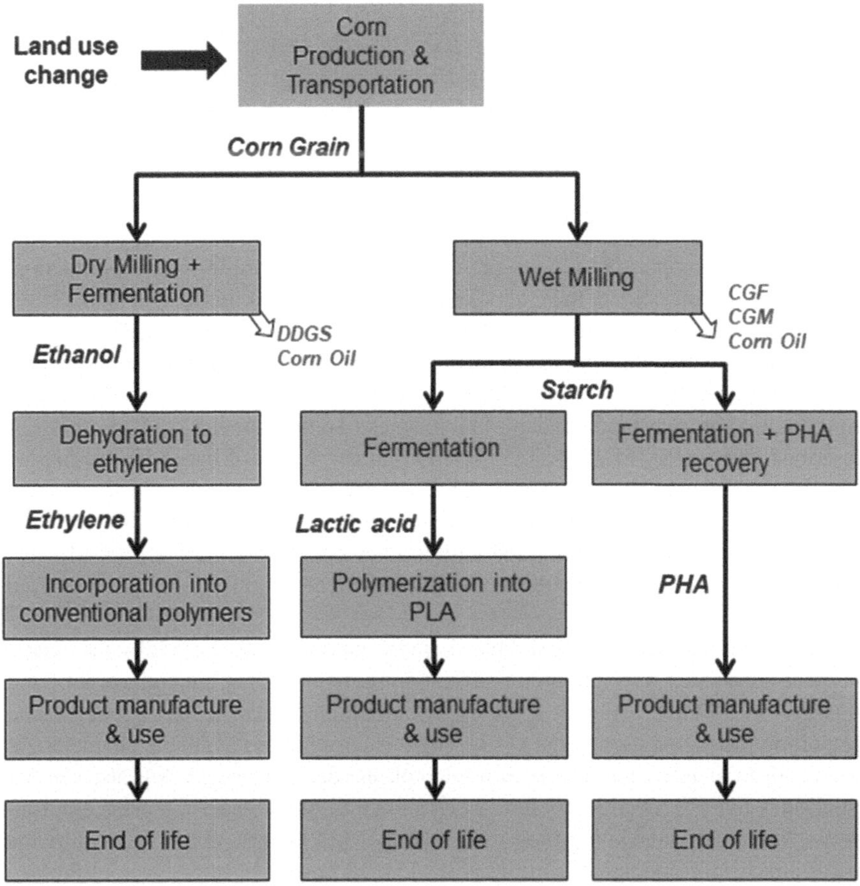


**Fig. 2** Biobased polymer production processes. Thick black arrows represent market-mediated effects, thin black arrows indicate main product flows, and white arrows indicate co-products: dried distillers grains with solubles (DDGS), corn oil, corn gluten feed (CGF), and corn gluten meal (CGM). Adapted from Posen et al. [99]

### 3.2.2 Treatment of Co-products

The availability of co-products is intrinsically linked to the choice of feedstock. For example, corn-based pathways tend to produce various animal feed co-products (see Fig. 2), while sugarcane and cellulosic pathways are often credited with surplus energy generation from the non-fermented portions. Whether and how the main plastic product is credited with emission reductions associated with these co-products is critical. For example, in their main scenarios, Posen et al. [99] estimate that full use of fermentation residues for energy production can reduce the mean life cycle GHG emissions from switchgrass-based PHA from +3 kg $CO_2$/kg PHA to −0.15 kg $CO_2$e/kg PHA. A similar, though less dramatic, reversal was observed when the co-products from corn-based PE were, respectively, ignored (resulting in life cycle emissions of 1.5 kg $CO_2$e/kg PE), credited with displacing other feed products (0.8 kg $CO_2$e/kg PE), or allocated a share of emissions in accordance with their mass (−0.15 kg $CO_2$e/kg PE).

### 3.2.3 End-of-Life

Although many bioplastic LCA studies focus primarily on cradle-to-gate production, end-of-life has been shown to be a major driver in life cycle emissions in some cases (e.g., Bohlmann [146], Gironi and Piemonte [147], Posen et al. [99], Hottle et al. [148]). Historically, the bioplastic market has been dominated by biodegradable plastics – including PHA, PLA, and TPS – though the market has shifted increasingly to durable (non-biodegradable) bioplastics like the bioethylene-based plastics described above (PE, PET, etc.) [135]. In contrast, the fossil-based plastic market includes small quantities of biodegradable plastics like polybutylene succinate and polyvinyl alcohol but is dominated by durable plastics like PE, PET, PS, PP, and PVC. Thus, even though biodegradability is a key consideration for end-of-life, it is only loosely associated with the comparison between fossil-based and biobased plastics.

The accumulation of long-lived plastic in the environment (especially the ocean) has been associated with a range of adverse impacts, including entanglement of and ingestion by wildlife, acting as a vector for other pollutants or invasive species, and despoiling natural environments [149]. There is currently no available characterization factor to quantify the impact of marine plastic debris using LCA [150]. Certain bioplastics like PHAs and TPS have been shown to biodegrade in marine environments, potentially alleviating these issues [151, 152]. Others, like PLA, require the use of industrial composting facilities [138, 151, 153]; such plastics can contribute to a reduction in landfill volumes but are less likely to have a direct impact on plastic accumulation in the environment.

In contrast to its impact on waste accumulation, biodegradation is often viewed negatively in terms of its impacts on GHG emissions [99, 154, 155]. Composting releases as $CO_2$ the carbon that might have otherwise been stored in the plastic, adding around 1.7 kg $CO_2$e/kg plastic in the case of PLA or 1.9 kg $CO_2$e/kg plastic in the case of PHA [99]. In certain cases, these emissions can represent a doubling of the life cycle emissions from the plastics in question, potentially leading them to having higher life cycle GHG emissions than their fossil counterparts. The situation can be even worse if the plastics are landfilled, potentially releasing their carbon as $CH_4$, an even more potent GHG. Landfilled PHA has been projected to add 3.4 kg $CO_2$e/kg plastic, while the question of whether PLA will degrade in a landfill remains controversial [156–158]. The issue is further complicated if energy recovery options are considered, with incineration tending to favor bioplastics like PLA and PHA over fossil ones (due to their lower carbon content) [99] and other methods like anaerobic digestion to produce biogas or renewable natural gas being possible only for the biodegradable plastics [159]. Due alternately to their mechanical properties or their low production volumes, bioplastics tend not to be recycled, adding further concerns to their environmental impact when compared to recycled conventional plastics. Exceptions to this observation are the bioethylene-based plastics, for which end-of-life is the same as for conventional ethylene plastics, since the biobased and fossil-based versions are chemically identical.

Generally, studies that include end-of-life effects within the scope see higher associated global warming potential impacts; however, this introduces a higher degree of uncertainty in the results, especially due to the lack of information on how these products will be disposed in real-world settings [138]. In a recent review of biopolymer end-of-life scenarios, Hottle et al. [148] show that global warming impacts vary substantially across scenarios; generally, the worst GHG impacts are seen for biobased plastics that degrade to $CH_4$ in landfill conditions. Where recycling was considered as an option (e.g., for bioethylene-based plastics), the ability to displace additional fossil plastics provided substantial benefit in terms of fossil fuel depletion, as well as carcinogenics and acidification in the case of bio-PET, though at the expense of smog creation and ozone depletion [148]. End-of-life was a relatively minor contributor to ecotoxicity, non-carcinogenic human toxicity, respiratory effects, and (with the exception of landfilled TPS) eutrophication.

### 3.2.4 Displacement of Conventional Products

Bioplastics also offer a prime example with respect to the complications associated with fossil product displacement. The market for plastics is wide and varied, with numerous potential "baseline" conventional products. Posen et al. [99] report mean emissions from common fossil thermoplastics in North America that range from around 1.5 kg $CO_2$e/kg plastic for plastics like PE and PP to a high of over 3 kg $CO_2$e/kg plastic for PS. Even within a product category, Posen et al. [59] likewise report 25% higher emissions from the production of petroleum (naphtha)-based fossil ethylene compared to natural gas ethane-based ethylene. This variability can lead to reversals in results, with bioplastics being more likely to reduce GHG emissions relative to PS than to PP and PE. Likewise biobased plastics are more likely to reduce GHG emissions relative to a naphtha-based ethylene plastic than to an equivalent ethane-based ethylene plastic.

Additional questions have been raised with respect to how to compare bioplastics to fossil plastics. Many studies (e.g., Kim and Dale [160], Groot and Borén [161], Vink and Davies [162], Posen et al. [99]) compare them simply on a mass basis (e.g. per kg), while others have investigated specific end products that differ in weight (e.g., Madival et al. [163], Suwanmanee et al. [164]) or have suggested general trends in substitution ratio – for example, the low density of TPS enabling it to displace greater quantities (on a mass basis) of conventional plastics as loose fill [165]. Whether bioplastics require greater, lesser, or equal mass to their fossil counterparts is generally application and plastic specific. If bioplastic products are shorter lived (e.g., single-use TPS bags vs reusable PP bags), this may also amplify their environmental impacts.

### 3.2.5 Indirect and Behavioral Effects

Few studies have investigated the indirect effects from bioplastics. As previously noted, the use of biodegradable plastics may facilitate diversion of food waste from landfill (eliminating the resulting $CH_4$ emissions from decomposition), by avoiding the need to separate organics from their packaging [68]. Some studies [59, 99, 166] have included emissions from indirect land use change in their assessment of bioplastics, generally finding these to be a major contributor to the life cycle emissions. Posen et al. likewise included a sensitivity analysis for indirect output use change associated with ethylene-based bioplastics (termed "indirect demand change" by the authors), concluding that this effect is less problematic than in the fuel industry due to the relatively lower per unit life cycle emissions of materials compared to fuels [59].

### 3.2.6 Trends in Life Cycle Results

In a recent review covering the environmental, economic, and social impacts of biobased plastics, Spierling et al. note that biobased plastic LCA results vary substantially between studies and are especially sensitive to the choice of feedstock and to a range of methodological factors such as the impact assessment method, allocation choices, co-product credits, and treatment of biogenic carbon [140]. Although the switching from fossil-based to biobased plastics by no means guarantees a reduction in GHG emissions, they are among the only products capable of exhibiting negative life cycle GHG emissions – particularly if they can act as a non-biodegradable carbon sink (e.g., bio-PE, potentially landfilled PLA) and are coupled with renewable energy sources in their production phase [42]. Done right, they can therefore contribute not only to reducing but actually reversing this particular environmental impact. It is worth noting that in some jurisdictions (e.g., Germany), only inert waste may be landfilled, and all municipal waste is to be incinerated beforehand. If biobased plastics were to be incinerated as part of such a system, they would lose the potential to be a carbon sink. Nevertheless, additional work is needed to understand the range of other environmental impacts of interest.

Hottle et al. [138] identified a lack of focus on non-GHG environmental aspects as a key shortcoming in the state of the science with respect to LCA of biobased plastics [138]. Based on a limited number of existing reviews [138, 142], PLA tends to exhibit higher acidification potential and eutrophication potential than fossil-based plastics; similar impacts on smog formation and human respiratory conditions; and lower impact on nonrenewable energy use and GHG emissions (only if carbon sequestration is assumed). PHA likewise exhibits higher acidification and eutrophication potential, with more variable results in terms of GHG emissions and energy use. Fewer studies are available for TPS and bioethylene plastics, which tend to score well on GHG emissions and nonrenewable energy use relative to fossil plastics, depending primarily on the feedstock source (e.g., see reviews in Posen et al. [59], Hottle et al. [138], Yates and Barlow [142]). Unsurprisingly, biobased materials in general – especially if using agricultural feedstocks – tend to perform poorly in terms of land use and water use impacts [167].

## 3.3 *Case Study 3: Enzyme Use in the Detergent Industry*

The BIO trade group reports that industrial biotechnology has allowed for product improvement simultaneously with reducing pollution impacts, from its earliest days, citing the example of how it addressed phosphate water pollution in the 1970s due to laundry detergents [1]. Phosphate-containing ingredients such as sodium tripolyphosphate (STPP) have been used in detergents in many regions across the world. When wastewater containing these phosphates is inadequately treated and released into the environment, water bodies may see eutrophication and subsequent algal blooms.

### 3.3.1 Displacement of Conventional Products

Phosphate is one of the primary components in detergent formulation, and it fulfills several functions including reducing water hardness, improving the efficiencies of surfactants, preventing redeposition of calcium and magnesium, emulsifying oily soils, stabilizing the alkalinity during the wash, and breaking and keeping the particles in suspension [168]. As phosphates have versatile functions in detergent, there is no single chemical alternative that can accomplish the full capacity of phosphates. The use of multiple substitutes to replace the key functions of phosphates is necessary to achieve satisfactory results in formulating a phosphate-free detergent, and enzymes are one of the viable substitutes. Enzymes such as protease, lipase, amylase, and cellulase [83] have been used to replace phosphates in detergents [169]. In 2013, the global market for enzymes used in detergents was valued at a $1 billion [170]. Since the detergent industry is part of a market strongly driven by consumer demand, the substitution of phosphates with enzymes was done in a way to ensure that product quality remained unaffected, for example, stain removal efficiencies and capabilities of enzyme-based detergents remained the same while retaining fabric color and quality.

### 3.3.2 Trends in Environmental Life Cycle Results

Enzyme-based detergents impact the environment across different dimensions:

- Generally, the environmental impacts of enzyme production are small, relative to the impact “savings” observed due to lower-energy and surfactant utilization in the use phase of the laundry life cycle [171].

- Enzymes allow for efficient washing at colder temperatures as well (relative to phosphates), thus reducing energy use, which typically results in GHG emissions savings. This also allows washing access to people with limited resources who may not have access to heated water [170]. Some studies assume a Northern European regional focus [171], where water is heated using electricity, and therefore, any avoided electricity use results in GHG savings (from a coal plant that is assumed to be on the margin). These emissions savings estimates may be different when considering a different region – for example, natural gas is used for water heating in many areas within the USA, which would instead be avoided with enzyme-based detergent use.
- Freshwater use and acidification impacts are also typically reduced, relative to phosphate-based detergents.
- To our knowledge, few LCA studies have considered how agricultural land use is affected due to increased enzyme production and use. For example, Nielsen and Schaetz [169] showed that agricultural land use ($m^2$) increases when carbonate or zeolite-based enzymatic ingredients are used instead of STPP (primarily because enzymes are dependent on agricultural production). However, they argue that these impacts are quantitatively "miniscule" compared to the environmental savings. We are not aware of any comprehensive analyses that have considered the global implications for land use change due to growth in enzymes for the detergent industry, among other industries.
- Similarly, Nielsen and Schaetz [169] also showed that the wash wastewater quality (measured by acute and chronic toxicity) may be affected negatively with the use of carbonate- or zeolite-based enzymatic ingredients compared to STPP. However, eutrophication impacts decrease, as a trade-off, relative to STPP use.

## 4 Emerging Trends in Industrial Biotechnology and Future Potential Environmental Impacts

Given the ubiquitousness of industrial biotechnology, it may be of interest to focus on a few emerging trends in this area and consider their environmental impacts.

### *4.1 Genetic Engineering/Modification*

These approaches have been used to improve resiliency and productivity of agricultural feedstocks used over the world on a large scale. While we do not address health risks in this chapter, one of the challenges associated with genetically modified crops is the potential proliferation of pesticide-resistant feedstocks [172]. As crops become resistant to mild versions of pesticides, stronger variants may be used that could potentially be more damaging to the environment, e.g., when they enter water systems. This may also translate more generally to crop pervasiveness or invasiveness [173] that may be a threat to native/wild species. Alternatively, modification that improves crop yields or resistance to pests may reduce land requirements and the need for agrochemicals, thereby reducing the environmental impact of agriculture [174, 175]. There is currently much debate around quantifying the environmental impacts of genetic modifications, and this will continue to be a growing area of research.

## 4.2 *Rapid Shift from Fossil Fuel Feedstock to Biobased Feedstock*

While we already see this shift beginning to occur in fuels production, this may extend to chemicals and products industries as well, especially given appropriate economic and policy signals. This will require greater quantities of biobased feedstock production in the future, potentially resulting in significant environmental impacts, as briefly discussed in Sect. 2. Increased demand for biobased feedstock may result in unprecedented land use change due to less than sustainable agricultural practices. It may increase demand for and divert land from other crops and/or protected areas while avoiding depleting resources typically considered non-renewable. In an effort to increase land productivity, intensive agricultural practices may be used (e.g., increasing nitrogen fertilizer, energy, irrigation water) which in turn impacts the environment.

## 4.3 *Evolving/Maturing Industry and Growing Need*

Industrial biotechnology is a rapidly growing area, and of major interest for efficiency and environmental reasons, as evidenced throughout this chapter. Technologies in some of these areas are emerging as robust alternatives to well-established industries (e.g., fossil fuels). While these incumbent industries have had, in some cases, decades of research and deployment, emerging industrial biotechnology processes are often less mature and may have greater potential for additional efficiency and yield improvements or may offer disruptive ways to improve upon these near-optimized incumbent systems. Industries that support these technologies are in turn moving toward lower environmental impacts, which have the potential to reduce life cycle impacts of biobased products and processes. Posen et al. [42] showed an example of this, demonstrating that bioplastics may become especially advantageous in terms of GHG emissions *after* low carbon energy technologies are more widely available/adopted to support their production. Industrial biotechnology will likely continue to play a key role in mitigating environmental concerns in the decades to come. For example, bioenergy coupled with carbon capture has been shown to be a key technology area in many decarbonization scenarios, one of the only category of technologies allowing for negative net carbon dioxide emissions (e.g., Venton [176]). Biobased processes may likewise play an increasingly important role, responding to progressively more costly and/or polluting competing fossil fuel sources (e.g., Wallington et al. [48]) and penetration limits of other renewable or low pollution technologies (e.g., Denholm and Margolis [177], Traut et al. [178], Clack et al. [179]). However, remaining challenges include costs, scale-up, as well as sustainable feedstock production to reduce potentially large-scale impacts such as land use change.

## 5 Conclusions

In this chapter, we presented a range of considerations for environmental impacts of industrial biotechnology. Although these processes are proposed or utilized as alternatives to conventional processes with the objective of reducing overall environmental impact, industrial biotechnology processes do have associated environmental impacts as well, which should be considered carefully. Life cycle assessment approaches offer comprehensive frameworks in considering these impacts, accounting for all supply chain stages where possible, and allowing for the accounting of multiple impact categories. A further discussion on sustainability assessments for industrial biotechnology is presented in Fröhling and Hiete [180]. Detailed economic modeling may also be required to understand certain unintended consequences of large-scale industrial biotechnology growth. One of the key takeaways from this chapter is that system boundaries should be defined carefully when comparing industrial biotechnology processes to conventional processes, in order to comprehensively account for relative environmental impacts.

While this chapter did not address societal/ethical aspects or risk assessment relating to industrial biotechnology, further discussion of these topics can be found in Osseweijer et al. [181] and Chen and Reniers [182], respectively.

## References

1. Biotechnology Innovation Organization (BIO) (2018) What is industrial biotechnology? https://www.bio.org/articles/what-industrial-biotechnology. Accessed 25 Jan 2018
2. House TW (2012) National bioeconomy blueprint, April 2012. Ind Biotechnol 8:97–102. https://doi.org/10.1089/ind.2012.1524
3. Australian Government Department of Industry Innovation and Science (2013) Industrial biotechnology and biomass industries. https://www.industry.gov.au/industry/IndustrySectors/nanotechnology/IndustrialBiotechnology/Pages/default.aspx. Accessed 25 Jan 2018

4. European Commission (2018) Biotechnology – growth. https://ec.europa.eu/growth/sectors/biotechnology_en. Accessed 25 Jan 2018
5. Verones F, Henderson AD, Lauren A, Ridoutt B, Ugaya C, Hellweg S (2016) LCIA framework and modelling guidance [TF 1 Crosscutting issues]. Global guidance for life cycle impact assessment indicators, vol 1. UNEP/SETAC Life Cycle Initiative, Paris, pp 40–57
6. Goedkoop M, Spriensma R, Effting S, Collignon M (2000) The eco-indicator 99: a damage oriented method for life-cycle impact assessment: manual for designers. PRé Consultants, Amersfoort
7. Guinée J (2002) Handbook on life cycle assessment: operational guide to the ISO standards. Springer, Berlin
8. Bare JC, Pennington DW, Thomas McKone GAN (2003) The tool for the reduction and assessment of chemical and other environmental impacts. J Ind Ecol 6:49–78
9. Jolliet O, Margni M, Charles R et al (2003) IMPACT 2002+: a new life cycle impact assessment methodology. Int J Life Cycle Assess 8:324–330. https://doi.org/10.1007/BF02978505
10. Frischknecht R, Steiner R, Jungbluth N (2009) The ecological scarcity method – eco-factors 2006. A method for impact assessment in LCA. Environmental Studies No. 0906. Swiss Confederation, Federal Office for the Environment FOEN, Bern
11. Huijbregts MAJ, Steinmann ZJN, Elshout PMF et al (2016) ReCiPe2016: a harmonised life cycle impact assessment method at midpoint and endpoint level. Int J Life Cycle Assess 22:138–147. https://doi.org/10.1007/s11367-016-1246-y
12. Vale A (2007) Methanol. Medicine 35:633–634. https://doi.org/10.1016/j.mpmed.2007.09.014
13. Lindner JP, Beck T, Bos U, Albrecht S (2018) Assessing land use and biodiversity impacts of industrial biotechnology. In: Fröhling M, Hiete M (eds) Sustainability and life cycle assessment in industrial biotechnology. Advances in biochemical engineering/biotechnology. Springer, Berlin, Heidelberg
14. Curran M, de Souza DM, Antón A et al (2016) How well does LCA model land use impacts on biodiversity? – A comparison with approaches from ecology and conservation. Environ Sci Technol 50:2782–2795. https://doi.org/10.1021/acs.est.5b04681
15. Broeren MLM, Zijp MC, Waaijers-van der Loop SL et al (2017) Environmental assessment of bio-based chemicals in early-stage development: a review of methods and indicators. Biofuels Bioprod Biorefin 11:701–718. https://doi.org/10.1002/bbb.1772
16. Park S, Croteau P, Boering KA et al (2012) Trends and seasonal cycles in the isotopic composition of nitrous oxide since 1940. Nat Geosci 5:261–265. https://doi.org/10.1038/ngeo1421
17. Ravishankara AR, Daniel JS, Portmann RW (2009) Nitrous oxide (N2O): the dominant ozone-depleting substance emitted in the 21st century. Science 326:123–125. https://doi.org/10.1126/science.1176985
18. Tilman D (1999) Global environmental impacts of agricultural expansion: the need for sustainable and efficient practices. Proc Natl Acad Sci U S A 96:5995–6000
19. Foley JA, DeFries R, Asner GP et al (2005) Global consequences of land use. Science 309:570–574. https://doi.org/10.1126/science.1111772
20. USDA ERS (2017) Irrigation and water use. https://www.ers.usda.gov/topics/farm-practices-management/irrigation-water-use.aspx. Accessed 4 Oct 2017
21. Qin Z, Dunn JB, Kwon H et al (2016) Influence of spatially dependent, modeled soil carbon emission factors on life-cycle greenhouse gas emissions of corn and cellulosic ethanol. GCB Bioenergy 8:1136–1149. https://doi.org/10.1111/gcbb.12333
22. Mullins KA, Griffin WM, Matthews HS (2011) Policy implications of uncertainty in modeled life-cycle greenhouse gas emissions of biofuels. Environ Sci Technol 45:132–138. https://doi.org/10.1021/es1024993

23. Warner E, Zhang Y, Inman D, Heath G (2013) Challenges in the estimation of greenhouse gas emissions from biofuel-induced global land-use change. Biofuels Bioprod Biorefin 8:114–125. https://doi.org/10.1002/bbb.1434
24. De Kleine RD, Anderson JE, Kim HC, Wallington TJ (2017) Life cycle assessment is the most relevant framework to evaluate biofuel greenhouse gas burdens. Biofuels Bioprod Biorefin 11:407–416. https://doi.org/10.1002/bbb.1752
25. The Federal Government (2012) Biorefineries roadmap as part of the German Federal Government action plans for the material and energetic utilisation of renewable raw materials. German Federal Government, Berlin
26. Saling P (2018) Assessing industrial biotechnology products. In: Fröhling M, Hiete M (eds) Sustainability and life cycle assessment in industrial biotechnology. Advances in biochemical engineering/biotechnology. Springer, Berlin, Heidelberg
27. Franklin Associates (2011) Cradle-to-gate life cycle inventory of nine plastic resins and four polyurethane precursors. Franklin Associates, Kansas
28. Venkatesh A, Jaramillo P, Griffin WM, Matthews HS (2011) Uncertainty analysis of life cycle greenhouse gas emissions from petroleum-based fuels and impacts on low carbon fuel policies. Environ Sci Technol 45:125–131
29. Venkatesh A, Jaramillo P, Michael Griffin W, Scott Matthews H (2012) Uncertainty in life cycle greenhouse gas emissions from United States coal. Energy Fuel 26:4917–4923. https://doi.org/10.1021/ef300693x
30. Choquette-Levy N, Zhong M, MacLean H, Bergerson J (2018) COPTEM: a model to investigate the factors driving crude oil pipeline transportation emissions. Environ Sci Technol 52:337–345. https://doi.org/10.1021/acs.est.7b03398
31. McKechnie J, Saville B, MacLean HL (2016) Steam-treated wood pellets: environmental and financial implications relative to fossil fuels and conventional pellets for electricity generation. Appl Energy 180:637–649. https://doi.org/10.1016/j.apenergy.2016.08.024
32. APEC Biofuels Taskforce (2011) Biofuel transportation and distribution options for APEC economies. Produced by BBI Biofuels Canada for APEC Secretariat, Singapore
33. Abrahams LS, Griffin WM, Matthews HS (2015) Assessment of policies to reduce core forest fragmentation from Marcellus shale development in Pennsylvania. Ecol Indic 52:153–160. https://doi.org/10.1016/j.ecolind.2014.11.031
34. Garg A, Kazunari K, Pulles T (2006) 2006 IPCC guidelines for national greenhouse gas inventories. Chapter 1: introduction. Intergovernmental panel on climate change, vol 2. IGES, Kanagawa
35. Curran MA (2012) Life cycle inventory modeling in practice. Life cycle assessment handbook: a guide for environmentally sustainable products. Scrivener Publishing/Wiley, Salem/Hoboken, pp 43–46
36. Menten F, Chèze B, Patouillard L, Bouvart F (2013) A review of LCA greenhouse gas emissions results for advanced biofuels: the use of meta-regression analysis. Renew Sust Energ Rev 26:108–134. https://doi.org/10.1016/j.rser.2013.04.021
37. DeCicco JM (2014) The liquid carbon challenge: evolving views on transportation fuels and climate. Wiley Interdiscip Rev Energy Environ 4:98–114. https://doi.org/10.1002/wene.133
38. Searchinger TD (2010) Biofuels and the need for additional carbon. Environ Res Lett 5:24007. https://doi.org/10.1088/1748-9326/5/2/024007
39. Wang M, Tyner WE, Williams D, Dunn JB (2015) Comments on and discussion of the liquid carbon challenge: evolving views on transportation fuels and climate. Argonne National Laboratory, Lemont
40. Wallington TJ, Anderson JE, Kurtz EM, Tennison PJ (2016) Biofuels, vehicle emissions, and urban air quality. Faraday Discuss 189:121–136. https://doi.org/10.1039/c5fd00205b
41. Bluhm K, Heger S, Seiler T-B et al (2012) Toxicological and ecotoxicological potencies of biofuels used for the transport sector-a literature review. Energy Environ Sci 5:7381–7392. https://doi.org/10.1039/C2EE03033K

42. Posen ID, Paulina J, Amy EL, Griffin WM (2017) Greenhouse gas mitigation for U.S. plastics production: energy first, feedstocks later. Environ Res Lett 12:34024
43. Smith P (2016) Soil carbon sequestration and biochar as negative emission technologies. Glob Chang Biol 22:1315–1324. https://doi.org/10.1111/gcb.13178
44. Vanholme B, Desmet T, Ronsse F et al (2013) Towards a carbon-negative sustainable bio-based economy. Front Plant Sci 4:174. https://doi.org/10.3389/fpls.2013.00174
45. Whitaker M, Heath GA, O'Donoughue P, Vorum M (2012) Life cycle greenhouse gas emissions of coal-fired electricity generation. J Ind Ecol 16:S53–S72. https://doi.org/10.1111/j.1530-9290.2012.00465.x
46. Kumar A, Schei T, Ahenkorah A et al (2011) Hydropower. IPCC special report on renewable energy sources and climate change mitigation. Cambridge University Press, Cambridge
47. Carnegie Endownment for International Peace (2018) Assessing global oils. http://oci.carnegieendowment.org/#total-emissions. Accessed 26 Jan 2018
48. Wallington TJ, Anderson JE, De Kleine RD et al (2016) When comparing alternative fuel-vehicle systems, life cycle assessment studies should consider trends in oil production. J Ind Ecol 21(2):244–248. https://doi.org/10.1111/jiec.12418
49. Anderson JE, DiCicco DM, Ginder JM et al (2012) High octane number ethanol–gasoline blends: quantifying the potential benefits in the United States. Fuel 97:585–594. https://doi.org/10.1016/j.fuel.2012.03.017
50. U.S. Department of Energy (2017) Alternative fuels data center: fuel properties comparison. https://www.afdc.energy.gov/fuels/fuel_properties.php
51. Fthenakis V, Kim HC (2009) Land use and electricity generation: a life-cycle analysis. Renew Sust Energ Rev 13:1465–1474. https://doi.org/10.1016/j.rser.2008.09.017
52. Trainor AM, McDonald RI, Fargione J (2016) Energy sprawl is the largest driver of land use change in United States. PLoS One 11:e0162269. https://doi.org/10.1371/journal.pone.0162269
53. Yeh S, Jordaan SM, Brandt AR et al (2010) Land use greenhouse gas emissions from conventional oil production and oil sands. Environ Sci Technol 44:8766–8772. https://doi.org/10.1021/es1013278
54. Bentsen NS (2017) Carbon debt and payback time – lost in the forest? Renew Sust Energ Rev 73:1211–1217. https://doi.org/10.1016/j.rser.2017.02.004
55. Fargione J, Hill J, Tilman D et al (2008) Land clearing and the biofuel carbon debt. Science 319:1235–1238. https://doi.org/10.1126/science.1152747
56. Gawel E, Ludwig G (2011) The iLUC dilemma: how to deal with indirect land use changes when governing energy crops? Land Use Policy 28:846–856. https://doi.org/10.1016/j.landusepol.2011.03.003
57. Rajagopal D (2013) The fuel market effects of biofuel policies and implications for regulations based on lifecycle emissions. Environ Res Lett 8:24013. https://doi.org/10.1088/1748-9326/8/2/024013
58. Bento A, Klotz R, Landry J (2012) Are there carbon savings from us biofuel policies? The critical importance of accounting for leakage in land and fuel markets. SSRN Electron J. https://doi.org/10.2139/ssrn.2219503
59. Posen ID, Griffin WM, Matthews HS, Azevedo IL (2015) Changing the renewable fuel standard to a renewable material standard: bioethylene case study. Environ Sci Technol 49:93–102. https://doi.org/10.1021/es503521r
60. Smeets E, Tabeau A, van Berkum S et al (2014) The impact of the rebound effect of the use of first generation biofuels in the EU on greenhouse gas emissions: a critical review. Renew Sust Energ Rev 38:393–403. https://doi.org/10.1016/j.rser.2014.05.035
61. Hill J, Tajibaeva L, Polasky S (2016) Climate consequences of low-carbon fuels: the United States Renewable Fuel Standard. Energy Policy 97:351–353. https://doi.org/10.1016/j.enpol.2016.07.035
62. Drabik D, de Gorter H (2011) Biofuel policies and carbon leakage. AgBioforum 14:103–110

63. Rajagopal D, Plevin RJ (2013) Implications of market-mediated emissions and uncertainty for biofuel policies. Energy Policy 56:75–82. https://doi.org/10.1016/j.enpol.2012.09.076
64. Thompson W, Whistance J, Meyer S (2011) Effects of US biofuel policies on US and world petroleum product markets with consequences for greenhouse gas emissions. Energy Policy 39:5509–5518. https://doi.org/10.1016/j.enpol.2011.05.011
65. Debnath D, Whistance J, Thompson W (2017) The causes of two-way U.S.–Brazil ethanol trade and the consequences for greenhouse gas emission. Energy 141:2045–2053. https://doi.org/10.1016/j.energy.2017.11.048
66. Seki SM, Michael GW, Chris H, Scott MH (2018) Refueling and infrastructure costs of expanding access to E85 in Pennsylvania. J Infrastruct Syst 24:4017045. https://doi.org/10.1061/(ASCE)IS.1943-555X.0000408
67. Girod B, De Haan P (2009) Mental rebound. Rebound Research Report Nr. 3. ETH, Zurich
68. Meeks D, Hottle T, Bilec MM, Landis AE (2015) Compostable biopolymer use in the real world: stakeholder interviews to better understand the motivations and realities of use and disposal in the US. Resour Conserv Recycl 105:134–142. https://doi.org/10.1016/j.resconrec.2015.10.022
69. Miller SA, Keoleian GA (2015) Framework for analyzing transformative technologies in life cycle assessment. Environ Sci Technol 49:3067–3075. https://doi.org/10.1021/es505217a
70. Brander M, Tipper R, Hutchison C, Davis G (2009) Consequential and attributional approaches to LCA: a guide to policy makers with specific reference to greenhouse gas LCA of biofuels. Econometrica, London
71. Curran MA, Mann M, Norris G (2005) The international workshop on electricity data for life cycle inventories. J Clean Prod 13:853–862. https://doi.org/10.1016/j.jclepro.2002.03.001
72. Earles JM, Halog A (2011) Consequential life cycle assessment: a review. Int J Life Cycle Assess 16:445–453. https://doi.org/10.1007/s11367-011-0275-9
73. Plevin RJ, Delucchi MA, Creutzig F (2014) Using attributional life cycle assessment to estimate climate-change mitigation benefits misleads policy makers. J Ind Ecol 18:73–83. https://doi.org/10.1111/Jiec.12074
74. Rajagopal D (2017) A synthesis of unilateral approaches to mitigating emissions leakage under incomplete policies. Clim Policy 17:573–590. https://doi.org/10.1080/14693062.2016.1150249
75. Rajagopal D (2017) A step towards a general framework for consequential life cycle assessment. J Ind Ecol 21:261–271. https://doi.org/10.1111/jiec.12433
76. Roos A, Ahlgren S (2018) Consequential life cycle assessment of bioenergy systems – a literature review. J Clean Prod 189:358–373. https://doi.org/10.1016/j.jclepro.2018.03.233
77. Weidema B (2003) Market information in life cycle assessment. Danish Environmental Protection Agency, Copenhagen
78. Zamagni A, Guinée J, Heijungs R et al (2012) Lights and shadows in consequential LCA. Int J Life Cycle Assess 17:904–918. https://doi.org/10.1007/s11367-012-0423-x
79. Gavrilescu M, Chisti Y (2005) Biotechnology – a sustainable alternative for chemical industry. Biotechnol Adv 23:471–499. https://doi.org/10.1016/j.biotechadv.2005.03.004
80. Abbasi T, Tauseef SM, Abbasi SA (2012) Anaerobic digestion for global warming control and energy generation – an overview. Renew Sust Energ Rev 16:3228–3242. https://doi.org/10.1016/j.rser.2012.02.046
81. Appels L, Lauwers J, Degrève J et al (2011) Anaerobic digestion in global bio-energy production: potential and research challenges. Renew Sust Energ Rev 15:4295–4301. https://doi.org/10.1016/j.rser.2011.07.121
82. Yoshida S, Hiraga K, Takehana T et al (2016) A bacterium that degrades and assimilates poly (ethylene terephthalate). Science 351:1196–1199. https://doi.org/10.1126/science.aad6359
83. Jegannathan KR, Nielsen PH (2013) Environmental assessment of enzyme use in industrial production – a literature review. J Clean Prod 42:228–240. https://doi.org/10.1016/j.jclepro.2012.11.005

84. Skals PB, Krabek A, Nielsen PH, Wenzel H (2008) Environmental assessment of enzyme assisted processing in pulp and paper industry. Int J Life Cycle Assess 13:124. https://doi.org/10.1065/lca2007.11.366
85. Nielsen PH, Kuilderd H, Zhou W, Lu X (2009) Enzyme biotechnology for sustainable textiles. In: Blackburn RS (ed) Sustainable textiles. Woodhead, Oxford, pp 113–138
86. Oxenbøll KM, Cowan D (2008) Enzymatic bioprocessing of oils and fats. Inf Int News Fats Oils Relat Mater 19:210
87. Oxenbøll K, Ernst S (2008) Environment as a new perspective on the use of enzymes in the food industry. Food Sci Technol 22:45–47
88. Henderson RK, Jiménez-González C, Preston C et al (2008) Peer review original research: EHS & LCA assessment for 7-ACA synthesis A case study for comparing biocatalytic & chemical synthesis. Ind Biotechnol 4:180–192. https://doi.org/10.1089/ind.2008.4.180
89. McKechnie J, MacLean HL (2014) Implications of emissions timing on the cost-effectiveness of greenhouse gas mitigation strategies: application to forest bioenergy systems. GCB Bioenergy 6:414–424. https://doi.org/10.1111/gcbb.12063
90. Fröhling M, Schweinle J, Meyer J, Schultmann F (2011) Logistics of renewable raw materials. Renewable raw materials. Wiley, Weinheim
91. Cherubini F, Peters GP, Berntsen T et al (2011) $CO_2$ emissions from biomass combustion for bioenergy: atmospheric decay and contribution to global warming. GCB Bioenergy 3:413–426. https://doi.org/10.1111/j.1757-1707.2011.01102.x
92. Levasseur A, Brandão M, Lesage P et al (2011) Valuing temporary carbon storage. Nat Clim Chang 2:6–8. https://doi.org/10.1038/nclimate1335
93. Almeida J, Degerickx J, Achten WMJ, Muys B (2015) Greenhouse gas emission timing in life cycle assessment and the global warming potential of perennial energy crops. Carbon Manag 6:185–195. https://doi.org/10.1080/17583004.2015.1109179
94. Schwietzke S, Griffin WM, Matthews HS (2011) Relevance of emissions timing in biofuel greenhouse gases and climate impacts. Environ Sci Technol 45:8197–8203. https://doi.org/10.1021/es2016236
95. McKechnie J, Zhang Y, Ogino A et al (2011) Impacts of co-location, co-production, and process energy source on life cycle energy use and greenhouse gas emissions of lignocellulosic ethanol. Biofuels Bioprod Biorefin 5:279–292. https://doi.org/10.1002/bbb.286
96. Lloyd SM, Ries R (2007) Characterizing, propagating, and analyzing uncertainty in life-cycle assessment: a survey of quantitative approaches. J Ind Ecol 11:161–179. https://doi.org/10.1162/jiec.2007.1136
97. Pfister S, Scherer L (2015) Uncertainty analysis of the environmental sustainability of biofuels. Energy Sustain Soc 5:30. https://doi.org/10.1186/s13705-015-0058-4
98. Plevin RJ, O'Hare M, Jones AD et al (2010) Greenhouse gas emissions from biofuels' indirect land use change are uncertain but may be much greater than previously estimated. Environ Sci Technol 44:8015–8021. https://doi.org/10.1021/es101946t
99. Posen ID, Jaramillo P, Griffin WM (2016) Uncertainty in the life cycle greenhouse gas emissions from U.S. production of three biobased polymer families. Environ Sci Technol 50:2846–2858. https://doi.org/10.1021/acs.est.5b05589
100. Sills DL, Paramita V, Franke MJ et al (2013) Quantitative uncertainty analysis of Life Cycle Assessment for algal biofuel production. Environ Sci Technol 47:687–694. https://doi.org/10.1021/es3029236
101. Spatari S, MacLean HL (2010) Characterizing model uncertainties in the life cycle of lignocellulose-based ethanol fuels. Environ Sci Technol 44:8773–8780. https://doi.org/10.1021/es102091a
102. Yan X, Boies AM (2013) Quantifying the uncertainties in life cycle greenhouse gas emissions for UK wheat ethanol. Environ Res Lett 8:15024. https://doi.org/10.1088/1748-9326/8/1/015024
103. Lloyd SM, Ries R (2007) Characterizing, propagating, and analyzing uncertainty in life-cycle assessment. J Ind Ecol 11:161–179

104. Fertitta-Roberts C, Spatari S, Grantz DA, Jenerette DG (2017) Trade-offs across productivity, GHG intensity, and pollutant loads from second-generation sorghum bioenergy. GCB Bioenergy 9:1764–1779. https://doi.org/10.1111/gcbb.12471
105. Sala S, Crenna E, Secchi M, Pant R (2017) Global normalisation factors for the environmental eootprint and life cycle assessment, EUR (28984). Publications Office of the European Union, Luxembourg. ISBN 978-92-79-77213-9, JRC109878. https://doi.org/10.2760/88930
106. Landis AE, Theis TL (2008) Comparison of life cycle impact assessment tools in the case of biofuels. 2008 IEEE international symposium on electronics and the environment, pp 1–7
107. Searchinger T, Heimlich R, Houghton RA et al (2008) Use of U.S. croplands for biofuels increases greenhouse gases through emissions from land-use change. Science 319:1238–1240. https://doi.org/10.1126/science.1151861
108. Cassidy E (2014) Ethanol's broken promise: using less corn ethanol reduces greenhouse gas emissions. Environmental Working Group, Washington
109. Pimentel D, Marklein A, Toth MA et al (2009) Food versus biofuels: environmental and economic costs. Hum Ecol 37:1–12. https://doi.org/10.1007/s10745-009-9215-8
110. Rathmann R, Szklo A, Schaeffer R (2010) Land use competition for production of food and liquid biofuels: an analysis of the arguments in the current debate. Renew Energy 35:14–22. https://doi.org/10.1016/j.renene.2009.02.025
111. Borrion AL, McManus MC, Hammond GP (2012) Environmental life cycle assessment of lignocellulosic conversion to ethanol: a review. Renew Sust Energ Rev 16:4638–4650. https://doi.org/10.1016/j.rser.2012.04.016
112. US Environmental Protection Agency (2010) Renewable fuel standard program (RFS2) regulatory impact analysis. EPA-420-R-10-006
113. Wiloso EI, Heijungs R, de Snoo GR (2012) LCA of second generation bioethanol: a review and some issues to be resolved for good LCA practice. Renew Sust Energ Rev 16:5295–5308. https://doi.org/10.1016/j.rser.2012.04.035
114. Searchinger T, Heimlich R (2015) Avoiding bioenergy competition for food crops and land. World Resources Institute, Washington
115. Energy Independence and Security Act of 2007 (EISA 2007). Public Law 110–140
116. Schwarzenegger A (2007) Executive order S-01-07: low carbon fuel standard. California Energy Commission, Sacramento
117. Morales M, Quintero J, Conejeros R, Aroca G (2015) Life cycle assessment of lignocellulosic bioethanol: environmental impacts and energy balance. Renew Sust Energ Rev 42:1349–1361. https://doi.org/10.1016/j.rser.2014.10.097
118. Gerbrandt K, Chu PL, Simmonds A et al (2016) Life cycle assessment of lignocellulosic ethanol: a review of key factors and methods affecting calculated GHG emissions and energy use. Curr Opin Biotechnol 38:63–70. https://doi.org/10.1016/j.copbio.2015.12.021
119. González-García S, Moreira MT, Feijoo G (2010) Environmental performance of lignocellulosic bioethanol production from Alfalfa stems. Biofuels Bioprod Biorefin 4:118–131. https://doi.org/10.1002/bbb.204
120. Spatari S, Bagley DM, MacLean HL (2010) Life cycle evaluation of emerging lignocellulosic ethanol conversion technologies. Bioresour Technol 101:654–667. https://doi.org/10.1016/j.biortech.2009.08.067
121. Scown CD, Nazaroff WW, Mishra U et al (2012) Lifecycle greenhouse gas implications of US national scenarios for cellulosic ethanol production. Environ Res Lett 7:14011. https://doi.org/10.1088/1748-9326/7/1/014011
122. McKechnie J, Pourbafrani M, Saville BA, MacLean HL (2015) Exploring impacts of process technology development and regional factors on life cycle greenhouse gas emissions of corn stover ethanol. Renew Energy 76:726–734. https://doi.org/10.1016/j.renene.2014.11.088
123. Kim S, Dale BE, Jenkins R (2009) Life cycle assessment of corn grain and corn stover in the United States. Int J Life Cycle Assess 14:160–174. https://doi.org/10.1007/s11367-008-0054-4
124. Yang B, Wyman CE (2008) Pretreatment: the key to unlocking low-cost cellulosic ethanol. Biofuels Bioprod Biorefin 2:26–40. https://doi.org/10.1002/bbb.49

125. MacLean HL, Spatari S (2009) The contribution of enzymes and process chemicals to the life cycle of ethanol. Environ Res Lett 4:14001. https://doi.org/10.1088/1748-9326/4/1/014001
126. Hong Y, Nizami A-S, Bafrani MP et al (2013) Impact of cellulase production on environmental and financial metrics for lignocellulosic ethanol. Biofuels Bioprod Biorefin 7:303–313. https://doi.org/10.1002/bbb.1393
127. Scown CD, Gokhale AA, Willems PA et al (2014) Role of lignin in reducing life-cycle carbon emissions, water use, and cost for United States cellulosic biofuels. Environ Sci Technol 48:8446–8455. https://doi.org/10.1021/es5012753
128. Pourbafrani M, McKechnie J, Shen T et al (2014) Impacts of pre-treatment technologies and co-products on greenhouse gas emissions and energy use of lignocellulosic ethanol production. J Clean Prod 78:104–111. https://doi.org/10.1016/j.jclepro.2014.04.050
129. Laure S, Leschinsky M, Frohling M et al (2014) Assessment of an Organosolv lignocellulose biorefinery concept based on a material flow analysis of a pilot plant. Cellul Chem Technol 48:793–798
130. Karavalakis G, Durbin TD, Shrivastava M et al (2012) Impacts of ethanol fuel level on emissions of regulated and unregulated pollutants from a fleet of gasoline light-duty vehicles. Fuel 93:549–558. https://doi.org/10.1016/j.fuel.2011.09.021
131. Geringer B, Spreitzer J, Mayer M, Martin C (2014) Meta-analysis for an E20/25 technical development study – task 2: meta-analysis of E20/25 trial reports and associated data. Vienna University of Technology, Institute for Powertrains and Automotive Technology, Wien
132. Patel M, Zhang X, Kumar A (2016) Techno-economic and life cycle assessment on lignocellulosic biomass thermochemical conversion technologies: a review. Renew Sust Energ Rev 53:1486–1499. https://doi.org/10.1016/j.rser.2015.09.070
133. Scown CD, Horvath A, McKone TE (2011) Water footprint of U.S. transportation fuels. Environ Sci Technol 45:2541–2553. https://doi.org/10.1021/es102633h
134. Wu M, Zhang Z, Chiu Y (2014) Life-cycle water quantity and water quality implications of biofuels. Curr Sustain Energy Rep 1:3–10. https://doi.org/10.1007/s40518-013-0001-2
135. Philp J (2014) OECD policies for bioplastics in the context of a bioeconomy, 2013. Ind Biotechnol 10:19–21. https://doi.org/10.1089/ind.2013.1612
136. Shen L, Worrell E, Patel M (2010) Present and future development in plastics from biomass. Biofuels Bioprod Biorefin 4:25–40. https://doi.org/10.1002/bbb.189
137. Mohammadi Nafchi A, Nafchi AM, Moradpour M et al (2013) Thermoplastic starches: properties, challenges, and prospects. Starch 65:61–72. https://doi.org/10.1002/star.201200201
138. Hottle TA, Bilec MM, Landis AE (2013) Sustainability assessments of bio-based polymers. Polym Degrad Stab 98:1898–1907. https://doi.org/10.1016/j.polymdegradstab.2013.06.016
139. Janssen LPBM, Moscicki L (2006) Thermoplastic starch as packaging material. Acta Sci Pol Tech Agrar 5:19–25
140. Spierling S, Knüpffer E, Behnsen H et al (2018) Bio-based plastics – a review of environmental, social and economic impact assessments. J Clean Prod 185:476–491. https://doi.org/10.1016/j.jclepro.2018.03.014
141. Bos H, Conijn S, Patel M (2010) Sustainability aspects of biobased applications. UR Food & Biobased Research, Wageningen
142. Yates MR, Barlow CY (2013) Life cycle assessments of biodegradable, commercial biopolymers – a critical review. Resour Conserv Recycl 78:54–66. https://doi.org/10.1016/j.resconrec.2013.06.010
143. Kim S, Dale BE (2005) Life cycle assessment study of biopolymers (polyhydroxyalkanoates) derived from no-tilled corn. Int J Life Cycle Assess 10:200–210. https://doi.org/10.1065/lca2004.08.171
144. Sainju UM (2016) A global meta-analysis on the impact of management practices on net global warming potential and greenhouse gas intensity from cropland soils. PLoS One 11 (2):1–26. https://doi.org/10.1371/journal.pone.0148527

145. Wightman JL, Duxbury JM, Woodbury PB (2015) Land quality and management practices strongly affect greenhouse gas emissions of bioenergy feedstocks. BioEnergy Res 4:1681–1690. https://doi.org/10.1007/s12155-015-9620-3
146. Bohlmann GM (2004) Biodegradable packaging life-cycle assessment. Environ Prog 23:342–346. https://doi.org/10.1002/ep.10053
147. Gironi F, Piemonte V (2010) Life cycle assessment of polylactic acid and polyethylene terephthalate bottles for drinking water. Environ Prog Sustain Energy 30:459–468. https://doi.org/10.1002/ep.10490
148. Hottle TA, Bilec MM, Landis AE (2017) Biopolymer production and end of life comparisons using life cycle assessment. Resour Conserv Recycl 122:295–306. https://doi.org/10.1016/j.resconrec.2017.03.002
149. Gregory MR (2009) Environmental implications of plastic debris in marine settings – entanglement, ingestion, smothering, hangers-on, hitch-hiking and alien invasions. Philos Trans R Soc Lond Ser B Biol Sci 364:2013–2025. https://doi.org/10.1098/rstb.2008.0265
150. Woods JS, Veltman K, Huijbregts MAJ et al (2016) Towards a meaningful assessment of marine ecological impacts in life cycle assessment (LCA). Environ Int 89–90:48–61. https://doi.org/10.1016/j.envint.2015.12.033
151. Emadian SM, Onay TT, Demirel B (2017) Biodegradation of bioplastics in natural environments. Waste Manag 59:526–536. https://doi.org/10.1016/j.wasman.2016.10.006
152. Tolinski M (2011) Polymer properties and environmental footprints. Plastics and sustainability: towards a peaceful coexistence between bio-based and fossil fuel-based plastics. Wiley, Hoboken
153. Martin RT, Camargo LP, Miller SA (2014) Marine-degradable polylactic acid. Green Chem 16:1768–1773. https://doi.org/10.1039/c3gc42604a
154. Levis JW, Barlaz MA (2011) Is biodegradability a desirable attribute for discarded solid waste? Perspectives from a national landfill greenhouse gas inventory model. Environ Sci Technol 45:5470–5476. https://doi.org/10.1021/es200721s
155. Sevenster MN (2015) Time-dependent life-cycle assessment of bio-based packaging materials. Sustainability assessment of renewables-based products. Wiley, Hoboken, pp 347–360
156. Kolstad JJ, Vink ETH, De Wilde B, Debeer L (2012) Assessment of anaerobic degradation of Ingeo™ polylactides under accelerated landfill conditions. Polym Degrad Stab 97:1131–1141. https://doi.org/10.1016/j.polymdegradstab.2012.04.003
157. Krause MJ, Townsend TG (2016) Life-cycle assumptions of landfilled polylactic acid underpredict methane generation. Environ Sci Technol Lett 3(4):166–169. https://doi.org/10.1021/acs.estlett.6b00068
158. Leejarkpai T, Mungcharoen T, Suwanmanee U (2016) Comparative assessment of global warming impact and eco-efficiency of PS (polystyrene), PET (polyethylene terephthalate) and PLA (polylactic acid) boxes. J Clean Prod 125:95–107. https://doi.org/10.1016/j.jclepro.2016.03.029
159. European Bioplastics (2015) Anaerobic digestion fact sheet. http://docs.european-bioplastics.org/publications/bp/EUBP_BP_Anaerobic_digestion.pdf. Accessed 10 Oct 2017
160. Kim S, Dale BE (2008) Energy and greenhouse gas profiles of polyhydroxybutyrates derived from corn grain: a life cycle perspective. Environ Sci Technol 42:7690–7695
161. Groot WJ, Borén T (2010) Life cycle assessment of the manufacture of lactide and PLA biopolymers from sugarcane in Thailand. Int J Life Cycle Assess 15:970–984. https://doi.org/10.1007/s11367-010-0225-y
162. Vink ETH, Davies S (2015) Life cycle inventory and impact assessment data for 2014 Ingeo™ polylactide production. Ind Biotechnol 11:167–180. https://doi.org/10.1089/ind.2015.0003
163. Madival S, Auras R, Singh SP, Narayan R (2009) Assessment of the environmental profile of PLA, PET and PS clamshell containers using LCA methodology. J Clean Prod 17:1183–1194. https://doi.org/10.1016/j.jclepro.2009.03.015
164. Suwanmanee U, Varabuntoonvit V, Chaiwutthinan P et al (2012) Life cycle assessment of single use thermoform boxes made from polystyrene (PS), polylactic acid, (PLA), and

PLA/starch: cradle to consumer gate. Int J Life Cycle Assess 18:401–417. https://doi.org/10.1007/s11367-012-0479-7
165. Shen L, Patel MK (2008) Life cycle assessment of polysaccharide materials: a review. J Polym Environ 16:154–167. https://doi.org/10.1007/s10924-008-0092-9
166. Piemonte V, Gironi F (2010) Land-use change emissions: how green are the bioplastics? Environ Prog Sustain Energy 30:685–691. https://doi.org/10.1002/ep.10518
167. Pawelzik P, Carus M, Hotchkiss J et al (2013) Critical aspects in the life cycle assessment (LCA) of bio-based materials – reviewing methodologies and deriving recommendations. Resour Conserv Recycl 73:211–228. https://doi.org/10.1016/j.resconrec.2013.02.006
168. Sekhon BS, Sangha MK (2004) Detergents – zeolites and enzymes excel cleaning power. Resonance 9:35–45. https://doi.org/10.1007/BF02837576
169. Nielsen AM, Schaetz T (2012) Taking steps towards a phosphate-free future. Househ Pers Care Today 2:13–16
170. Sarmiento F, Peralta R, Blamey JM (2015) Cold and hot extremozymes: industrial relevance and current trends. Front Bioeng Biotechnol 3:148. https://doi.org/10.3389/fbioe.2015.00148
171. Nielsen PH, Skagerlind P (2007) Cost-neutral replacement of surfactants with enzymes. Househ Pers Care Today 4:3–7
172. Union of Concerned Scientists (2017) Genetic engineering risks and impacts. http://www.ucsusa.org/food_and_agriculture/our-failing-food-system/genetic-engineering/risks-of-genetic-engineering.html. Accessed 11 Oct 2017
173. Tsatsakis AM, Nawaz MA, Kouretas D et al (2017) Environmental impacts of genetically modified plants: a review. Environ Res 156:818–833. https://doi.org/10.1016/j.envres.2017.03.011
174. Brookes G, Barfoot P (2017) Environmental impacts of genetically modified (GM) crop use 1996–2015: impacts on pesticide use and carbon emissions. GM Crops Food 8:117–147. https://doi.org/10.1080/21645698.2017.1309490
175. Phipps R, Park J (2002) Environmental benefits of genetically modified crops: global and European perspectives on their ability to reduce pesticide use. J Anim Feed Sci 11:1–18. https://doi.org/10.22358/jafs/67788/2002
176. Venton D (2016) Core concept: can bioenergy with carbon capture and storage make an impact? Proc Natl Acad Sci U S A 113:13260–13262. https://doi.org/10.1073/pnas.1617583113
177. Denholm P, Margolis RM (2007) Evaluating the limits of solar photovoltaics (PV) in electric power systems utilizing energy storage and other enabling technologies. Energy Policy 35:4424–4433. https://doi.org/10.1016/j.enpol.2007.03.004
178. Traut EJ, Cherng TC, Hendrickson C, Michalek JJ (2013) US residential charging potential for electric vehicles. Transp Res Part D Transp Environ 25:139–145. https://doi.org/10.1016/j.trd.2013.10.001
179. Clack CTM, Qvist SA, Apt J et al (2017) Evaluation of a proposal for reliable low-cost grid power with 100% wind, water, and solar. Proc Natl Acad Sci 114:6722–6727. https://doi.org/10.1073/pnas.1610381114
180. Fröhling M, Hiete M (2018) Sustainability and life cycle assessments in industrial biotechnology: a review of current approaches and future needs. In: Fröhling M, Hiete M (eds) Sustainability and life cycle assessment in industrial biotechnology. Advances in biochemical engineering/biotechnology. Springer, Berlin, Heidelberg
181. Osseweijer P, Posada Duque J, Asveld L (2018) Societal and ethical aspects of industrial biotechnology. In: Fröhling M, Hiete M (eds) Sustainability and life cycle assessment in industrial biotechnology. Advances in biochemical engineering/biotechnology. Springer, Berlin, Heidelberg
182. Chen C, Reniers G (2018) Risk assessment of processes and products in industrial biotechnology. In: Fröhling M, Hiete M (eds) Sustainability and life cycle assessment in industrial biotechnology. Advances in biochemical engineering/biotechnology. Springer, Berlin, Heidelberg

Adv Biochem Eng Biotechnol (2020) 173: 121–142
DOI: 10.1007/10_2019_100

Published online: 22 July 2019

# Societal and Ethical Issues in Industrial Biotechnology

**Lotte Asveld, Patricia Osseweijer, and John A. Posada**

## Contents



**Abstract** In this chapter we aim to give an overview of the main societal and ethical issues that are currently voiced around industrial biotechnology. We will illustrate this with some recent cases, such as the development of synthetic artemisinin, synthetic vanillin and vegetable oil produced by engineered algae. We show that current societal and ethical issues in industrial biotechnology centre on the following five themes: sustainability, naturalness, innovation trajectories, risk management and

L. Asveld (✉), P. Osseweijer, and J. A. Posada
Biotechnology and Society Group, Delft University of Technology, Delft, The Netherlands
e-mail: l.asveld@tudelft.nl

economic justice. In each of these themes, clashing public opinions fuel the public debate on the acceptability of new industrial biotechnology. In some cases this has led to the failure of otherwise promising innovations. In the last part, we provide suggestions on how to deal with these ethical and societal aspects based on the approach of Responsible Research and Innovation (RRI).

**Keywords** Economic justice, Ethical and social issues, Naturalness, Responsible research and innovation, Sustainability

## 1 Introduction

Until now industrial biotechnology has not received the same kind of public scrutiny as plant biotechnology has, especially not in relation to genetic modification [1]. The societal and ethical issues which do emerge around industrial biotechnology are more broadly oriented to its role as enabling technology with its many claims on applications aimed at sustainability, such as biofuels and biochemicals. With industrial biotechnological applications becoming more abundant, these public concerns do increase within the broader debate on bio-based economy, sustainable development goals and climate change, but also towards specific technological issues.

In this chapter we aim to give an overview of the main societal and ethical issues that are currently voiced around industrial biotechnology. We will illustrate this with some recent cases, such as the development of synthetic artemisinin (see also Schürrle, this volume), synthetic vanillin and vegetable oil produced by engineered algae. We do not include a case on biofuels, because the public debate on biofuels has already been documented extensively [2, 3]. Where relevant we will refer to this debate. We did not include pharmaceutical products, because that would make the chapter too wide ranging. We mainly focus on the development of a bio-based economy (or bioeconomy), here understood as an effort to derive high-quality and highly sustainable products from biomass [2]. We will also consider the wider societal debate on the bioeconomy as can be found in public reports, newspapers and websites.

We claim that current societal and ethical issues in industrial biotechnology centre on the following five themes: sustainability, naturalness, innovation trajectories, risk management and economic justice. In each of these themes, clashing public opinions fuel the public debate on the acceptability of new industrial biotechnology. These clashes in the public opinion bring out salient ethical and societal aspects for industrial biotechnology. In the last part, we provide suggestions on how to deal with these ethical and societal aspects based on the approach of Responsible Research and Innovation (RRI).

## 2 Some Recent Controversies in Industrial Biotechnology

Recent controversies in industrial biotechnology are here shortly described by using three examples, namely, the production of artemisinin, vanillin and algae-based oil. These cases can be taken as indicators for the social and ethical issues that are relevant for industrial biotechnology in general. They point out clashes in perceptions and underlying values. Based on the issues that are central in these cases, we arrive at five general societal themes relevant to industrial biotechnology, namely: (1) What is sustainability and how can it be measured? (2) What is natural? (3) How should risks of emerging industrial biotechnologies be managed? (4) How will industrial biotechnological trajectories develop? and (5) Who benefits from these new technologies?

In the cases mentioned, criticism is most clearly articulated by the ETC Group,[1] a Canadian NGO opposing specific technologies such as synthetic biotechnology (synbio). Although this might be seen as only one actor, the ETC Group often represents a broad group of NGOs and thereby a widely shared societal perspective. For instance, in the Ecover case discussed below, the ETC Group started a petition against the company Ecover which was signed by 17 other NGOs.

This is not to say that the ETC Group represents a view shared by all environmental NGOs. Other environmental NGOs often express a more nuanced view on new technologies, i.e. they do not categorically condemn technologies such as synthetic biology but remain open to see if they could possibly produce benefits and, if so, under what conditions. The England, Wales, and Northern Ireland division of Friends of the Earth (FoE EWNI) is an example of such an NGO. Greenpeace, however, is often taking a position comparable to that of the ETC Group, i.e. categorically rejecting genetic technologies.

The position of the ETC Group is very interesting because it represents a very outspoken position, diametrically opposed to those in favour of industrial biotechnology. Many other perspectives on synthetic biology and green chemistry can be expected to be somewhere in between those strongly in favour of industrial biotechnology and those vehemently opposing it. Focussing on these two positions as indicative for societal concerns brings out the most well-articulated concerns, assumptions and beliefs.

### 2.1 *The Case of Synthetic Artemisinin Production*

In 2005, supported by funds from the Bill & Melinda Gates Foundation, the US-based company Amyris achieved a 'breakthrough'[2] by developing artemisinic acid, a precursor to artemisinin, the main ingredient for antimalaria drugs. Malaria

[1] www.etcgroup.org.

[2] http://investors.amyris.com/news-releases/news-release-details/amyris-scientists-describe-breakthrough-development-anti.

treatments are currently mainly based on the Chinese sweet wormwood plant, from which artemisinin is extracted. However, this process is costly and time-consuming, according to Amyris. The semi-synthetic artemisinin (SSA) provides a viable and cost-effective alternative, as the company states. The platform for producing SSA is yeast whose metabolic pathways have been engineered. Amyris partnered with the pharmaceutical company Sanofi to produce 'cost-effective malaria treatments', which became available in 2013.

At the time, SSA was hailed as the first triumph for synthetic biology. It showed it was possible to provide a viable alternative to naturally occurring substances, with great promise for combatting a persisting global health issue: malaria. However, not everybody was convinced of the merits of this innovation. The most vocal of the opponents of SSA is the Canadian-based ETC Group. Their main objection is that the production of SSA will undermine the agricultural production of wormwood, thereby undermining the livelihoods of farmers growing the wormwood. Sanofi can keep their prices low because of the Gates Foundation support and thereby undercut the competition. Moreover, this technology allows for the concentration of economic power in the hands of one company at the expense of many small producers. Their criticism is supported by the Dutch Royal Tropical Institute who state in a 2006 report:

> The advantage of synthetic artemisinin is the combination of its predictability and, eventually, cheap production. Pharmaceutical companies will be able to enhance their control over the production process and will not have to depend on numerous supply chain actors, such as thousands of individual producers and local extractors. Long transportation distances across multiple borders will be replaced by on-the-spot production and manufacturing. However, there are also disadvantages: pharmaceutical companies will accumulate control and power over the production process; artemisia producers will lose a source of income; and local production, extraction and (possibly) manufacturing of ACT (Artemisinin based Combination Therapy, LA) in regions where malaria is prevalent will shift to the main production sites of Western pharmaceutical companies. (Heemskerk et al. [4], p. 51)

Proponents of SSA such as Amyris and Sanofi state that SSA is not intended to replace agricultural production of artemisinin, but as a supplement to reduce volatility of supply and prices (ibid). The worries as expressed by ETC Group and the Dutch Royal Tropical Institute are therefore unnecessary.

As it turned out, the supply of artemisinin has indeed been volatile over the years; however, the availability of SSA does not seem to have had a big impact on that, although it does appear to have helped stabilise the prices [5]. Sanofi has in any case not increased its production of SSA, because prices of naturally derived artemisinin are too low for them to compete with and because the demand has plateaued due to better diagnostics (ibid).

This case shows that where industrial biotechnology offers an alternative to agriculturally produced substances, questions arise about who benefits from the new production method and who is in disadvantage. It also shows that the consequences as well as the uptake of an innovation can be unpredictable. We will come back to this point later when we discuss strategies to deal with societal concerns in Sect. 4.

## 2.2 *The Case of Vanillin Production*

In 2011 the Swiss-based company Evolva[3] developed a synthetic version of vanillin, produced by yeast whose metabolic pathways had been reprogrammed, comparable to the Amyris platform. Vanillin is the most prominent ingredient of vanilla. Most of the vanillin on the market is produced through petrochemical or chemical processes. This synthetic vanillin is much cheaper than the natural vanilla derived from the vanilla orchid, which makes up less than 1% of the vanillin used today [6].

Evolva believes that the vanillin they produce through the yeast platform is natural and more sustainable and offers a higher quality than the other artificial vanillin. It is more sustainable because it does not rely on fossil resources or paper pulp, such as other artificial vanillin. The quality is better because it comes closer chemically to natural vanilla. It is natural because it is produced through fermentation, which under EU and US law is considered a traditional or natural food production process (ibid).

However, again the ETC Group, along with other environmental organisations such as Friends of the Earth USA,[4] opposed this innovation and the associated claims [7]. They do not dispute the quality of this product, but they state that this form of vanillin is not sustainable or natural because it has been produced with the use of highly engineered organisms. Furthermore, they fear this vanillin will undermine the livelihoods of vanilla farmers, who do produce in a sustainable manner, with respect for their direct natural environment. These two environmental organisations claim that industrial biotechnological processes are far from respecting local ecology since they could turn any crop into a feedstock for their processing facilities, while the 'natural' value chain needs to be attuned to the local ecology and support it in order to ensure ongoing production. Hence, according to the environmental organisations, this traditional value chain is better for the conservation of fragile rain forests.

## 2.3 *The Case of Algae-Based Oil Production*

When Ecover,[5] a Belgian company producing sustainable cleaning products, announced a change to one of the ingredients in its basic cleaning formula, it suddenly found itself under attack from a coalition of environmental organisations whose members used to be among Ecover's most loyal customers, with the international ETC Group prominent among them. The new ingredient which invoked all these criticisms was vegetable oil produced from genetically engineered algae [8], a procedure developed by the US-based company Solazyme.[6]

---

[3]www.evolva.com.

[4]https://foe.org.

[5]https://www.ecover.com/nl/.

[6]http://solazymeindustrials.com/.

As far as Ecover was concerned, this ingredient did not fundamentally differ from anything it had used before. In their detergents Ecover had used enzymes produced by genetically modified bacteria for years, as most companies in this area do and had hardly received any criticism for it. In the eyes of its critics, however, the oil produced by engineered algae does represent something fundamentally different. To these critics, the engineered algae symbolise a socio-technological system that is inherently unsustainable because it reinforces existing economic inequalities. The controversy led Ecover to stop using the algae-produced oil and to reflect on its strategy as a company seeking to be a front-runner in the field of sustainable innovations as Ecover describes itself [9].

In an open letter in *The Ecologist*, Jim Thomas, ETC Group's spokesperson on this matter, condemned Ecover for using what he considered to be synthetic biology which he refers to as 'extreme genetic engineering'. In his letter that was signed by 17 other NGOs, Thomas voiced concerns about the safety risks associated with this technology as well as possible socio-economic effects such the displacement of income for small farmers that depend on coconut oil [8]. Coconut oil could also provide a sustainable alternative to palm oil, one that is far less disruptive, in the eyes of Thomas and the wider environmental coalition.

This response took Ecover and Solazyme completely by surprise, as Tom Domen, long-term innovation manager at Ecover, described in an interview [9]. Both companies considered the algae technology to be in line with the technologies that were already widely used, such as enzymes derived from genetically engineered bacteria, the so-called white or industrial biotechnology. Ecover has been using such enzymes in their detergents for a long time. There is hardly any opposition against this white biotechnology because they are kept in containment in industrial plants, which minimises the risk of escape and contamination of the environment. To Ecover and Solazyme, the algae-based oil production was just another variation on an existing theme, one that is supposed to solve a pressing sustainability issue, namely, the problematic production of palm oil, the demand for which continues to grow.

## 3 Five Social and Ethical Issues in Industrial Biotechnology

These cases indicate some common societal themes that are brought forwards by innovations in industrial biotechnology. As stated above these are:

(a) Sustainability
(b) Naturalness
(c) Risk management
(d) Innovation trajectories
(e) Economic justice

These themes are interesting and relevant to the advancement of industrial biotechnology because different perspectives exist on how we should deal with them. These perspectives can be related to different values, assumptions and beliefs among different actors as described below.

### 3.1 Sustainability

Whether an application of industrial biotechnology can be considered sustainable can be determined by measuring the quantifiable impact of that application, such as contribution to $CO_2$ reduction, use of resources such as water, release of toxic substances and so forth [10, 11]. However, as emerges from the cases discussed above, other nonquantifiable factors may also play a role in assessing sustainability. These factors may be difficult to quantify because they are very complex and surrounded by many uncertainties and/or difference in perspectives, for instance, the effect on the social well-being of people working in a bio-based value chain such as for biofuels. Although efforts are ongoing to measure this, well-being is notoriously hard to define because of the many different ways to operationalise it [12]. Nevertheless, literature on assessment of social sustainability for bioeconomy is becoming more frequent, and several social issues like employment, working condition, labour right, gender equality, social development and food security have been discussed [13–15].

Some other aspects of sustainability may be hard to quantify because their assessment is very ideological and relates to preferred societal structures and possible future effects of a specific technology. This may be the case when a biotechnological application competes with other applications that may be considered more natural or the production thereof requires more attention for preserving local ecosystems. This is, for instance, one of the issues that emerges in the vanillin case, where the production of natural vanilla is tied up closely with local ecosystems and local traditional farming practices. Proponents of the synbio vanillin will say that it does not compete with traditional vanilla but with chemical production of artificial vanillin, compared to which it can be considered more sustainable because it needs less land. However, many actors oppose genetic engineering in any form because they deem it inherently unsustainable, while more ecologically sound technologies, in their view, are available. They perceive genetic engineering as enabling economic monopolies and as introducing unnecessary risks [16].

The advancement of industrial biotechnology and many of its products depends on a reliable and widely supported system for sustainability assessment. Sustainability is not something that can be directly witnessed. Potential consumers and society at large need reliable indicators to show them which products are sustainable and which aren't. Such indicators will only be considered reliable when people feel that they reflect the concerns they have about sustainability and hence serve the wider public good and not a particular interest. When designing indicators for the sustainability assessment of products from industrial biotechnology, it is important to acknowledge the different views on sustainability [17, 18]. These different views will be further explicated in the following subthemes which all relate to the overarching theme of sustainability.

## 3.2 Naturalness

The issue of sustainability is closely linked to that of naturalness. The concept of naturalness raises two distinct issues. Firstly, there is the question of what natural is exactly and which products can be termed natural and which cannot. Secondly, there is the issue of what nature is and how we should relate to nature.

The first question pops up around the labelling of products derived from industrial biotechnology. If a fragrance or flavour is made through microbial or enzymatic processes, it can be labelled natural under both US and European laws. However, if this production relies on engineered organisms under industrial conditions, it might not fit with commonly held conceptions of natural which mostly refer to something existing in or produced by nature, as it is described in a dictionary [19]. To the vanillin produced with the help of engineered organisms, natural might suggest that it is actually derived from the vanilla orchid [20]. Environmental organisations are calling to make the distinction between these two types of products clearer [21].

The second issue, to which the first is connected, is a more fundamental one that originates from differences in values, beliefs and convictions. For some people (such as environmental activists), nature is something fragile that should be treated with the utmost care, so as not to upset vulnerable ecological balances. For other people (such as some working in industrial biotechnology or in high-risk investment), nature is essentially a resource, which has many wonderful things to offer and can provide viable solutions to pressing problems [22].

Different worldviews in which nature plays a pivotal part are depicted below. These are adapted from cultural theory (cf. [16, 23]). The little ball represents nature and the position in which nature is supposed to be, i.e. it sits in a precarious balance (vulnerable nature) or it is safely contained and can take a hit (nature as resource). These two positions are usually the most outspoken ones in discussions on genetic technologies (Fig. 1).

The perspective described as 'controllable nature' can be considered a midway position between vulnerable nature and nature as a resource, i.e. nature is considered to be relatively robust, but risks to the 'natural balance' are also acknowledged. Within this perspective, (global) regulation is considered essential to avoid any disastrous effects. Policymakers are typically put in this perspective, but also some environmental NGOs fit here, like FoE EWNI.

The capricious nature perspective is mostly associated with groups that have little political power and little influence on the economic conditions of their lives. Nature is considered to be something capricious and uncontrollable, comparable to many other aspects of live. Smallholders in developing countries may, for instance, be placed in this quadrant.

From the perspective that nature is essentially something vulnerable, approaching living organisms as entities that can be controlled and designed is a seriously flawed misconception about how we should deal with living organisms. In this perspective, living organisms are inherently unpredictable and should be treated as such. Pursuing a strategy in which living organisms are treated as predictable and controllable is

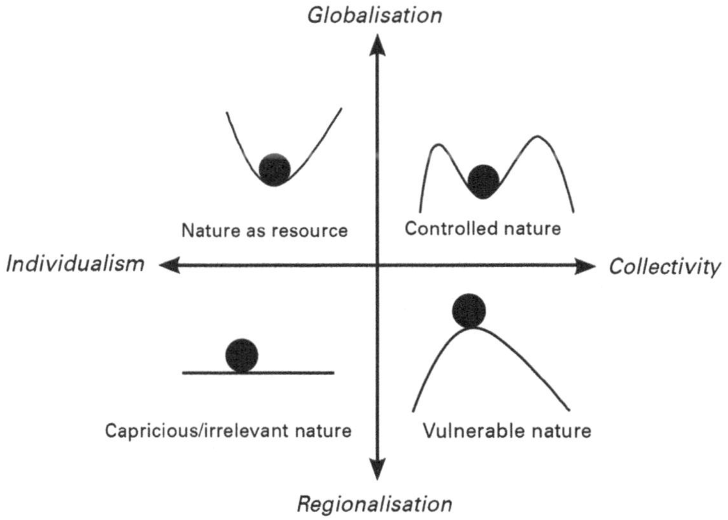


**Fig. 1** Worldviews

therefore basically a mistake which diverts money and other resources away for more viable and truly sustainable solutions such as community based, organic farming practices [7, 9, 24].

From the perspective that nature is a resource that is essentially robust, industrial biotechnology is an excellent opportunity to find optimal solutions to pressing problems such as climate change and scarce resources. What's more, we cannot afford to forego the many possibilities that nature has to offer via industrial biotechnology if we want to achieve a sustainable world. This position is diametrically opposed to that of the perspective of nature as ultimately fragile [9].

The above-described perspectives are two extreme positions. There are many other positions possible that may be somewhere in between these two, or totally different altogether, such as a position that is more or less indifferent about nature and does not see any way humans could control nature. However, the positions explicated above very clearly represent a source of conflict about the acceptability of biotechnology and are therefore relevant. In considering how to assess industrial biotechnology, it can help to keep these two positions in mind to assure a complete picture of possible societal and ethical issues.

## 3.3 *Risk Management*

These differing perspectives on naturalness give rise to differing perspectives on the management of risk. Risk management refers to the identification (e.g. is there a risk?), estimation (e.g. how big is the risk?) and evaluation of the risk (e.g. how acceptable is the risk?) [25]. For many people, the risks of industrial biotechnology are small, because the organisms are kept in closed vats. The chances of any of them escaping are low, and even if they would escape, they probably would not survive outside of the industrial conditions.

However, for other actors such as Friends of the Earth, USA, and the ETC Group, the risks of applications such as synbio are both undeniable and considerable:

> While other types of pollution such as synthetic chemicals break down over time and do not breed, synthetic biological creations are designed to self-replicate and once released into the environment they would be impossible to stop and could wipe out entire species. This type of pollution, known as genetic pollution, can be devastating since it cannot be cleaned up. (FoE [26], p. 9)

These environmental NGOs think that those working in industrial biotechnology should be very careful with engineered organisms because basically they cannot be controlled. Since they are living organisms, they might escape from any setting, adapt to their environment and disturb it.

Experts in the field agree that there is no way to contain synthetic or genetically engineered organisms – particularly algae. According to Lissa Morganthaler-Jones, CEO and co-founder of Liverfuels Inc., a small number of genetically engineered algae have already leaked from the lab into the environment. 'They have been carried out on skin, on hair and all sorts of other ways, like being blown on a breeze out the air conditioning system', she said. (FoE [26], p. 9)

To other actors, the increased sophistication in biotechnology, the advancement to synthetic biology, indicates higher levels of safety. Because it is possible to control living organisms to an ever-increasing extent, the risks become smaller and smaller. Safety switches can be built in, for instance [9]. Safety switches are traits that ensure that living organisms can only survive within a specific, controlled environment. They might need a specific substance, for instance, that is only available in a laboratory. If the organisms leave the controlled environment, they will not be able to survive [27].

From the above we can conclude that there is a difference in the way the risks of industrial biotechnology may be identified and estimated. Moreover, there is also a difference in the way the risks are assessed. To critics of industrial biotechnology, even if the risks are small, which they do not think is the case anyhow, they would not evaluate these risks as acceptable because they think there are other, better alternatives to achieve a sustainable society or to create high-quality products, mostly through sustainable eco-agricultural practices, such as natural vanilla in the vanillin case or sustainably sourced coconut oil in the Ecover case.

## *3.4 Innovation Trajectories*

Another important societal and ethical question is what kind of innovation trajectory industrial biotechnology is supporting. In the Ecover case, for instance, some of the critics thought that the engineered algae would lead to a technological lock-in. They saw the algae as a platform that enables an optimally rationalised management of biomass, which serves the interest of large industry, but which does not stimulate sustainable agriculture. Additionally, these critics state that there are more viable opportunities to achieve a sustainable agriculture which are being foregone by concentrating on engineered algae. In this case, sustainably sourced coconut oil was proposed as an alternative.

However, other actors have a different view on the future innovative potential of the algae. The sustainability manager of Ecover, for instance, considered the use of engineered algae as a first stepping stone towards even more sustainable applications, such as algae that do not have a large need for sugar as a feedstock, but instead rely on sunlight and water. Additionally the plants for these algae can be distributed in a decentralised manner thereby avoiding concentration of knowledge and power [9]. This actor envisioned a totally different innovation trajectory connected to the algae.

Such clashes in expectations about innovation trajectories also emerge in the vanillin case and the artemisinin case, where the developers of the product claim that their application will not compete with the plant-derived alternative, but instead will serve to stabilise the market (artemisinin) and/or compete with the less sustainable petrochemical version. Opponents instead think that the innovation will compete with their natural counterparts and will only serve the interests of specific companies.

A similar conflict in expectations can be seen around biofuels. When the first-generation biofuels were being put to use, many supporters of this technology claimed that these first-generation biofuels would provide a stepping stone for more sustainable second- and third-generation biofuels. However, many critics feared that the first-generation biofuels would turn out to be a technological lock-in, implying that once all the investments in first-generation biofuels had been made, there was no incentive for the industry to switch to more sustainable next-generation biofuels [2].

The first-generation biofuels did indeed prove to be somewhat of a lock-in [28]. It turned out to be difficult for the EU to lower the cap for first-generation biofuels in the directive for sustainable transport due to resistance from the first-generation biofuels industries [29]. However, while more sustainable second- and third-generation biofuels haven't become available in large quantities yet, first-generation biofuels have become more sophisticated, and there is evidence suggesting that the first generation is as sustainable as the second generation, thereby questioning the need to look to second- and third-generation biofuels as sustainable solutions [30].

The articulation of a future innovation trajectory can serve as a legitimisation for a specific application. Even if the direct benefit of a specific application is not immediately clear, it can still seem desirable because of the future innovations the application enables [31]. The environmental life cycle assessment (LCA) of the engineered algae in the Ecover case, for instance, did not show a huge improvement in comparison to alternatives such as palm oil [16]. However, because Ecover perceived the algae as a contribution to a potentially more sustainable technology, they embraced the algae nonetheless. The same thing can be said to apply to first-generation biofuels, which received governmental support, at least partly, because of their potential contribution to more sophisticated second- and third-generation biofuels [2].

Because these expectations about the future of innovation trajectories can have a considerable impact on the shaping of technology and society alike, it is important that they are open for input by a wide range of actors to ensure a democratic development of technology [32]. For example, some relevant questions are: What purpose do we expect an innovation to serve? Do we pursue an innovation simply because it is technologically possible? What are possible alternatives to the innovation trajectory? Wide-ranging input to such questions is not only desirable for democratic purposes but also for instrumental ones. If the envisioned innovation trajectory has wide societal support, the technology can be expected to disseminate more successfully and to reach societally desirable goals such as sustainability more easily.

## 3.5 Economic Justice

Another important issue for industrial biotechnology is that of economic justice, or put differently, who benefits from this technology? Many critics state that applications of industrial biotechnology lead to a concentration of knowledge and power in the hands of a few companies. This has been said for biofuels [24, 26, 33, 34] as well as for more speciality chemicals ([4, 24]; ETC Group 2016):

> What is being sold as a benign and beneficial switch from black carbon to green carbon is in fact a red hot resource grab (from South to North) to capture a new source of wealth. If the grab succeeds, then plundering the biomass of the South to cheaply run the industrial economies of the North will be an act of 21st century imperialism that deepens injustice and worsens poverty and hunger. Moreover, pillaging fragile ecosystems for their carbon and sugar stocks is a murderous move on an already overstressed planet. (ETC Group [24])

Each of the cases described above features a prominent concern for the fate of small-scale farmers producing the natural substance for which a synthetic alternative is produced. Such concerns also extend to farmers producing biomass for biofuels, or farmers who might be forced to abandon their land in favour of large biofuel producers.

In contrast with this concern, many authors point out that if the production of biomass for bio-based products is managed under the right conditions, biofuels can have beneficial, sustainable effects for both small-scale farmers and society as a whole [35]. Such conditions encompass good governance [36], an appropriate division of responsibilities [33], investments in agricultural innovations and stable price regime [37] and the inclusion of a wide range of stakeholders in the design of a sustainable bioeconomy (ibid; [33, 38, 39]).

Overall the value of economic justice is widely shared. The main challenge here is how to bring it about effectively in relation to industrial biotechnology. Some actors claim that industrial biotechnology should be largely abandoned because they see it as inherently tied up with unjust economic systems. Others see many promising opportunities offered by industrial biotechnology [17]. The call for wide-ranging participation is one we will pick up on here as a means of dealing with the many societal and ethical issues surrounding industrial biotechnology.

# 4 Responsible Research and Innovation for Industrial Biotechnology

As we have shown, there are many societally intricate issues related to industrial biotechnology. To support societally intricate technological trajectories, the approach of Responsible Research and Innovation (RRI) has been proposed. RRI has been defined as:

> A transparent, interactive process by which societal actors and innovators become mutually responsive to each other with a view to the ethical acceptability, sustainability and societal desirability of the innovation process and its marketable products (in order to allow a proper embedding of scientific and technological advances in our society). (Von Schomberg [40], p. 9)

The approach of Responsible Research and Innovation (RRI) can be understood as an attempt to align new technologies with societal concerns and needs. RRI is intended to help designers and manufacturers of new technologies identify and accommodate public concerns when developing a new technology by engaging with a wide range of relevant actors [41]. As such RRI can be considered as a tool to answer questions about the direction in which we would want to use available scientific and technical knowledge (cf. [32]).

RRI has four dimensions, namely, anticipation, reflexivity, inclusion and responsiveness [42]. We will consider in turn how these might play out a role in industrial biotechnology.

## *4.1 Anticipation*

> Anticipation involves systematic thinking aimed at increasing resilience, while revealing new opportunities for innovation and the shaping of agendas for socially-robust risk research. (Stilgoe et al. [42])

Aside from promising visions on sustainability, new industrial biotechnologies also bring about new uncertainties. Questions arise about the exact environmental impacts of new technologies and about how to control new potential risks. And questions arise about what sustainability amounts to and what innovation trajectories should be instigated to achieve sustainability. Well-informed anticipation that includes a variety of perspectives may potentially substantially reduce uncertainty and thereby prevent the occurrence of unwanted consequences. RRI asks innovators to consider the possible effects of their innovation in an integral and structured way.

However, anticipation is always necessarily limited. Unexpected consequences may emerge even after the most thorough and inclusive anticipation efforts. As van de Poel [43] shows, this is due to the complex epistemological uncertainties that surrounds a new technology, i.e. technologies can have impacts on so many different levels that it is often impossible to predict all of them correctly, also because many of these effects depend on how individuals will eventually apply a new technology and what (moral) meanings they associate with a technology.

Still it is important to try and foresee the possible effects an innovation might have and to try and accommodate this, as much as possible. When an alternative for a naturally occurring substance is produced by means of industrial biotechnology, it makes sense to think about the effects the new product may have on the existing value chain. Moreover, it is also possible to anticipate uncertainty and try to design an innovation in such a way that it can be adapted if unforeseen, unwanted effects occur (ibid). With the production of biofuels, for instance, it can be advisable, for example, to set up production systems that are flexible in terms of feedstock [44]. If new insights emerge that indicate that a particular feedstock might not be so sustainable after all, a flexible production system allows to switch to another, more sustainable feedstock.

## *4.2 Reflexivity*

RRI asks for reflexivity in actors implying that they critically assess their own preconceptions. When different actors have different perceptions about the desirability of a technology, it can be possible for actors to construct a compromise or even a shared perspective on that technology. However, such a shared perspective or compromise requires a willingness to reconsider one's own position and the associated preconceptions [16, 45]. If environmental activists are unwilling to reconsider their preconception that *all* genetic engineering is unsustainable, it will be hard to achieve a common vision with other people who are convinced that genetic engineering is essential to achieve a sustainable society. However, possibly these groups can find a common ground, such as agreeing that instead of genetic engineering, directed evolution as a means to achieve sustainable applications, is considered acceptable by all groups involved. To achieve such a common ground, each of the actors involved will need to carefully consider their values, beliefs and convictions to see what kind of innovations are compatible with it or what kind of compromises might be acceptable.

## 4.3 Inclusion

Inclusion implies that a wide range of people and perspectives is taken into account when developing a new technology [42]. In the bio-based economy, inclusion already takes place in many shapes, such as for the certification and monitoring of sustainable biomass [46]. There is an ongoing international debate on the norms for sustainable biomass as well as on the quality of the labels monitoring these norms [17]. A wide variety of actors takes part in these debates, such as NGOs, companies and governmental organisations. Their views are incorporated in schemes for the certification and monitoring of the sustainability. The criteria by which sustainability is determined are hence done in an inclusive, participatory manner.

According to RRI such inclusion should go beyond the formulation of criteria and extend to the actual design choices that are being made when developing a new technology. There are of course many practical and institutional barriers to actually implement a wide variety of perspectives into the R&D phase of innovation, such as confidentiality issues, a balanced division of responsibility [47] and stakeholders who might not be willing to get involved [48] or are unable to get involved due to geographical or time-management reasons [49].

Even if these barriers occur, there are still actions that companies and other innovators can undertake to assure a representation of a wide variety of perspectives. One option is to learn from other related cases what are relevant concerns from stakeholders. The cases described above can, for instance, serve as guide for the kind of societal concerns that might affect comparable industrial biotechnology products [16]. Also, stakeholders that are at a given time unavailable might be represented by other parties that are available such as academic experts or NGOs [49].

Once a wide variety of perspectives has been identified, either indirectly as described above or directly through interviews or workshops, they can be used to inform the design choices made in an innovation trajectory. Such choices can, for instance, concern the choice for a particular feedstock, for a particular kind of technology or for centralised or decentralised production facilities [44].

## 4.4 Responsiveness

The last dimension of RRI is that of responsiveness and this might be seen as an overarching attitude for which the other three strategies are essential conditions. Responsiveness is the action that is taken after innovators have anticipated possible effects of their innovation, have been reflective and have included a wide variety of perspectives [32, 49].

Stilgoe et al. [42] define responsiveness as a willingness to change an innovation when it becomes clear that it conflicts on crucial issues with values of other stakeholders: 'Responsible innovation requires a capacity to change shape or direction in response to stakeholder and public values and changing circumstances'.

Especially in a setting that is continuously evolving and where the learning curve on social values is steep such as in the bio-based economy, a responsive attitude is crucial to achieve societally robust innovations. The public outcry over using food crops for fuel has, for instance, intensified the policy support for fuels from nonedible parts of crops and algae. This can be considered a responsive attitude towards societal concerns of biotechnological innovations.

It can be difficult to be responsive because innovations are sometimes locked into their own trajectory. It is possible to change the policy surrounding biofuels, but it may also be difficult to change the production platforms of the biofuels. Responsiveness is hence not always technologically possible without abandoning existing production facilities. Therefore responsiveness can be a costly affair if not managed properly. According to RRI philosophy, companies should always be aware of the need for responsiveness and try to incorporate it in their innovation strategies by ensuring some form of flexibility. For the biotech companies described in the beginning, such responsiveness seems available to different degrees because they can switch to different end products if needed without changing the core of their business, namely, the production platform (engineered micro-organisms).

Responsiveness does not necessarily always imply a change in course of those developing an innovation. Although they had the option to change to other end products, all three biotech companies that faced societal criticism (e.g. Amyris, Evolva and Solazyme) are actually still producing as they were before. They have considered the societal criticism, and they did not deem it necessary to change their innovations. The management of Solazyme started a new company focussing on health foods based on non-engineered algae, named TerraVia, thereby expanding their portfolio. They are also still producing oil from engineered algae. However, the end-user of the algae-based oil that Solazyme produced, Ecover, has stopped using the oil as ingredient for their detergents, even though they still deem it a desirable innovation. The innovation manager, Tom Domen, thought the company needed to reconsider its communication and engagement strategies before continuing with a controversial innovation such as the algae-based oil [9].

These different responses to societal criticism show that responsiveness does not necessarily imply the same course of actions for each company. Ultimately, the response to other stakeholders' values and concerns needs to be in line with the innovators' own values and concerns; otherwise it becomes a hollow public relations exercise in which the company or the innovators lose track of their own moral compass and motivation. Responsiveness does not imply a blind catering to societal concerns; it does, however, imply a reflection on one's own motives and values in light of such societal concerns. This might lead to an adaptation of the technology, but it should, at minimum, lead to a better articulation and explication of the reasons and values behind choices made in a particular innovation process.

## 5 Reflecting Social Issues in Sustainability Assessment for Industrial Biotechnology

Sustainability was already introduced in Sect. 3.1 as one of the five social ethical issues in relation to potential social, environmental and economic impacts that may derive from the application of novel biotechnologies. Such impacts should in principle be qualitatively or quantitatively assessed to determine whether a process, product or service should be realised or not. In other words, knowing the potential positive or negative consequences (in terms of social, environmental and economic impacts) of a novel biotechnology-related project – throughout the entire value chain – is a valuable instrument to motivate and support debates among the five social ethical issues (see Sect. 3) and allows a structured reflection on the four dimensions of RRI (see Sect. 4).

In the case of economic and environmental impacts assessment, multiple methodologies are already available and have extensively been described in literature [50]. However, in the case of social sustainability assessment, literature is scarce since methodologies are still under development, being social impact assessment (SIA) and social life cycle assessment (SLCA) the most commonly used approaches [13, 51]. The former considers on-site specific impacts, while the latter accounts for the entire life cycle. In the case of the SLCA methodology, a twofold classification of social impacts is considered, i.e. stakeholder categories and impact categories, and such social impacts are subdivided into social, socio-economic and geographical subcategories. These subcategories deal with 31 aspects of the entire value chain, such as working conditions and employment, health and safety (H&S) aspects (at different levels), access to resources (material and immaterial), contribution to economic and technology development and corporate responsibility, among others [52].

In the particular case of industrial biotechnology for biofuels and biochemicals production, some of the most critical social issues are related to food security, land use, water availability, energy security, rural and social development, employment, working conditions and health and safety impacts [50]. For instance, land expansion for industrial biotechnology applications had raised concerns in the last decade on food security and land competition for food production, especially in view of the increasing food demand of a constantly growing population. Although this connection has been in public scrutiny as the food-vs-fuel debate, there is evidence that the effect of biofuels production on food prices is limited as compared to the effects from the oil prices [35]. Another concern from biomass production and expansion for industrial biotechnology applications is that such projects may significantly affect water availability and quality for other basic uses like sanitation or food production. Although this concern highly depends on contextual features (e.g. geographical location, crop type, cultivation practice and agricultural practices, among others), it has also been demonstrated that water consumption per ton of bio-based feedstock can significantly be decreased due to technological improvements in water recovery and recycling by using closed-loop water cycles and municipal wastewaters.

However, it is also acknowledged that further water stress would raise from growing biomass production demands [50]. On the other hand, industrial biotechnology-related projects, like biofuels and biochemicals production, have shown positive social effects, at local and global scales, in terms of employment creation (e.g. over 3.5 million (direct and indirect) jobs globally in the bioenergy transportation sector by 2010 [53]), value generation (e.g. increase in the municipal GDP per capita, regional tax income and poverty reduction [54]), infrastructure investments and social services contributions.

## 6 Conclusion

Industrial biotechnology carries the promise of sustainable solutions based on natural resources. However, some issues invoke societal criticism, showing that different actors have different perspectives on salient issues. These issues include sustainability, naturalness, risk management, innovation trajectories and economic justice. To achieve societally robust innovations, innovators can learn from actors who have a different view on a specific application. The framework of Responsible Research and Innovations offers guidelines to organise such learning. These guidelines are based on the principles of anticipation, inclusion, reflexivity and responsiveness. The outcome of such a learning process might be that an innovation trajectory is adapted, or at minimum that the innovators are aware of possible objections to their innovation and can widen their understanding of their own motivation for pursuing that innovation. Currently efforts are on the way to integrate social and economic aspects into LCA, but these are still in their infancy.

## References

1. Gaskell G, Stares S, Allansdottir A et al (2006) Eurobarometer 64.3: Europeans and biotechnology in 2005: patterns and trends, a report to the European Commission's Directorate-General for Research, Brussels
2. Asveld L, Qv E, Stemerding D (eds) (2011) Getting to the core of the bio-economy: a perspective on the sustainable promise of biomass. Rathenau Institute, The Hague
3. Osseweijer P, Ammann K, Kinderlerer J (2010) Societal issues in industrial biotechnology. In: Soetaert W, Vandamme EJ (eds) Industrial biotechnology: sustainable growth and economic success, 1st edn. Wiley, Oxford, pp 457–481
4. Heemskerk W, Schallig H, De Steenhuijsen Piters B (2006) The World of Artemisia in 44 questions. The Royal Tropical Institute, The Hague
5. Peplow M (2016) Synthetic biology's first malaria drug meets market resistance. Nature 530:389–390
6. Bomgartner M (2016) The problem with vanilla. Chem Eng News. 94(36):38–42. http://cen.acs.org/articles/94/i36/problem-vanilla.html. Accessed 19 Sept 2017

7. ETC group (2013) Vanilla and synthetic biology. A case-study. In: ETC group SynBio case studies. http://www.ETCgroup.org/files/Vanilla_SynBio_case_study_Oct2013.pdf. Accessed 22 Sept 2017
8. Thomas J (2014) Ecover pioneers 'synthetic biology' in consumer products. In: Ecologist. https://theecologist.org/2014/jun/16/ecover-pioneers-synthetic-biology-consumer-products. Accessed 10 Nov 2014
9. Asveld L, Stemerding D (2016) Algae oil in trial. Conflicting views on technology and nature. Rathenau Instituut, The Hague
10. Fröhling M, Hiete M (2018) Sustainability and life cycle assessments in industrial biotechnology: a review of current approaches and future needs. In: Fröhling M, Hiete M (eds) Sustainability and life cycle assessment in industrial biotechnology, advances in biochemical engineering/biotechnology. Springer, Berlin
11. Venkatesh A, Posen ID, HL ML, Chu PL, Griffin WM, Saville BA (2019) Environmental aspects of biotechnology. In: Fröhling M, Hiete M (eds) Sustainability and life cycle assessment in industrial biotechnology, advances in biochemical engineering/biotechnology. Springer, Berlin
12. van der Deijl W (2017) Are measures of well-being philosophically adequate? Philos Soc Sci 47 (3):209–234
13. Macombe C (2019) Social Life Cycle Assessment (sLCA) for Industrial Biotechnology. In: Fröhling M, Hiete M (eds) Sustainability and life cycle assessment in industrial biotechnology, advances in biochemical engineering/biotechnology. Springer, Berlin
14. Martínez SH, van Eijck J, da Cunha MP et al (2013) Analysis of socio-economic impacts of sustainable sugarcane–ethanol production by means of inter-regional input–output analysis: demonstrated for Northeast Brazil. Renew Sustain Energy Rev 28:290–316
15. Wicke B, Smeets E, Tabeau A et al (2009) Macroeconomic impacts of bioenergy production on surplus agricultural land – a case study of Argentina. Renew Sustain Energy Rev 13 (9):2463–2473
16. Asveld L, Stemerding D (2018) Social learning in the bioeconomy: the case of Ecover. In: Van de Poel I, Asveld L, Mehos D (eds) Experimentation beyond the laboratory: new perspectives on technology in society. Routledge, London, pp 103–124
17. Asveld L, Ganzevles J, Osseweijer P, Landeweerd L (2014) Naturally sustainable: the social aspects of the transition to a sustainable bio-economy. Delft University of Technology, Delft
18. Parada MP, Osseweijer P, Posada JA (2017) Sustainable biorefineries, an analysis of practices for incorporating sustainability in biorefinery design. Ind Crop Prod 106:105–123
19. Merriam-Webster (n.d.) Natural. In the dictionary of Merriam-Webster. Retrieved from https://www.merriam-webster.com/dictionary/natural. Accessed 22 Sept 2017
20. Gallage NJ, Møller BL (2015) Vanillin-bioconversion and bioengineering of the most popular plant flavor and its de novo biosynthesis in the vanilla orchid. Mol Plant 8(1):40–57
21. Friends of the Earth (2013) Extreme genetic engineering in your ice cream? In: Friends of the earth finance & economic systems issue. https://foe.org/2013-08-extreme-genetic-engineering-in-your-ice-cream. Accessed 22 Sept 2017
22. Douglas M, Wildavsky A (1983) Risk and culture: an essay on the selection of technological and environmental dangers. University of California Press, Berkeley
23. Thompson JB (1990) Ideology and modern culture: critical theory in the era of mass communication. Polity Press, Cambridge
24. ETC group (2011) The new biomasters. Synthetic biology and the next assault on biodiversity and livelihoods. ETC Group Communiqué # 104, Ontario
25. Chen C, Reniers G (2018) Risk assessment of processes and products in industrial biotechnology. In: Fröhling M, Hiete M (eds) Sustainability and life cycle assessment in industrial biotechnology, advances in biochemical engineering/biotechnology. Springer, Berlin
26. Friends of the Earth (2010) Synthetic solutions to the climate crisis: the dangers of synthetic biology for biofuels production. Friends of the Earth, Amsterdam
27. Robaey ZH (2018) Dealing with risks of biotechnology: understanding the potential of Safe-by-Design. Report commissioned by the Dutch Ministry of I&W, The Hague

28. Asveld L (2016) The need for governance by experimentation: the case of biofuels. Sci Eng Ethics 22:815–830. https://doi.org/10.1007/s11948-015-9729-y
29. Sharman A, Holmes J (2010) Evidence-based policy or policy-based evidence gathering? Biofuels, the EU and the 10% target. Environ Policy Gov 20(5):309–321
30. Dammer L, Carus M, Piotrowski S et al (2017) Sustainable first and second generation bioethanol for Europe. Nova-Institute, Hürth
31. Borup M, Brown N, Konrad K, Van Lente H (2006) The sociology of expectations in science and technology. Tech Anal Strat Manag 18(3–4):285–298
32. Owen R, Macnaghten P, Stilgoe J (2012) Responsible research and innovation: from science in society to science for society, with society. Sci Public Policy 39:751–760
33. Balkema A, Pols A (2015) Biofuels: sustainable innovation or gold rush? Identifying responsibilities for biofuel innovations. In: Koops BJ, Oosterlaken I, Romijn H, Swierstra T, Van den Hoven J (eds) Responsible innovation 2: concepts, approaches, and applications. Springer, Dordrecht, pp 283–303
34. Borras Jr SM, McMichael P, Scoones I et al (2010) The politics of biofuels, land and agrarian change: editors' introduction. J Peasant Stud 37(4):575–592
35. Souza GM, Victoria RL, Joly CA et al (2015) Bioenergy & sustainability: bridging the gaps. SCOPE72, Paris, Sao Paulo
36. Diaz-Chavez R, Morese MM, Colangeli M et al (2015) Social considerations. In: Souza GM, Victoria RL, Joly CA, Verdade M (eds) Scientific committee on problems of the environment (SCOPE) report: bioenergy & sustainability: bridging the gaps. SCOPE72, Paris, Sao Paulo, pp 529–552
37. Kline KL, Msangi S, Dale VH et al (2017) Reconciling food security and bioenergy: priorities for action. GCB Bioenergy 9:557–576. https://doi.org/10.1111/gcbb.12366
38. Levidow L, Birch K, Papaioannou T (2012) EU agri-innovation policy: two contending visions of the bio-economy. Crit Policy Stud 6(1):40–65
39. Schuurbiers D, Osseweijer P, Kinderlerer J (2007) Future issues in industrial biotechnology. Biotechnol J 2007(2):1112–1120
40. Von Schomberg R (2011) Towards responsible research and innovation in the information and communication technologies and security technologies fields. Publication Office of the European Union, Luxembourg
41. Van den Hoven JVD, Jacob K (2013) Options for strengthening responsible research and innovation. European Commission, Brussels
42. Stilgoe J, Owen R, Macnaghten P (2013) Developing a framework for responsible innovation. Res Policy 42(9):1568–1580
43. Van de Poel I (2017) Society as a laboratory to experiment with new technologies. In: Bowman D, Stokes E, Rip A (eds) Embedding new technologies into society: a regulatory, ethical and societal perspective, 1st edn. Pan Stanford Publishing, Singapore, pp 61–68
44. Parada MP, Asveld L, Osseweijer P et al (2018) Setting the design space of biorefineries through sustainability values, a practical approach. Biofuels Bioprod Biorefin 12(1):29–44
45. Roling N (2002) Beyond the aggregation of individual preferences: moving from multiple to distributed cognition in resource dilemmas. In: Leeuwis C, Assen RP (eds) Wheelbarrows full of frogs: social learning in rural resource management. Koninklijke Van Gorcum, pp 25–47
46. Van Dam J, Junginger M, Faaij A et al (2008) Overview of recent developments in sustainable biomass certification. Biomass Bioenergy 32(8):749–780
47. Blok V, Lemmens P (2015) The emerging concept of responsible innovation. Three reasons why it is questionable and calls for a radical transformation of the concept of innovation. In: Koops BJ, Oosterlaken I, Romijn H et al (eds) Responsible innovation 2: concepts, approaches, and applications. Springer, Dordrecht, pp 19–35
48. Noorman M, Swierstra T, Zandbergen D (2017) Questioning the normative core of RI: the challenges posed to stakeholder engagement in a corporate setting. In: Asveld L, Van dam-Mieras MEC, Swierstra T et al (eds) Responsible innovation 3: a European agenda. Springer, New York, pp 231–239

49. Sonck M, Asveld L, Landeweerd L et al (2017) Creative tensions: mutual responsiveness adapted to private sector research and development. Life Sci Soc Policy 13(1):14
50. Posada JA, Osseweijer P (2016) Socio-economic and environmental considerations for sustainable supply and fractionation of lignocellulosic biomass in a biorefinery context. In: Mussatto S (ed) Biomass fractionation technologies for a lignocellulosic feedstock based biorefinery. 1st edn. Elsevier, Amsterdam pp 611–631
51. Lehmann A, Russi D, Bala A et al (2011) Integration of social aspects in decision support, based on life cycle thinking. Sustainability 3:562–577
52. UNEP/SETAC (2013) (Pre-publication version) the methodological sheets for subcategories in social life cycle assessment (S-LCA). UNEP/SETAC Life Cycle Initiative, Gothenburg
53. IRENA (2011) Renewable energy jobs: status, prospects and policies. IRENA Working Paper, Abu Dhabi
54. Satolo LF, Bacchi M (2013) Impacts of the recent expansion of the sugarcane sector on municipal per capita income in São Paulo state. ISRN Economics 2013:1–14

Adv Biochem Eng Biotechnol (2020) 173: 143–204
DOI: 10.1007/10_2020_122

Published online: 30 March 2020

# Sustainability and Life Cycle Assessment in Industrial Biotechnology: A Review of Current Approaches and Future Needs

**Magnus Fröhling and Michael Hiete**

## Contents



M. Fröhling (✉)
Technical University of Munich (TUM), TUM Campus Straubing for Biotechnology and Sustainability, Straubing, Germany
e-mail: magnus.froehling@tum.de

M. Hiete
Ulm University, Department of Business Chemistry, Ulm, Germany
e-mail: michael.hiete@uni-ulm.de

**Abstract** The development and implementation of industrial biotechnology (IB) is associated with high expectations for reductions of environmental impacts and risks, particularly in terms of climate change and fossil resource depletion, positive socioeconomic effects, hopes for new competitive products and processes, and development in rural areas. However, not all products and processes are really advantageous with regard to sustainability criteria, and not all are economically successful and accepted by stakeholders. Sustainability and life cycle assessment can play an important role to assess IB products and processes, often accompanying development processes from the early stages onwards. Such assessments can identify key factors regarding sustainability criteria, enable a determination of both product and process performance, or aid in prospectively estimating such performance and its consequences. Thus, development processes, investment decisions, policymaking, and the communication with stakeholders can be supported. This contribution reviews the field of sustainability and life cycle assessment in IB. We explore relevant literature from a methodical and application perspective and categorise suitable methodologies, methods, and tools. We characterise IB from an assessment perspective and indicate challenges, discuss approaches to address these, and identify possible fields of future research. Thus, students, researchers, and practitioners in the field of IB will obtain an up-to-date overview, references to relevant fields of literature, and guidance for own studies in this important and fast-emerging topic.

**Graphical Abstract**

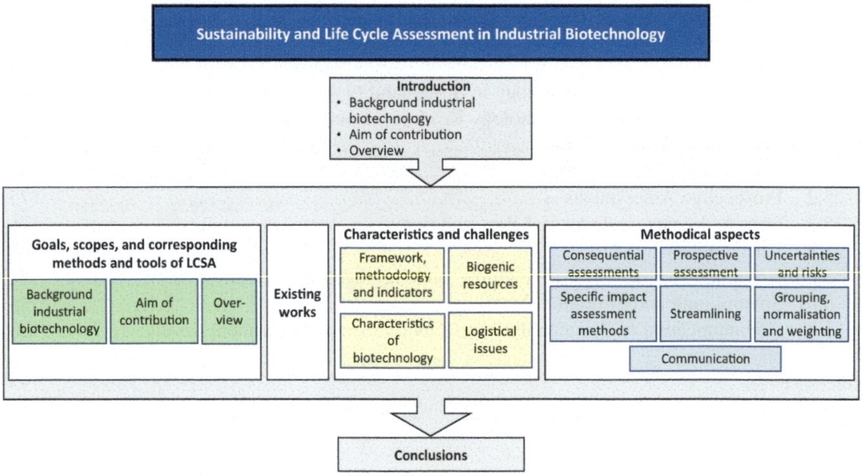


**Keywords** Bioeconomy, Biogenic resource, Consequential assessment, Prospective assessment, Renewable raw material, Streamlined assessment

# 1 Introduction

In industrial biotechnology (IB), typically biogenic resources are transformed using microorganisms or substances thereof to provide food and fodder, substances, materials, as well as energy. IB is promoted as a promising key technology in numerous fields of business, politics, research, and society as it contains the promise of revolutionising or at least complementing chemical production with nature as a model. As IB utilises biogenic resources, it aims at avoiding the depletion of non-renewable fossil resources and the associated greenhouse gas (GHG) emissions. The mild reaction conditions of biological processes are believed to reduce environmental impacts and risks, and the supply of biogenic resources is expected to have positive effects for socioeconomic development, particularly in rural areas. The promise of IB goes beyond merely offering an alternative to chemical production: IB enables the production of substances which would have otherwise had to be extracted from natural resources in costly and complex processes, if at all; moreover, it also enables the development of new substances and materials with targeted characteristics. However, the advantages in terms of sustainability must not be taken for granted (e.g. [1, 2], both this volume, for environmental, health, and land use impacts; [3, 4] for social impacts; and [5] for risks), and case-by-case assessments are necessary to identify whether a chemical or biotechnological process is more sustainable to produce a certain substance or provide its function [6]. Finally, as IB is, with few exceptions, still in its infancy, a favourable policy framework is required to overcome market entry and other barriers [7, 8] as well as reduce risks for investors.

Against this backdrop, sustainability and life cycle assessment may be used to ascertain the environmental, social, and economic performance of IB products, processes, and concepts and highlight the needs and starting points for further improvements and optimisation. Moreover, a comparison with "conventional" products, processes, and concepts may serve as a basis for decision-making for investors, funding agencies, and political decision-makers. The results of the assessment may indicate, for example, promising investment options or a need for policy support or funding to avoid or overcome weaknesses of the emerging technology of IB via research and development. Finally, the assessment results may also provide a profound basis for the discussion and communication of such concepts with stakeholders to achieve an early involvement in the development and broader acceptance in stakeholder management processes.

Numerous approaches and tools have been developed for sustainability and life cycle assessment, which largely differ in terms of pursued goals, input information required, and level of detail of the results (e.g. [9–11]). The aim of this contribution is to review the field of life cycle and sustainability assessment for IB by (1) sketching – with a clear focus on industrial biotechnology – a conceptual framework for sustainability and life cycle assessment, (2) providing an overview of current approaches and tools as well as their applications as guidance for both academia and practitioners, and (3) identifying current challenges and future research needs.

In order to achieve this aim, the contribution is structured in the following manner: following this introduction, Sect. 2 explores the field of sustainability and life cycle assessment for IB. Further, it categorises the methodologies, methods, and tools as well as presents the general aspects guiding the appropriate choice for the intended purpose. Section 4 provides a brief review of existing works in this field. On that basis, Sect. 3 discusses the characteristics and challenges of sustainability and life cycle assessment in IB. This includes the identification of suitable approaches for a given purpose, characteristics of the use of biogenic resources, logistical and biotechnological issues, as well as challenges induced when applying these methods at lower technology readiness levels. Section 5 deals with methodical aspects in the assessment of IB, encompassing consequential and prospective assessments which play an important role. The section discusses challenges, mainly for environmental impact assessment methods, arising inter alia from risks and uncertainties as well as the optional steps of grouping, normalisation, weighting, and aggregation. Finally, it describes the efforts to streamline and simplify such assessments and approaches to present and communicate the results. Section 6 closes the contribution with conclusions and an outlook for future research and applications.

# 2 Goals, Scopes and Corresponding Methodologies, Methods, and Tools

This section aims to provide a general overview of different methodologies, methods, and tools for sustainability and life cycle assessment and guidance for their selection to deal with a problem at hand. Before that, a few important terms are briefly defined. This is necessary, as terms used in this field are not always self-explanatory and are occasionally used interchangeably; in certain cases, authors use the same term in different contexts, thereby causing difficulties.

## *2.1 Terminology*

In the following account, sustainability is understood in the sense of the Brundtland Commission, which has defined sustainable development as "a development that meets the needs of the present without compromising the ability of future generations to meet their own needs" [12]; moreover, it is also included in the interpretation of the three pillars conception of sustainability with interconnected social, economic, and environmental dimensions [13]. The definition of the Brundtland Commission emphasises not only intragenerational (e.g. industrial and developing countries) and intergenerational (e.g. current vs. future generations) justice but also the dynamic process inherent in the quest for sustainability as well as the existence of environmental limitations, also termed biophysical limits or planetary boundaries (cf. [14]). Therefore, sustainable development is a complex, value-laden concept with numerous perceptions [15], as is – consequently – sustainability assessment.

The aim of sustainability assessment is to "measure or predict the sustainability of a [...] system" [16]. Sustainability assessment may support decision- and policymaking towards sustainable development [15] and is, thus, solution-oriented. Sala et al. [15, 17] have developed a conceptual framework for sustainability science and assessment at three levels:

- Ontological level – that is, the concepts underlying sustainability, such as the concept of capitals (natural, human, social, manufactured, financial), precautionary principle, biophysical limits, dematerialisation, etc., and the relationships at different hierarchical levels, for example, from the individual to all of humanity.
- Epistemological level – that is, the development and foundations of scientific methods and knowledge. For sustainability science, this implies a shift to post-normal science, which implies that science is participative, uncertain, and exploratory as well as value-laden and related to cultural perspectives.
- The methodological level – that calls for "multi-inter-trans-disciplinary" methods to find solutions for sustainability problems.

Although this framework sounds rather theoretical, it is helpful to understand the particularities and problems inherent in sustainability assessment. For example, although life cycle assessment (LCA) is regarded as a science-based methodology, conducting an LCA in practice may require to make value-based decisions – for example, when weighting different impact indicators or when selecting a life cycle impact assessment (LCIA) method, such as IMPACT (World+), ReCiPe, CML, TRACI, or Eco-Indicator 99. Moreover, the sustainability principles underlying all sustainability targets are the translation of values [15]. The conceptual framework also highlights the relevance of uncertainties, the management of which represents a major challenge in all sustainability assessment methodologies and methods.

Following [17], the following hierarchy of terms is used for sustainability assessment (top-down) in this chapter:

- *Framework*: The framework represents the highest level. It can be conceptual and/or structural, thereby enabling the integration of different methods or results thereof. For example, life cycle sustainability assessment (LCSA) is often interpreted as a framework integrating (environmental) LCA, social life cycle assessment (sLCA), and life cycle costing (LCC).
- *Methodology*: LCA, sLCA, and LCC are examples of methodologies. They provide general descriptions of how an assessment can be conducted. In contrast with methods such as ReCiPe or CML, they are not sufficiently concrete to allow an actual assessment.

- *Method*: A method is a "set of models, tools, and indicators" [17].
  - The *model* enables the calculation of the impacts of an intervention. Different types of models may be applied. For quantitative results, mathematical models are required.
  - A *tool* implies the implementation of the model as software, application, or database.
  - *Indicators* are used to describe a concrete impact. Different indicators may exist for the same impact; these indicators differ in terms of their suitability to describe the impact, their complexity, and difficulty in determining the indicator value, etc. Indicators may be quantitative, semi-quantitative, or qualitative.

Often the terms "analysis" and "assessment" are used interchangeably. However, the terms "life cycle sustainability assessment" (LCSA (assessment)) and "life cycle sustainability analysis" (LCSA (analysis)) may have a different meaning (cf. [18] with a detailed comparison). LCSA (assessment) aims at product assessments and is often translated as LCSA = LCA + sLCA + LCC [19], whereas LCSA (analysis) is broader in scope and includes coverage of, for example, dynamic effects, feedback, etc. as well as in scale, which may range from micro (product) to meso (e.g. sector) and macro (economy-wide) [18]. Therefore, LCSA (analysis) is less standardised and uses integrated models.

Further misunderstandings may arise from the terms "integrated assessment" and "impact assessment". In sustainability assessment, "integrated" ideally implies a transdisciplinary, intersectoral, and participatory approach [15]; however, integrated may also be used to emphasise that several sectors are considered, or that more than one dimension of sustainability is considered, or that the problem is approached in an inter- or multidisciplinary manner. Ostensible "integrated assessment models" (IAM) are complex and tailor-made models to assess policy options – for example, the Regional Air Pollution Information and Simulation model (RAINS model) and its successor, the Greenhouse Gas and Air Pollution Interactions and Synergies model (GAINS model), are used to assess and refine clean air and climate mitigation policies in the UNECE [20]. On the one hand, "impact assessment" is called the third phase in life cycle assessment (as life cycle impact assessment, LCIA). In the LCIA phase, inputs such as biomass and outputs such as $CO_2$ to air produced over the life cycle of a product and determined in the life cycle inventory phase are assigned to impact categories such as biotic resource depletion, acidification, or human toxicity. On the other hand, the term impact assessment is used in combinations of "x impact assessment" where "x" may stand for "environmental" (EIA), "social" (SIA), "health" (HIA), "gender", "human rights", etc. (cf. International Association for Impact Assessment (IAIA, https://www.iaia.org)). These must not be confused with LCIA, as these provide a procedure to determine the impacts in a certain field of interest arising from a *concrete* action, such as a planned infrastructure project.

Further, strategic environmental assessment (SEA) is similar to EIA, but refers to the assessment of policy, plans, and programs proposed by public organisations – for example, a land use plan. These types of impact assessments are often legally required – for example, as laid down in the EIA Directive of the European Union – if a certain project qualifies for it due to its nature or size. Numerous countries have equivalent laws. However, terms may vary among countries – for example, in the UK, "sustainability appraisal" is used, and in the USA, "environmental assessment" (EA) is used.

Overall, this short overview shows that a variety of terms are used, mostly for a concrete context, but occasionally also interchangeably. For a novice, a problem arises from the fact that terms sound rather similar and from the large number of methods and tools whose goals do not, in all cases, become evident from their names. The following section provides an overview and a categorisation of the tools.

## 2.2 Categorisation of Methodologies, Methods, and Tools

Several attempts have been made to provide overviews and categorise methodologies, methods, and tools for sustainability assessment as well as for life cycle assessment ([9, 10, 15, 17, 18, 21], e.g. [22–26]). Though categorisation is also a value as such, its main goal is to guide potential users (e.g. [27]). In the EU project *Sustainability Assessment Methods and Tools to Support Decision-Making in the Process Industries (SAMT)* (2015–2016), 51 methodologies/methods and 38 tools were reviewed with respect to 6 items (essence, scope, relevance, requirements, outcome, and information) and assigned to 1 of 7 method categories and 6 tool categories, respectively [10]. The large numbers might initially function as a deterrent, but LCA methodologies and methods account for 19 of the 38 tools and 9 of the 51 methods [10], which considerably reduces the complexity. The resulting review of methodologies and methods is highly useful to gain a quick but in-depth overview of the various methodologies, methods, and tools that exist. The focus on process industries makes it even more interesting for assessments in IB.

The main question that arises is which methodology/method and tool is "best suited" for a question at hand.

This comprises various aspects such as (cf. also [22]):

- The appropriateness for the *purpose* which can be, for example, decision support when choosing between product systems or processes, identification of sustainability deficits over the life cycle of a product, accounting, reporting, monitoring, or certifying that certain requirements are met, better understanding – for example, the energy and material flows in a system as a base for future optimization, ex-post assessments utilising existing data or prospective assessment for technologies in development for which major assumptions must be made;
- The *concept of sustainability* taken as base for the evaluation; in the concept of strong sustainability, for example, the ecological dimension takes a particular role as the presence of planetary boundaries is accepted.

- The *aspects and impacts considered*: certain methodologies, methods, and tools focus on single dimensions of sustainability or even selected aspects within one dimension – for example, the carbon footprint on greenhouse effect or material input per service unit (MIPS) on material use; certain methods and tools focus on material or energy flows, others on impacts on one of the areas of protection that is the endpoints: (1) impacts on human health, (2) impacts on natural environment, and (3) scarcity of natural resources.
- The *object of analysis* which can be, for example, a product, process, technology, or system but also an organisation, a project, a plan, or policy as well as a substance or material but also a geographical area such as a region or a nation; most objects can be either generic, for example, a reusable glass bottle for one litre of water in Western Europe, or specific, for example, a biogas plant at a specific location as in environmental impact assessment.
- The *complexity and procedural quality of the assessment* affecting input data needs and sophistication of methodologies, methods, and tools, for example, to what extent consequences in the system arising from the item to be assessed, are accounted for, but also whether, for example, the dynamics of the system, uncertainties in the data, parameters, and assessment method are addressed or whether impacts are localised.
- *Target group* which can be academia, industry, customers, legal authorities, the broader public, etc.

Proposals for categorisation consequently strongly vary between authors and [23] alone make four proposals to support tool selection:

1. *Perspective of the assessment*
   Two broad perspectives may be distinguished: a *biophysical perspective* which aims at quantifying resource consumption and a *monetary perspective* which aims at measuring, for example, human well-being, economic efficiency, growth, and welfare. Indicator-based approaches are positioned as those adopting a more comprehensive view.

2. *Coverage of five main desirable features*

   (a) Consideration of economic, environmental, and social issues and including interrelations (triple bottom line, integrated assessment)
   (b) Inclusion of future impacts (prospective, predictive, ex ante assessment)
   (c) Accounting for inherent uncertainties (risk aversion and precautionary assessment)
   (d) Allowance for inter- and intragenerational *equity*, not just efficiency (distributional assessment)
   (e) Inclusion of stakeholder expectations (participatory assessment)

3. *Criterion for acceptability*
   From a sustainability perspective, the methodologies/methods can be grouped according to three types of acceptance criteria [28] in the following categories:

(a) Baseline-led or environmental impact assessment-led assessment, which implies that the assessed object must not be less sustainable than a predefined baseline; a worse performance in one criterion may be overcompensated for by a better one in another criterion; in case of compensation between the three dimensions of sustainability, this corresponds to the concept of weak sustainability.
(b) Objectives-led assessment, which implies that trade-offs like those in the baseline-led assessment must be avoided; instead, a sustainable solution leads to a win-win situation.
(c) "Assessment for sustainability" approaches, which assess against a definition of sustainability and corresponding criteria and not as in the previous one against a relative target, such as the current situation, other options, etc.

4. *Stakeholder values*
   Criteria for considering stakeholder requirements and views in the tool selection.

This brief overview indicates the complexity and difficulties when selecting an appropriate methodology/method and tool. An overview of selected clusters for methods and tools, respectively, identified in the EU-SAMT project [10] is presented in Table 1 (note that these authors do not distinguish methodologies and methods). This overview highlights that over the years, a variety of methodologies/methods and tools were developed.

Central to the determination of suitable methods or methodical tool sets is the rationale against which the assessment is conducted. For a holistic sustainability assessment, most studies break this assessment down to the three dimensions of sustainability [29]. However, in many cases, the analyses focus on single dimensions or even single criteria, like the contribution to global warming via the emission of GHG in the carbon footprint method.

As mentioned above, the life cycle approach and (environmental) LCA hold a key position within sustainability assessment. LCA focusses on environmental aspects and impacts of products throughout their life cycle [30]. The definition of product is very broad, encompassing both goods and services – that is, material and immaterial objects. LCA follows a system analysis approach. The product's life cycle is considered as a product system fulfilling a function. This product system consists of unit processes – that is, the system is disaggregated in the smallest considered units. The definition of these processes is subject to the definition of the LCA study. The guiding principle for the identification of processes is their function. It can be as detailed as a single chemical reaction or unit operation in the process chain or an aggregation of larger units, like a process chain, and go up to policy options or entire countries (cf. [31]). Even though LCA adopts a product-centric approach, processes can be assessed by this functional definition. Consequently, one can find studies considering products of biotechnological processes (cf., e.g. [32, 33]), biotechnological processes (cf., e.g. [34, 35]), or feedstock provision (cf., e.g. [36, 37]). In order to attain comparable results, the LCA methodology has been standardised within the ISO14000 standard family (cf. esp. DIN EN ISO 14040:2006 [30] and DIN EN ISO 14044:2018-05 [38]).

However, with regard to the social and economic dimension of sustainability – for example, in techno-economic assessments or social assessments – there are widely used and accepted methodologies, methods, and tools, even though the procedures in this field are not as standardised as for the environmental dimension. These methodologies, methods, and tools often employ similar approaches such as a holistic life cycle approach and are based (particularly in the techno-economic field) on the same or very similar material and energy flows. For the social dimension, the activities underlying the assessment are often different, for example, the working hours. Holistic approaches to sustainability assessment aim at integrating all three dimensions. Thus, under the umbrella of LCSA, a coherent network compatible with ISO standards is proposed, combining (environmental) LCA with life cycle costing (cf. [39]) and social LCA (cf. [4, 40]).

Table 1 further shows that numerous methodologies/methods and tools focus on single dimensions or even single aspects. For the novice in sustainability assessment, Table 1 might provide a wrong impression. Even though the list of examples is well filled for hybrid and integrated methodologies, methods, and tools, in industrial practice, hybrid methodologies, methods, and tools play only a small role due to their high complexity. In fact, even in academia, only a few specialised research institutes utilise these methodologies, methods, and tools. Moreover, for integrated methodologies, methods, and tools, it must be emphasised that their degree of popularity and prevalence in practice is largely varying – whereas, for example, eco-efficiency analysis (EEA) has been even standardised in ISO 14045:2012, others are still in development or barely known.

## 2.3 *Framework for Sustainability and Life Cycle Assessment of Industrial Biotechnology*

The described broad background and large variety of causes and goals for sustainability and life cycle assessments mentioned above impose several requirements and challenges. Not surprisingly, the goals – such as identification of sustainability problems of existing products or decision aid in the choice of materials or provision of information for a stakeholder dialogue – determine not only the scope of the study but also which methodologies and methods are best suited, the data requirements, etc. (cf., e.g. [41]).

### 2.3.1 Sustainability Assessment: A Multicriteria Problem

In LCA, three areas of protection (natural environment, human health, and resources) are considered, each with several impact categories. In the ILCD[1]

[1]International Reference Life Cycle Data System.

**Table 1** Selected clusters of methodologies/methods and tools for sustainability assessment (in accordance with [10])

| Type | Cluster | Description | Examples |
|---|---|---|---|
| Methodologies/ methods | Life cycle methodologies/ methods | • Based on life cycle thinking (cradle to grave of product systems) to avoid transfer of impacts to other life cycle stages<br>• Pure assessment<br>• Largely applied and partially standardised, particularly environmental LCA<br>• Increased use of footprint indicators focusing on single aspects or impacts | Environmental<br>• Life cycle assessment (LCA)<br>• Ecological footprint (EF)<br>• Carbon footprint (CF)<br>• Water footprint (WF)<br>• Material input per service unit (MIPS)<br>• Cumulative energy demand (CED)<br>• Exergetic LCA (E-LCA)<br>• Emergy analysis (EA)<br>Social<br>• Social LCA (sLCA)<br>Economic<br>• Life cycle costing (LCC)<br>• Environmental life cycle costing (E-LCC)<br>Sustainability<br>• Life cycle sustainability assessment (LCSA) |
| | Hybrid methodologies/methods | • Here, hybrid implies that sub-methods are "computed together", not in parallel or one after the other as in integrated methods<br>• Originally, "hybrid" LCA referred to the combination of input-output models and LCA to be able to model consequences of a decision in other areas (consequential LCA). This is the case in EE-IOA/LCA, LCA/PEM but not in LCO and LCAA | • Hybrid environmentally extended input-output analysis and LCA (EE-IOA/LCA)<br>• HybridLCA + partial equilibrium model (LCA/PEM)<br>• (HybridEE-IOA/LCA + general equilibrium model (EE-IOA/LCA/GEM)<br>• Life cycle optimisation (LCO)<br>• Life cycle activity analysis (LCAA) |
| | Integrated methodologies/ methods | • Assessment data in several sustainability dimensions are integrated to support decision-making. This requires, for example, normalising, weighting, aggregation, or benchmarking | • Eco-efficiency analysis (EEA),<br>• Socio-eco-efficiency analysis (SEEbalance®)<br>• Product sustainability assessment (PROSA)<br>• Integrated method for calculation/measurement of resource efficiency (ESSENZ)<br>• Sustainable value (SustV)<br>• Ecodesign (EcoD)<br>• Product-oriented |

(continued)

**Table 1** (continued)

| Type | Cluster | Description | Examples |
|---|---|---|---|
| | | | environmental management system (POEMS)<br>• Life cycle index (LInX)<br>• Composite Sustainable Development Index (CSDI)<br>• COMbining environmental Performance indicators, LIfe cycle approach and Multicriteria to assess the overall ENvironmental impact (COMPLIMENT) |
| | Methodologies/ methods specific to the chemical industry | • Particular relevance for chemical industry, for example, due to focus on chemical processes or chemicals | • Atom economy<br>• Environmental factor (E-factor)<br>• Process mass intensity (PMI)<br>• Reaction mass efficiency (RME)<br>• IChemE Sustainable Development Progress Metrics (IChemE)<br>Although eco-efficiency analysis (EEA) and socio-eco-efficiency analysis (SEEbalance®) have their origins in the chemical industry, they are rather general methods and thus appear under integrated methods |
| Tools | Full LCA tools | • Tools that enable the calculation of environmental life cycle inventory (LCI) and subsequently perform a life cycle impact assessment (LCIA)<br>• A few tools extend to social LCA and hybrid approaches | GaBi, openLCA, SimaPro, TEAM, Umberto, SULCA, CMLCA, RangeLCA, EASETECH, Brightway2 |
| | Simplified/ streamlined LCA tools | • Tools enabling a quick analysis of adaptations, for example, in design<br>• Non-experts as a user group<br>• Simplification from using precalculated, generic data and from restriction to selected impacts | CCaLC, ECO-it, Ecolizer, InstantLCA, EcoFly, EcoTransIT, Bilan-Produit, GaBi Envision, SimaPro Compact, Umberto NXT $CO_2$ |

(continued)

**Table 1** (continued)

| Type | Cluster | Description | Examples |
|---|---|---|---|
| | Integrated tools | • Inputs and outputs in the three dimensions of sustainability<br>• Concrete tools are available<br>• Mainly for corporate sustainability management | • Global Environmental Management Initiative Sustainable Development Planner (GEMI SD Planner)<br>• Sustainability Balanced Scorecard (SBSC)<br>• Sustainable Innovation through Design tools (SInnDesign)<br>• Future-Fit Business Benchmark (FutureFitBB)<br>• Sustainability, Competency, Opportunity, and Reporting, Evaluation (S-CORE) |
| | Tools specific to the chemical industry | • Tools with particular relevance for the chemical industry, that is, with a focus on chemical processes or chemicals | • Environmental Assessment Tool for Organic Syntheses (EATOS)<br>• EcoScale for choosing organic preparations<br>• EcoSolvent tool for assessing waste-solvent treatment<br>• iSustain Green Chemistry Index for scoring chemical products and processes (iSustain)<br>• Persistent Bioaccumulative and Toxic profiler (PBT profiler)<br>• Waste Reduction Algorithm (WAR)<br>• Chemical Screening Tool For Exposures and Environmental Releases (ChemSTEER)<br>• Green Chemical Alternatives Purchasing Wizard<br>• Solvent selection guides<br>• Process Mass Intensity Calculator (PMIC)<br>• Green Chemistry Assistant (GCA) |

**Table 2** Impact categories in LCA [42]

| Baseline impact categories | Additional impact categories |
|---|---|
| Depletion of abiotic resources | Impacts of land use<br>• Loss of life support function<br>• Loss of biodiversity |
| Impacts of land use<br>• Land competition | Ecotoxicity<br>• Freshwater sediment ecotoxicity<br>• Marine sediment ecotoxicity |
| Climate change | Impacts of ionising radiation |
| Stratospheric ozone depletion | Odour<br>• Malodorous air<br>• Malodorous water |
| Human toxicity | Noise |
| Ecotoxicity<br>• Freshwater aquatic ecotoxicity<br>• Marine aquatic ecotoxicity<br>• Terrestrial ecotoxicity | Waste heat |
| Photo-oxidant formation | Casualties<br>• Lethal<br>• Non-lethal |
| Acidification | Depletion of biotic resources |
| Eutrophication | Desiccation |

handbook on LCIA methodologies and methods [42],[2] baseline and additional impact categories are defined (cf. Table 2) which highlight the multicriteria aspect of assessment. In fact, impacts may be aggregated up to the level of the areas of protection based on scientific findings, for example, *disability-adjusted life years lost* (DALYs) for damage to human health, the *potentially disappeared fraction of species* (PDF) for damage to ecosystems, or *surplus energy* for damage to natural resource availability as in the ReCiPe2016 LCIA method [43]. Beyond this, however, aggregation must be based on values. The most prominent approach is the one employed in ReCiPe2016, which relies on the cultural theory of risk and defines three main perspectives [43]:

- An *individualistic perspective* based on short-term interests, technological optimism, negligence of disputed impact categories
- A *hierarchical perspective* as a balanced position
- An *egalitarian perspective* adopting a rather precautionary position with a long-term perspective

The three perspectives are then operationalised for several of the impact categories (but not all due to a lack of a sound scientific base to do so) [43].

[2]The ILCD handbook series was developed by the European Commission to harmonise LCA studies funded by the Commission and provides an excellent detailed overview of LCA.

In social LCA, for each stakeholder group (workers, local community, value chain actors, consumers, and society), approximately four to nine subcategories are distinguished (cf. also [44] with a recent review of indicators used) (Table 3).

Finally, in life cycle costing, all aspects are combined into one common currency in order to make aggregation simpler. However, questions regarding how to aggregate costs arising for different stakeholder groups, in different currencies, or at different times as well as internal and external costs, can be complicated.

To conclude, in LCA and social LCA, as well as when aggregating the three dimensions in LCSA, aggregation approaches are needed if the amount of information can be reduced to a few numbers only. Thus, two challenges arise. The first challenge is the choice of criteria or indicators which are relevant for the study – for example, choosing from the list of additional impact categories in Table 2. The second challenge is to consider the multiple criteria adequately. On the top level, when aggregating environmental, social, and economic impacts, the sustainability

**Table 3** Categories in social LCA [45]

| Stakeholder category | Subcategory |
|---|---|
| Worker | Freedom of association and collective bargaining<br>Child labour<br>Fair salary<br>Working hours<br>Forced labour<br>Equal opportunities/discrimination<br>Health and safety<br>Social benefits/social security |
| Consumer | Health and safety<br>Feedback mechanism<br>Consumer privacy<br>Transparency<br>End-of-life responsibility |
| Local community | Access to material resources<br>Access to immaterial resources<br>Delocalisation and migration<br>Cultural heritage<br>Safe and healthy living conditions<br>Respect for indigenous rights<br>Community engagement<br>Local employment<br>Secure living conditions |
| Society | Public commitments to sustainability issues<br>Contribution to economic development<br>Prevention and mitigation of armed conflicts<br>Technology development<br>Corruption |
| Value chain actors (excluding consumers) | Fair competition<br>Promotion social responsibility<br>Supplier relationships<br>Respect of intellectual property rights |

principle followed (strong or a weak sustainability, cf. Sect. 2.2) is decisive, as it determines whether compensation among the three dimensions is appropriate. In LCA and sLCA, the aggregation is usually achieved through normalisation, weighting, and aggregation. Note that in product assessment, normalisation of indicator values, for example, by country population – that is, per capita – does not permit a subsequent aggregation, as no one would propose to add the indicator values on a country-scale either, which would be an equivalent case. Thus, normalisation must be used only to "inform about the relative magnitude of each of the characterised scores for the different impact categories" [46]. In case of strong sustainability, approaches based on the concept of satisficing could be a solution (cf. [47]).

Overall, this leads to common problems associated with multicriteria decision analysis (MCDA). Making use of MCDA methodologies and methods can be helpful to achieve weightings that correspond better to the preferences of decision-makers.

### 2.3.2 Decision Context of the Assessment

Sustainability assessments are often used as a base for decision-making. If the analysed object – for example, a new biotechnological process – is expected to result in far-reaching qualitative or quantitative changes in industry, it is consequent to also account for these changes in the assessment, which are also termed structural changes. Based on the questions of whether or not major changes in the background system are expected and whether or not other systems are affected and included in the modelling approach, Björn et al. [48] distinguish the following three decision contexts:

- *Micro-level decision support*: No structural changes are expected from the decision made for which the assessment serves as a base. For example, the assessment of an enzyme is not expected to lead to (relevant) replacement of existing processes and equipment. A typical application case is the assessment for the improvement of existing products and processes. An object of analysis at the micro level could be, for example, a product, a process, or a technology as long as only minor structural changes are expected.
- *Meso- and macro-level decision support*: Structural changes in one or several systems are expected as a result of the decision based on the assessment's outcomes due to interactions with the system under assessment. A typical case would be decision-making to support policy, for example, a regulation for a higher minimum blending of biodiesel in diesel oil with impacts on, for example, necessary infrastructure for biodiesel production as well as agriculture and possibly the petroleum industry. The meso level considers local communities up to subnational regions [49], whereas the macro level adopts a national or international view ([40], cf. [48, 50]).
- *Accounting purposes*: No decision is made based on the assessment, for example, because the decision has already been made. Then, the assessment serves to

assess and document what has already happened or may still happen. Thus, the assessment is purely descriptive. However, changes in other systems due to interaction with the one under assessment still need to be accounted for. An example of this could be a scientific analysis of the impacts of Transport Biofuels Directive 2003/30/EC.

The purpose of the study and the level of the analysis determine the necessary degree of detail in consideration, whether specific or average data can be used, and which methodologies and methods can or must be employed. The ranges in the degree of detail extend from aggregate black box models – particularly for macro-level studies – to in-depth process modelling of processes or even single chemical reactions, for example, in the process of material development. The challenge here is to find a level of detail which is appropriate for the problem, in view of the complexity of the assessment and the necessary data quality and the financial and temporal efforts of the assessment.

### 2.3.3 Attributional and Consequential Assessment

Taking up the issues described in the previous subsection, we describe in the following the differences between the methodological approaches of attributional and consequential assessments.

In an attributional assessment, environmental aspects and impacts of the object under study – mostly a product throughout its life cycle – will be identified, quantified, and assessed. This implies that the entire product system is considered with input and output flows, and the impacts of these are determined as attributes of this system [48]. This can be used to compare different alternatives and scenarios for that system. However, the focus is on that system and assuming that no major changes occur outside the system under study.

In the application of LCA and further scientific discussion, the notion grew that such an attributional and static comparison of alternatives is often not sufficient for decision-making, since a decision will lead to changes. This was the starting point for the second line of development, the so-called "consequential" assessments. These focus on the evaluation of the impacts ("consequences") of decisions or actions. For such analyses, the scope of considerations must be broadened to also encompass aspects and related systems which are not covered in attributional approaches (cf. [40]). This requires a different modelling of the data as well as the impact assessment methods. Consequential assessments consider marginal data in order to derive the particular effects. For example, a consequential LCA would, when assessing a new biotechnological process using a biogenic feedstock, account for the increased demand for the biogenic and the decreased demand for the fossil feedstock. Thus, the system boundaries and models must be expanded in order to assess the effects on connected systems. This often includes going methodically beyond the usually employed methodologies and methods and include, for example, economic models, land use models, or others in an LCA study. Both approaches can lead to different results. Searchinger et al. [51] demonstrated that, while an attributional assessment of biofuel production shows the oft-cited GHG emission reductions of 20%, a consequential assessment taking into account land use change leads to an emission increase of almost 50%. This result must not be interpreted in the sense that results of LCA are arbitrary or that one of the two types for assessment is superior to the other. Instead, both concepts must rather be considered complementary [41], and the choice of either one has to be made depending on the aim and scope of the study.

### 2.3.4 Retrospective and Prospective Assessment

Somehow related, but not congruent, with the distinction between attributional and consequential studies is the differentiation between retrospective and prospective assessments. Prospective assessments have a broader temporal horizon. They anticipate the possibility of larger changes in the object of study and its surroundings over time. Therefore, they examine these changes in two areas: the first one is the so-called foreground system on which the product or technology under study may impose larger changes over longer time horizons. The second one is the so-called background system, where the changes in the boundary conditions – that is, the societal, economic, political, technological, and ecological environment – are systematically analysed [52]. For the foreground system, often the scale-up from an early development and low volume level – for example, lab or bench scale – to the intended industrial operation at full scale raises methodical challenges for the assessment (cf. Sect. 4.5). The changes in the background system – for example, the national power supply providing the electricity for the operation of the plant – often add further major sources of uncertainty to the analyses. Thus far, standardised methods in this field are missing (cf. [53]).

In addition, there exists a certain fuzziness regarding the terminology. Terms such as *ex ante* or anticipatory assessments are also used in relation to prospective assessments. Despite a few approaches for individual definitions (cf., e.g. [54]), we do not distinguish these further in this chapter and refer only to the term "prospective assessment".

On the other hand, retrospective studies deal with current or previous results or situations for products or technologies [53] – that is, situations in which the foreground and background may be considered as time-invariant. Both retrospective and prospective approaches can be applied in attributional and consequential assessments [41]. The application of the approaches and the use of corresponding methods are to be defined in the initial stages of the assessment – that is, in the goal and scope definition (cf. [31]). Most LCA studies are retrospective, which is particularly justified for short-lived products or snapshot assessments. However, even products with longer lifetime or processes in development are assessed retrospectively, mainly because of the lack of data on future developments and the uncertainties inherent in these. Whereas it is common in economic assessments to anticipate future improvements from learning, utilising the so-called learning curve concept, this is largely neglected for environmental impacts [55]. This is expected to result in a systematic underestimation of the environmental performance of new products, processes, and technologies that could not have benefitted as strongly as their established counterparts from learning.

The fulfilment of the purpose of the assessment also requires an adequate definition of the functions of the considered product systems and the unit in which this function is measured. Further, data needs and issues such as a critical review of the assessments by external experts need to be decided on [41]. A key role is taken by the functional unit and with it the reference flow, which is the amount of product required to fulfil the desired function. Determining an appropriate functional unit is of utmost importance when comparing different products. Although these aspects are critical for performing an assessment study, they are not further considered here, as their importance is similar in other IB assessment studies and the principal requirements in this field do not differ from common practice. Thus, we refer the interested reader to LCA textbooks such as [56–59] for further reading.

## 3 Current Use of Assessment Approaches

As shown above, a large variety of methodologies, methods, and tools is available for sustainability assessment or for single dimensions or parts thereof. Consequently, the question that arises is to what extent are these methodologies, methods, and tools used in an industrial biotechnology context. To answer this question, we conducted a (limited) literature review. We decided to use Scopus database (www.scopus.com) for this analysis, as the Scopus database is wider than, for example, Web of Science, which is more focused on high-quality academic journals, but less broad than, for example, Google Scholar, which contains more grey literature. It is obvious that searching all three databases – and others such as the those from major publishers – and synthesising the results would be best; however, since our aim is to provide an overview only, we decided to restrict ourselves to a search in Scopus while bearing in mind that results might change slightly when using other databases.

As a search strategy, we chose to identify relevant work without being overwhelmed by the number of documents found. For example, the Boolean search string ALL ("life cycle" OR "sustainability") AND ("analysis" OR "assessment" OR "management") AND ("biotechn*" OR "biochem*" OR "enyzm*") in the title, abstract, or keywords (Title-ABS-KEY) (no restriction on age) resulted in 6,300 documents in September 2019. As our main interest is in papers that focus on sustainability assessment or parts thereof, we narrowed down the results by requiring that the title must contain either "life cycle" or "sustainability" as well as "analysis" or "assessment" or "management", along with "biotechn*" or "biochem*" or "enzym*" in the title, abstract, or keywords, thereby resulting in 280 documents.[3] Eliminating irrelevant documents – for example, those focusing on biochemistry or

[3]The exact search string was TITLE ("Life Cycle" OR "Sustainability") AND TITLE ("Analysis" OR "Assessment" OR "Management") AND TITLE-ABS-KEY ("Biotechn*" OR "Biochem*" OR "Enzym*").

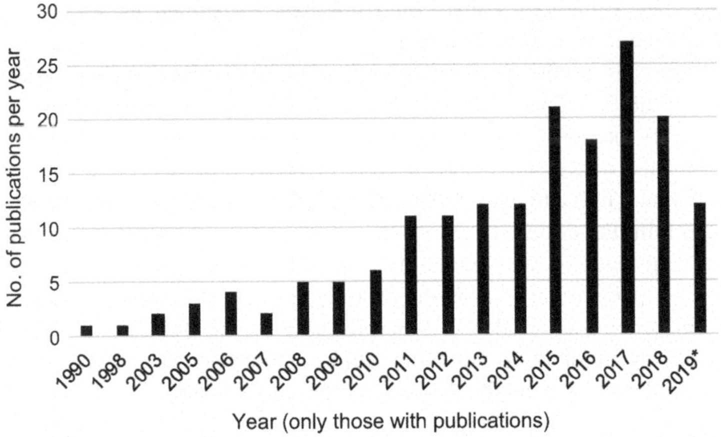


**Fig. 1** Number of identified publications per year (see text for details); ∗search was conducted in September 2019

waste management – resulted in a sample of 157 documents. From complementary searches, we found 6 and 4 relevant documents when replacing ("life cycle" OR "sustainability") by "social" and "energ∗"in the title, respectively, and another 10 when replacing ("life cycle" OR "sustainability") by "eco-efficiency" in Title-ABS-KEY, of which 16 were added as there were 4 doublets, thereby resulting in a total of 173 publications. The low number of documents found on social assessment is in line with the findings of Macombe in this book [4]; however, she also included biofuels in a further step to enlarge the basis. Of the 173 publications, there are 149 journal articles, 9 contributions in proceedings, 6 book chapters, and 9 books.

A temporal analysis of the identified publications emphasises that sustainability assessment in the field of industrial biotechnology is still a rather young and highly dynamic field, which accelerated in the 2010s (Fig. 1). Interestingly, the number of publications remained at a fairly constant level over the last 5 years, with approximately 20 publications per year. More enlightening is a look at the journals in which research was published (Table 4). A clear division becomes obvious. Approximately 40% of the publications appeared in journals focusing on biomass use, like *Bioresource Technology*, which leads the ranking with 25 publications. At ranks 2–4, there are journals which can be considered as typical publication outlets of the life cycle and sustainability assessment community: *International Journal of Life Cycle Assessment*, *Journal of Cleaner Production*, and *Environmental Science and Technology* (together 22%); the *Journal of Industrial Ecology* with two publications and possibly *Science of the Total Environment* with four publications also belong to this group. Several journals are purely dedicated to energy issues such as *Applied Energy*, *Biotechnology for Biofuels*, *Energy*, *Energy Conversion and Management*, *Fuel*, *Journal of Renewable and Sustainable Energy*, and *Renewable and Sustainable Energy Reviews* (a total of 18 articles) with additional journals with a partial focus on bioenergy, such as *Biomass and Bioenergy*, *Biofuels*, and *Bioproducts and*

**Table 4** Ranking of journals with highest number of identified publications (only those with two or more publications) (see text for details)

| Journal | No. of publications |
|---|---|
| Bioresource Technology | 25 |
| International Journal of Life Cycle Assessment | 14 |
| Journal of Cleaner Production | 13 |
| Environmental Science and Technology | 6 |
| Applied Energy | 4 |
| Biomass and Bioenergy | 5 |
| Biofuels, Bioproducts and Biorefining | 4 |
| Science of the Total Environment | 4 |
| ACS Sustainable Chemistry and Engineering | 3 |
| Biotechnology for Biofuels | 3 |
| Biotechnology Progress | 3 |
| Chemical Engineering Transactions | 3 |
| Energy | 3 |
| Energy Conversion and Management | 2 |
| Environmental Science and Pollution Research | 2 |
| Fuel | 2 |
| Journal of Biotechnology | 2 |
| Journal of Industrial Ecology | 2 |
| Journal of Renewable and Sustainable Energy | 2 |
| Renewable and Sustainable Energy Reviews | 2 |
| Waste Management and Research | 2 |

*Biorefining* highlighting the fact that biomass is still above all considered as an energy source. Interestingly, dedicated biotechnology journals are missing in the list, thereby indicating a divide between biotechnology R&D and sustainability assessment outlets. One might speculate whether this also hinders exchange between these groups. Overall, the identified outlets show that authors of sustainability assessments have to decide whether they wish to emphasise the methodological part and publish in, for example, *International Journal of Life Cycle Assessment* or *Journal of Cleaner Production*, or focus on the application case and choose a biotechnology or bioenergy-related journal. For the reader, we recommend to use searches in databases covering both fields in order to identify relevant literature, as there are no leading journals in this intersection topic.

An analysis of the publications' titles provides insights into the topics covered. Methodology-wise, 46 of 173 publications have "life cycle" in their title, which corresponds to 27%. The majority of them (34 or 20%) include "life cycle assessment". The search string "cost" yields 26 counts (15%), which is equally frequent as the string "sustaina", thereby indicating a high importance of cost analyses, and even more so as another five publications have "econ" in their titles. In contrast, there are only five publications (3%) with "social" in their titles and one with "life cycle sustainability assessment" and one with "streaml" for streamlined LCA (cf. Sect. 5.6).

Assessments are often performed to compare different processes, products, or raw materials and, thus, as support for decision-making, which is revealed by the high count for "compar" (22 publications or 13%). Reviews are rather scarce; there are only three publications – one review on sustainability assessment of integrated biorefineries from 2015 and two on assessing algae-based biofuels from 2011 and 2015.

Examining what is assessed also provides various insights – that is, the raw materials, the processes, and their purpose. Fuel and energy purposes are most frequent (22 or 13% and 19 or 11% contain "fuel" and "energy", respectively), whereas only four have "polymers" in their titles and three "chemicals". Further, 16 publications (9%) have "biorefin" in their titles, that is, they address biorefineries.

In terms of raw material, with four counts (2%) each equally rare are "corn", "straw", and "sugar". For "cellulos", there are 19 (11%) and for "wood" there are three publications highlighting the strong interest in the use of lignocellulosic biomass, which is likely to avoid direct competition with food production. When examining what is assessed, the most frequent is "ethanol" (42 counts, 24%), followed by "alga" (20 counts, 12%) and "enzym" (10 counts, 6%).

Please note that combinations of word occurrences are possible and that the inclusion of ethanol in the title could have various implications, such as ethanol being used as feedstock, product, etc. With regard to processes, five titles contain "ferment".

Overall, this brief analysis reveals the strong focus on energy use and biofuel production, whereas the production of chemicals and materials is rather seldom assessed. In order to avoid competition with food and use thus far un(der)used feedstock, lignocellulose is often considered as a raw material. Finally, biorefineries as complex systems are apparently more frequently assessed than single processes or enzymes.

## 4 Characteristics of Industrial Biotechnology for Assessments

After sketching the field of sustainability and life cycle assessment in general in the previous section, we characterise the development and implementation of industrial biotechnology in the following sections and also highlight particularities and challenges.

### *4.1 Identification of Relevant Framework and Indicators*

Following the goals of an assessment in industrial biotechnology, it is necessary to determine fundamental characteristics of the study – that is, the functional unit,

**Table 5** Important sustainability categories of industrial biotechnology (after [60], modified and extended aspects in italics)

| Environmental | Economic | Social |
|---|---|---|
| GHG emissions | *Investment* | Human health |
| Land use | Feedstock costs | Labour rights |
| Land use change | *Operational costs* | Working conditions |
| Biodiversity | Local income generation | Land ownership |
| Water consumption | Subsidies | Food security |
| Emissions to water and soil | – | Rural development |
| *Emissions to air* | – | *Risks* |
| *End-of-life related aspects* | – | *Indigenous people* |
| – | – | *Traditional supply chains* |

system boundaries, considered life cycle stages [41] – as well as adequate (relevant) dimensions and impact categories for the assessment. Therefore, the first question is which sustainability dimensions must be covered. Thus, a suitable assessment framework is determined. As described above, monodimensional studies considering either environmental, (techno-)economic, or social criteria as well as two- or three-dimensional studies can be included, such as techno-economic and environmental aspects in ostensible eco-efficiency or whole life cycle sustainability assessments (cf. [40]).

Which sustainability impacts must be covered needs to be determined for each sustainability dimension. A set of particularly relevant categories for IB is provided in Table 5. The chapters by Venkatesh et al. [1], Festel [8], and Asveld et al. [3] in this book treat environmental, economic, societal, and ethical aspects of industrial biotechnology in detail. Note that ethical aspects have been barely covered thus far.

Within the selected framework, suitable operationalisations – that is, methodologies, methods, tools, and indicators – must be selected to assess the sustainability categories in meaningful measures. This selection must be performed in a context-specific manner: (1) for simple questions, an analysis of one or few selected categories may suffice. These must be selected following a prioritisation and hierarchisation. (2) For example, for a comparison of two or more items, different technically suitable processes and all relevant categories of each item must be taken into account, and the consistency of the methodical approaches, system boundaries, and data must be ensured. (3) A full sustainability assessment must cover all relevant sustainability impacts [60] as well as positive ones in social LCA.

Many of the sustainability impacts of industrial biotechnology are associated with feedstock provision, logistics, size and capacity, processing characteristics, and associated risks in comparison to the petrochemical industry. Sections 4.2–4.5 deal with these characteristics and challenges.

Further chapters of this book cover specific methodologies, methods, tools, and indicators for the sustainability categories of Table 5. Lindner et al. [2] deal with land use, land use change, and biodiversity issues, Chen and Reniers [5] with risk assessment, Macombe [4] with social life cycle assessment, and Saling [6] with the sustainability assessment of biotechnology products.

## 4.2 Utilisation of Biogenic Resources

The production of the renewable raw materials plays a crucial role in the sustainability and life cycle assessment of biomass-based processes. From an economic perspective, the raw material supply at predictable and reasonable costs is one of the biggest challenges. Simultaneously, the production of the feedstock contributes to a large extent to the environmental impacts, particularly through land and fertiliser use. Moreover, depending on the region in which the feedstock is produced, significant social issues may also arise. These can either be evoked by the production of the feedstock or arise from the replacement of a natural production technique by biotechnological processes [3].

An important factor for the assessment of industrial biotechnology processes based on renewable biogenic materials is that these are grown in a natural environment. Renewable raw materials result from recent biological production processes of the present natural environment.[4] For biogenic raw materials produced in agriculture and forestry, natural factors such as climate, seasons, weather, topography, soil, size, and spatial distribution of agricultural land and forests determine the decision regarding which feedstock to grow with which machinery, thereby influencing achievable yields [62]. Further, the feedstock differ in terms of the characteristics, both within and among species, as illustrated by the large improvements made in plant breeding. The feedstock characteristics of a species in general determine the possible utilisation pathways. However, the plant species differ not only in terms of their characteristics – for example, hardwood or softwood – but also different types of hardwood. Thus, attention and flexibility are required to account for these characteristics during utilisation.

In the assessments, it must be considered that the feedstock has thus far been bred mainly for utilisation as food and fodder or for energetic purposes, thereby leaving room for improvements for feedstock use in IB. When estimating utilisation potentials, new plants – from breeding or biotechnological modifications – can achieve higher yields in terms of the envisaged components or tailor further characteristics to the utilisation. Such productivity gains often lead to better results in the assessments but must not be taken for granted, particularly due to increased stress from climate change. However, successful breeding may also lead to higher yields or changing characteristics of other feedstock. Therefore, a static consideration of these competing feedstocks is problematic. A transparent solution would be to analyse potential effects on all potential input materials in a scenario analysis (cf. Sect. 5.4).

Potential competition with other utilisations of the feedstock must also be accounted for. With regard to economic aspects, existing usages often have to be considered, thereby limiting the potentially usable amounts in an area and prolonging the distances from which the biomass must be transported to a conversion plant [62, 63]. In addition, many of the biomass-based value chains which are

[4] "Recent" refers to the so-called short carbon cycle. Included is a formation of the organic matter of plants, algae, maritime organisms, woods, microorganisms, animals, and organic residues from households, agriculture, and animals from the food and fodder industry (cf. VDI 6310 [61]).

currently under discussion and in development aim at the use of the same potentials [64, 65], thereby causing direct raw material competition if these value chains are implemented. Using wood and residues as feedstock avoids direct competition with food and fodder chains, although the potential of these resources is limited [64, 65]. Nevertheless, an increasing demand for biogenic raw materials will lead to a competition with regard to land use, both nationally and internationally; this is likely to lead to indirect competition with food and fodder chains and nature protection of rain forests, savannah, and grass lands in particular [2, 66]. The direct and indirect competition leads to price effects on food and fodder, which can as of today already be observed (e.g. [67]) with social implications.

Apart from the biogenic raw materials from agriculture and forestry, large quantities of organic residues – such as sewage sludge, residues from households, and from food and fodder chains – can be used. While the price of such residues is often attractive, available amounts, regular supply, changing compositions, and efforts for catchment and provision of these residues are often limiting factors for their utilisation.

## *4.3 Logistical Issues Relevant for Sustainability Assessment*

Logistical issues arise along all bio-based supply chains and, thus, for numerous applications of industrial biotechnology. Biogenic raw material production and provision to the industrial plant come into focus, as biogenic raw materials have higher O:C ratios and, thus, low specific calorific values on mass basis in comparison to fossil raw materials, such as coal or natural gas. This makes longer transports unfavourable, from both an economic and an ecological perspective, and transport a relevant process in all sustainability assessments. Therefore, numerous logistical configurations of new bio-based supply chains aim to source the raw material regionally and avoid long-distance transport of (untreated) raw materials. Nevertheless, transport modes and economies of scale do matter, particularly for utilisation/conversion plants requiring large amounts of biomass. Increasing the supplied amount of biomass requires extending the radius of supply, thereby leading ceteris paribus to an overproportional increase in transport costs even when price increases and decreasing yields are not considered [68]. For bulk materials, the geographical proximity to transportation means – that is, rail and inland or naval shipping ports – is important to supply large-scale plants with bulk materials over longer distances. The selection of a transport mode is not only relevant in cost terms but also in terms of the other two dimensions of sustainability [69, 70]. Moreover, seasonal accruement of raw materials can demand for large processing capacities, thereby also reducing the annual degree of capacity utilisation. The suitability for storage may also be limited, for example, if the storage capacity is too small or if the raw material needs to be processed soon after harvest. Lastly, the harvested amounts vary interannually.

Certain multistage conversion processes provide the possibility of a spatial and/or temporal decoupling of processing steps to overcome the trade-off between transportation costs and economies of scale as well as the limited storability of raw materials. The prepared and conditioned raw material or the intermediate obtained from primary conversion can have better transportation and storage properties. In a spatial decoupling instead of one large integrated plant covering all processing steps, a number of smaller processing units are built up and decentralised in close proximity to the raw materials. The intermediates can then be shipped with better transportation properties to a larger central plant where additional processing steps are implemented. This has consequences regarding the economic and ecological parts of the assessment [71, 72].

In a temporal decoupling, better storage properties of intermediates after a pretreatment, conditioning, or primary conversion in comparison to raw materials are used to arrive at smaller capacities for further conversion steps. The first conversion steps of a process chain are built with large capacities to cope with large amounts of seasonally accruing raw materials, which is often prone to microbial attack and other degradation, thereby making a fast first processing indispensable. The intermediate is then stored and processed more continuously throughout the year. Examples of such configurations can be found, for example, in sugar mills. Nevertheless, it requires individual considerations for different utilisation chains and to take into account the regional conditions to identify the best suited logistical concept for each chain.

This also holds for the location and network strategy. In order to achieve economies of scope, the integration of one or several biotechnological processes into an existing plant site or supply chain offers various potentials for economic and environmental advantages. Outside battery limits, investment can be reduced by using existing infrastructure, facilities, and logistical processes. Operating costs may be lowered through lower plant-specific overheads and lower prices for, for example, steam, electricity, heat, and cooling. Process integration can lead to environmental advantages, for example, from exchanges between heating and cooling demands of processes. However, this poses challenges regarding the assessment and allocation of impacts [73]. Constructing new plants or supply chains provides the possibility of implementing ideal solutions for the particular supply chain at the cost of omitting the described potentials of economies of scope [62].

## *4.4 Characteristics of Biotechnology in Sustainability Assessment*

Each technology has its specifics, and greater attention must be paid to these during assessment. This subsection aims to provide a brief overview of the impacts often considered as relevant when assessing IB as well as of indicators specifically developed for (mostly) chemical processes.

The use of microorganisms or enzymes associated with further specifics is characteristic for IB and IB processes: a substrate, mostly a biogenic raw material or a product made thereof, is needed; it is converted using biochemical processes to a desired product under mostly mild conditions in terms of temperature, pressure, pH level, etc. in aqueous media to ensure that microorganisms survive and enzymes are not denatured. The reaction typically yields a product in low concentrations, such that often additional processing steps for concentration and cleaning are needed. As the microorganisms and enzymes are typically more targeted than chemical reactions, co-products are less common. Moreover, several classes of chemicals that are associated with environmental problems – for example, halogenated organic substances with the widespread problem of persistence in the environment – are unlikely for IB reactions. Unlike catalysed chemical reactions, IB also does not rely on scarce metals with a large environmental footprint. In their review, Jegannathan and Nielsen [74] summarise that enzymes have benefits in terms of milder reaction conditions, biodegradability, and reduced or no toxicity and can be produced "with less raw material, chemical, water and/or energy consumption and with less problematic waste generation than traditional processes". Although this view of the two authors working for Novozymes appears reasonable in general, there are products which can be better produced chemically. An example of this is synthetic astaxanthin [6]. In other cases, the overall picture is less clear, as indicated by the example of vitamin B2 provided by Saling in this book [6], where the biotechnological process is advantageous in several categories but worse in land use. Undoubtedly, such findings are only snapshots, and ongoing optimisation of the different possibilities of producing a substance of interest can quickly change the picture. This must be kept in mind, particularly for biotechnology processes which are generally less optimised yet than their chemical counterparts and likely also more susceptible to optimisation. For example, Ma et al. [75] were able to improve the volumetric productivity per mass of catalyst in a concrete case by a factor of 2,500.

Overall, Renner and Klöpffer [76] conclude that biotechnological processes and products are not, per se, more (environmentally) sustainable than their conventional counterparts for the following reasons:

- Agricultural feedstock production typically involves a number of resource- and emission-intensive steps (cf. also Sects. 5.3.2 and 5.3.3) and corresponding impacts. The most important impacts among these are land use, including biodiversity impacts (cf. [2]), soil degradation, eutrophication (both nitrogen and phosphorus), ecotoxicity from the use of crop protection products, and GHG emissions. Impacts arising in the agricultural upstream chain may largely dominate the entire LCA.
- Low yields and output concentrations require intensive post-processing steps. This may result in high energy demand (in spite of the low temperatures); high water consumption, which might be of high relevance in regions with water scarcity; and emissions in receiving water bodies if wastewater is not appropriately treated as well. Post-processing may result in high costs.
- IB may utilise genetically modified organisms (GMOs) (cf. Sect. 5.3.1). In IB, in contrast to green biotechnology, release of GMOs can be considered unintentional "biological emissions" from inappropriate waste disposal or accidents.

For assessing chemical reactions, several (simple) indicators have been developed with the aim of providing a quick screening to identify more and less promising processes.

Due to the strong similarities between chemical production and IB production, metrics used for assessing chemical processes and synthesis routes are presented in the following account. A sound overview of LCA of chemicals and chemical products is provided by Fantke and Ernstoff [77]. The intersection of LCA and chemistry is also addressed by other authors – for example, [78–82] – and the assessment of IB processes with a focus on benchmarking biocatalytic and fermentation processes in [83]. While certain assessment tools can be applied once the chemical or biochemical reaction is known – that is, in silico, for example, atom economy and carbon mass economy (see below) – the application of other indicators requires knowledge of the processes and is thus used during reaction development like the E-factor [83].

Fantke and Ernstoff [77] analyse the conceptual relationships between sustainability assessment, product and material design, and regulatory chemicals management, each with their own but also (partially) shared tools. To the group of sustainability assessment tools, they assign LCA, green and sustainable chemistry tools, as well as (in part) chemical alternatives assessment. Green (and sustainable) chemistry has its own metrics, e.g. [84–86] (see below). Chemical alternatives assessment (CAA) serves to identify and evaluate alternative solutions for hazardous chemicals [77]. Since chemicals are used for the production of almost all industrial products, in most LCAs also chemicals are assessed from a life cycle perspective. Though eco- and human toxicity are in general important impact categories for chemicals, others may be similarly important or even more important. It is also not possible to identify a life cycle stage that is in general more important than others. Apart from the production of a chemical, also its use (in particular, if contained in consumer products) or its end-of-life may be highly relevant. LCAs of chemicals are conducted to compare different chemical products, reactions, synthesis routes, or feedstock, to determine the environmental profile of a chemical and to identify relevant life cycle stages and process steps, often with unexpected outcomes due to shifting of impacts to other impact categories, life cycle stages, etc. [77]. This highlights the value of LCA as a decision support tool.

However, several challenges arise when conducting LCAs of chemicals. Whereas multifunctionality and the setting of appropriate system boundaries are also challenges for LCA of chemicals, the need to allocate to co-products, complex LCI emission pathways, or incomplete emission inventories and lack of data needed for LCIA are more specific to chemical production and chemicals. An additional issue is the toxicity of the mixtures of chemicals [87], which is in contrast to the assumption of linearity and additivity of the effects of LCA.

Several indicators are typically used by chemists and chemical engineers to assess the performance of a chemical reaction, mostly for screening purposes – for example, (cf. [88] and literature cited therein)

$$\text{Conversion} = \frac{(\text{Reactant consumed in the reaction})}{(\text{Reactant supplied to the reaction})} \quad (1)$$

$$\text{Selectivity} = \frac{(\text{Amount produced of desired product})}{(\text{Reactant consumed in the reaction})} \times (\text{Stoichiometric factor}) \quad (2)$$

$$\text{Yield} = \frac{(\text{Amount produced of desired product})}{((\text{Limiting})\ \text{Reactant supplied to the reaction})} \times (\text{Stoichiometric factor}) \quad (3)$$

Due to limitations inherent in these indicators – for example, the yield ignores that reactants are typically added in excess to drive the reaction as well as catalysts, solvents, non-limiting reactants, etc. – further indicators at the process and product levels were proposed [89]. One of them is atom economy, which is a measure that indicates the extent to which, according to the reaction equation, the reactants are incorporated in the desired product [85]

$$\text{Atom economy} = \frac{(\text{Molecular weight of desired product})}{(\text{Molecular weight of reactants supplied to the reaction})} \quad (4)$$

Atom economy can be calculated also for a series of reactions. Solvents, reagents, and materials used in catalytic quantities only are disregarded in atom economy, and the reaction needs to be correctly balanced [85]. Though adding information to identifying green reactions, atom economy alone is not a sufficient criterion for greenness as, for example, energy demand is disregarded as well as whether inputs or outputs and reaction conditions are environmentally sound and inherently safe. Also processing efforts of the products are in general not included and, most importantly, information about the yield.

In order to account for these deficits, the indicator "reaction mass efficiency" (RME) was proposed, integrating yield and atom economy [85] where

$$\text{Reaction mass efficiency (RME)} = (\text{Yield}) \times (\text{Atom economy}) \times \frac{1}{(\text{Stoichiometric factor})} \quad (5)$$

Thus, reaction mass efficiency measures the extent to which supplied reactants are incorporated in the product as mass ratio – that is, accounting for incomplete reactions. For further details and variants of the RME, the reader is referred to [85, 90].

Another popular indicator is the E-factor, which is simply the total amount of waste generated for a product – that is, including solvents (but excluding water), losses, process aids, and even products from energy production – relative to the amount of that product [91, 92]. As such, the E-factor also accounts for the yield and is easily applicable for multistep processes and entire sectors [92].

$$\text{E-factor} = \frac{\text{Amount of waste produced}}{\text{Amount of desired product}} \tag{6}$$

The indicator "process mass intensity" (PMI) strongly resembles the E-factor [85]. Further screening indicators are carbon mass efficiency (mainly for biocatalytic processes) applicable for route feasibility evaluation and solvent and water intensity for reaction development [83].

A broader view on process metrics can be found in [93]. Differentiating between indicators for materials, equipment, operability, environment, health, safety (EHS) risk, and quality, these authors present a variety of indicators, for example, related to safety, maintenance, scalability, renewability, economics, etc. In particular, chemical safety is addressed in [94].

## 4.5 Challenges Induced by Technology Readiness Levels

The development stages of IB-based products and processes differ. Like other process developments, they can be differentiated roughly into three major stages – that is, *lab scale*, *pilot scale*, and *industrial scale*. These usually have ascending capacities and reach from works on the underlying basic principles – for example, single chemical reactions or an ideal process and its validation in the laboratory over pilot demonstrations – with the aim of being qualified for an industrial realisation to a successful practical implementation in industry. In order to scale up a new process from a basic idea to such an industrial implementation requires numerous tasks. Therefore, the development stages are differentiated further. Based on an approach of the US Department of Defense [95], a high-level expert group (HLEG) established by the European Commission proposed a nine-stage scale of *technology readiness levels (TRL)* for key enabling technologies [96], which is illustrated in Fig. 2 and has been taken up, for example, to describe the development stages of different biorefineries [97]. It must be noted that the first industrial realisation of a process chain is also subject to further learning and optimisation, and the following

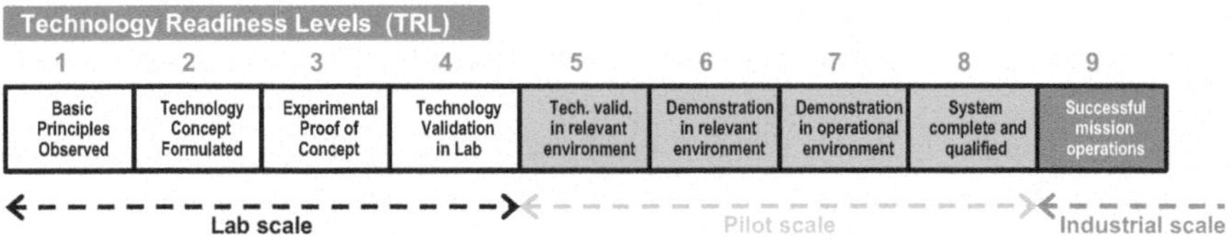


**Fig. 2** Technology readiness levels (adapted from [96])

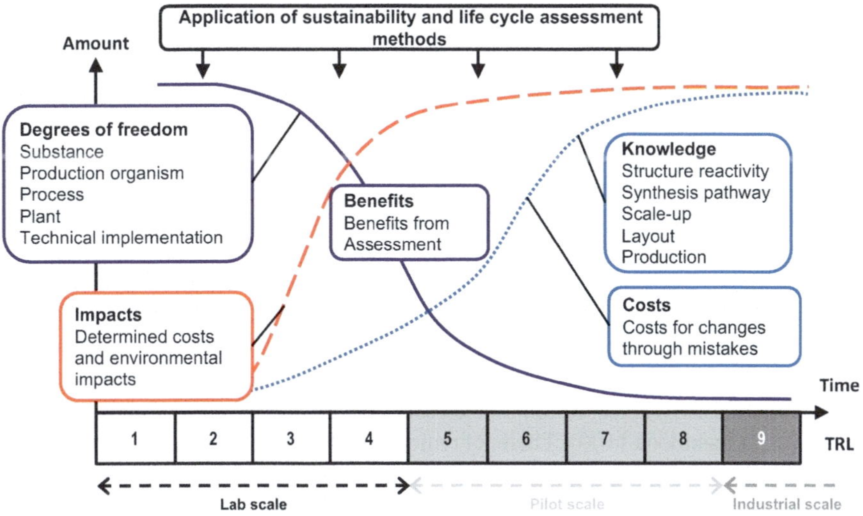


**Fig. 3** Potential impact of analysis and assessment and achievable accuracy for biotechnological processes (schematic, adapted from [98])

implementations might significantly reduce investments and operation costs due to learning and experience effects.

In most cases, the analysis and assessment of new IB processes and materials involve projections from the individual stage of development to an industrial-scale realisation. The potential benefits for an application of analysis and assessments of new bio-based supply chains as well as the achievable accuracy of such works evolve with contrary signs during the development process (see Fig. 3). With increasing TRL, the share of *determined* costs and environmental impacts increases. Following this, the degrees of freedom for the further development and the potential benefits from applied analysis and assessment methods on development decrease, although the accuracy of these works increases [99].

## 5 Methodical Aspects

After having characterised IB and the challenges associated with these characteristics for sustainability assessment in Sect. 4, this section focuses on how these can be addressed by utilising the methods and methodologies outlined in Sect. 2. In addition, further methodical aspects of specific relevance for IB assessment are discussed.

## *5.1 Consequential Assessments*

A first topic of specific relevance is the field of consequential assessments. Attributional approaches are well suited and recommended to compare different (biogenic) feedstocks and different processing routes and products in the sustainability and life cycle assessment approaches for IB [100]. However, consequential approaches are indicated in numerous cases: IB is often believed to substitute existing "conventional" chemical processes and biogenic feedstocks to substitute non-renewable ones, and the further development and implementation of bioeconomy concepts aims at larger system shifts. Thus, assessments as a means of decision support must account for the consequences and follow a consequential approach. Following this argumentation, consequential life cycle sustainability assessment (CLCSA) must be applied to two of the three use cases of LCA proposed in the ILCD handbook [100]. There, (1) micro-level decisions (products), (2) meso-macro-level decisions (policy), and (3) accounting (products and processes) are distinguished (cf. Sect. 2.3.2). Ekvall et al. [101] conclude that consequential approaches are the methodology of choice for the first two.

Accordingly, the young stream of CLCSA already encompasses a large number of works which are related to IB and the use of biogenic raw materials. This is evident in the works covered in the early review of Zamagni et al. [102] and several actual works such as [103, 104].

With regard to methodological aspects, Zamagni et al. [102] consider two important aspects of a consequential assessment. They first state that CLCSA affects the nature of modelling and second that the system boundaries must be set accordingly. By the nature of modelling, they imply "analysing principles, analytical techniques used and their limitations, all characteristics that define what the model can (and cannot) do". They conclude that with the exception of one publication [105], these aspects have not been dealt with in detail. With regard to system boundaries, the inclusion of economic aspects is particularly important. Despite a few works in this field, these efforts are mainly limited to the inclusion of basic effects – for example, by price elasticities – and a few markets. In particular, the interplay between different markets is omitted. Where considered, these effects are mostly modelled as exogenous parameters and are not endogenised [102]. Despite the age of the referred publication, to the best of our knowledge, the situation has not changed to date. In order to address this, a more intense application of scenario and sensitivity analyses [102] and a close link or integration to and with integrated assessment models are solution approaches. The latter one would be a particularly valuable methodical contribution in this field.

## 5.2 Prospective Assessments

As described above (particularly Sects. 4.5 and 5.1), the application of sustainability and life cycle assessment in the field of IB is often prospective, since it has to anticipate the further development of technologies and products towards an industrial realisation and their diffusion in the market.

Research focusing on prospective assessments for IB are rather rare. However, a recent review by Arvidsson et al. [52] summarises recommendations for prospective LCA studies of emerging technologies in comparison to retrospective studies. They divide these recommendations into three aspects: (1) the consideration of alternatives, (2) the data for the foreground system, and (3) the data for the background system.

With regard to the alternatives to consider, the challenge is to select those relevant for the future – that is, the considered point of time. Thus, currently available benchmark technologies may no longer play a role, while others – which are currently not considered to be competitive or under development – may become relevant. Since it is difficult to provide a valid prognosis about the eventually relevant technologies, Arvidsson et al. recommend to include as many alternatives that provide a similar function as possible and/or to conduct cradle-to-gate studies for promising emerging technologies in order to use them as building blocks in assessments [52]. What is challenging is that not all functions of new technologies or products are comparable and that the spectrum of by-products and their utilisation may differ. Thus, it is recommended to define multiple functional units, communicate assumptions transparently, and utilise scenarios as well [106].

Further, both for the foreground and background systems, data availability is a challenge in prospective assessments. For new processes, materials, and products, a lack of inventory data is likely; data quality may be an issue, for example, due to spatial or temporal validity and uncertainty due to the TRL [106]. A central requirement for valid assessments is that available alternatives and foreground and background systems consider the same point in time and a "mismatch" of the temporal assumptions must be avoided [52]. Thus, future developments must be considered not only for the foreground system but also for competing existing and mature technologies and the background system. For both, learning curves for efficiencies and cost reductions as well as technology diffusion must be taken into account [54].

For an assessment of the foreground system, the situation of the system at the end of the development process and at a reasonable industrial scale is considered. Although primary data of an industrial-scale implementation would be preferable [106], this data is, naturally, mostly not available. Here, scientific articles, patents, expert knowledge, research reports, and process simulations provide valuable sources that can be used [52]. For a scale-up, it is an approach to identify "hotspots" of sustainability impacts and focus on the assessment and development of them [106]. Works in the field of IB with a techno-economic and environmental focus deal, for example, with lignocellulose biorefineries [35] or chemo-enzymatic epoxidation of seed oil [34].

For the background system, the use of average data or representative proxy data is indicated [106]. However, one is interested in the situation of that system at a future point in time. Here, as well, scenario analyses have proven their worth. If the modeller regards a few of the scenarios as more probable, one can use these as predictive scenarios. If this is not possible, taking account of uncertainties and the elaboration of range scenarios to encompass a broad spectrum of possible futures (e.g. optimistic or pessimistic ones, baseline) is recommended. If scenarios become too numerous, efforts must be made for either aggregating them or selecting the most meaningful ones (cf. [47]).

This consideration neglects a joint consideration of consequential and prospective assessment. The work by Arvidson et al. [52] focuses on attributional assessments only. However, the focus on scenarios also enables consequential assessments, although, here again (cf. Sect. 5.1), as exogenous parameters. As described for consequential assessments, a modelling of the mutual influence between consequences and particularly the background system – for example, by an endogenisation of market mechanisms – would be a fruitful direction of research [54].

Moreover, what is often not considered is the step of impact assessment. The factual and normative relevance of impact categories may change over time. Issues which are nowadays in focus may lose their importance due to environmental protection efforts or common technological developments. Others may come into focus due to new scientific insights or system changes [52]. Such focus shifts can also be evoked through normative changes that influence the goal and scope definition as well as the interpretation step.

## 5.3 *Specific Impacts in Industrial Biotechnology*

The characteristics of IB lead also to some particularities to be considered in impact assessments. The most important aspects, in our view, are described in the following subsections.

### 5.3.1 Impact Assessment for Genetically Modified Organisms (GMOs)

Industrial biotechnology may utilise GMOs in two main areas: (1) for the production of substrates – for example, genetically modified maize for sugar production, and (2) to produce enzymes for or directly in biotechnological processes. With respect to GMOs, possible negative impacts on the environment and human health may be direct and indirect as well as with high and low probability of occurrence. Impacts on biodiversity from an herbicide resistant plant are typically indirect (e.g. [107] with a recent review of environmental impacts of genetically modified plants). As explained in Sect. 5.4, LCA as a tool focuses on rather average rather than extreme conditions [108] and, thus, largely neglects any risks – that is, any negative effects with low probability of occurrence which are also likely those effects which are most intensively discussed in the public. Apart from the "objective" risk perspective, there is also a perspective that discusses ethics (cf. [3]). Here, a distinction is made between processes resembling breeding – for example, by making use of chemical and physical agents to artificially induce mutations [109] in a rather uncontrolled manner – on the one hand and targeted genetic engineering using techniques of synthetic biology like CRISPR-Cas on the other hand. Although IB experts prefer genetic engineering techniques – as the results are foreseeable and, thus, considered as less risky – the wider public tends to prefer techniques resembling other well-known techniques, even if they might result in the production of unforeseeable products and are less efficient.

As Seager et al. [110] indicate, GMOs are different from the material and energy flows typically covered in LCA, since in case of GMOs, *information flows* must be addressed, and these flows cannot be captured in a life cycle inventory. Moreover, once in the environment, the information could, in principle, be replicated if self-reproducing organisms are released or if gene flows to other self-reproducing species were produced. Seager et al. [110] further argue that LCA has problems with the blurring of ecosphere and technosphere in case of GMOs. To conclude, LCA is not the methodology of choice to encompass all relevant aspects related to GMOs and must be complemented by other tools. Hence, when reading an LCA study involving GMOs, the reader must keep an eye on which aspects were covered in the LCA study and which were neglected and, thus, are not part of the results.

### 5.3.2 Climate Change Impacts Considering Biogenic Carbon and Temporal Effects

A key potential advantage of bioeconomy concepts is that through the use of biogenic feedstock, significant GWP reductions up to a carbon neutrality can be achieved. The classical argumentation – for example, in the biofuel field – is that biofuel combustion only emits carbon that has been bound in the production of the feedstock before. As this feedstock production takes place with a, in comparison to fossil fuels, short time horizon, numerous studies do not consider the $CO_2$ emissions originating from this so-called biogenic carbon and consider their use as carbon neutral. This argumentation is followed in numerous impact assessments, particularly with regard to bioenergy/biofuel research. However, this non-consideration may be an oversimplification [111].

In order to ensure for adequate accounting, for example, for emission trading, it is reasonable to account for both the initial sequestration and the positive or negative changes of carbon stock [112] and to model the carbon cycle on a systemic level. In addition, it must be considered whether the carbon is not emitted as $CO_2$ or as $CH_4$ [111].

The timing of sequestration and emission and the temporary storage of carbon also influence climate change impacts. Even if the uptake of carbon and the emissions during or at the end of a product life are accounted for, the effect of delaying the emission is neglected [112, 113]. For example, if a forest is cut to produce biofuels, the stored carbon is converted to carbon emissions. A reforestation for a reuptake of the emitted carbon will take decades. Thus, such a use leads to a temporary increase of the carbon in the atmosphere. In turn, the use of short rotation/ annual crops is compensated for through a regrowth in much shorter times [114].

Lastly, the time horizon which is considered also plays a role. Unlike most other environmental impact categories, the commonly used GWP considers a 100-year period instead of an infinite time horizon. This is against the intergenerational justice principle of sustainable development, since effects beyond the time horizon are not accounted for. However, the rationale underlying this is to have a planning horizon that is relevant for humans in order to avoid prolongations into the too far and, thus, too uncertain future on the one hand but to avoid an overweighting of the too near future on the other hand. Thus, most methods for LCIA regarding climate change consider the 100-year time horizon [112]. However, there are ongoing discussions regarding whether to split the GWP impact category to account for both short-term (20 years) and long-term (100 years) climate changes [115]. A reason for this distinction is that though climate change is disastrous in itself, a *fast* climate change is even worse, as nature and humans lack the time for adaptation in such a case.

Several recent works have studied the effects of the aspects described above for different products and application fields. For example, Garcia and Freire [113] investigated the impact of different carbon footprint methods on the assessment of results of particle cardboard. Other studies deal with timber-framed buildings [116, 117], buildings in general [117], or biofuels [118]. A very recent and comprehensive study by Brandão et al. compares LCIA approaches for bioenergy systems in the context of climate change [114]. These encompass the frequently used $GWP_{100}$ and $GTP_{100}$ (Global Temperature Change Potential (GTP)) as well as several methods which focus specifically on the above-mentioned aspects. Certain methods use $CO_2$ decay curves; others employ the avoidance of radiative forcing or time-averaged carbon stocks. Further methods employ time correction factors and time-adjusted warming potentials. Others discount emissions, use specific characterisation factors for $CO_2$ emissions from combustion for temporary biogenic carbon releases that are compatible with commonly known LCIA approaches, or employ climate tipping potentials. Brandão et al. apply 15 methods to 3 bioenergy case studies – that is, bioenergy use of forest grown for 25 years to replace a carbon-rich mature forest after usage, a similar forest grown on pasture land, and an annual energy crop. The results reveal that from the first to the third case study, the differences among the methods from the LCIA are drastically reduced. In the "forest after forest" case study, negative assessments occur – that is, negative impacts on climate change through the bioenergy in comparison to fossil fuels. The "forest after pasture" case study reveals only positive effects despite a few substantial differences between the methods. For the energy crops, the difference is rather small. It can be concluded that each method has its specific application field and that when a comprehensive assessment is needed, decision-makers must be aware of the importance of the mentioned temporal aspects and choose suitable methods to consider these. A consideration of the biogenic carbon and temporal effects is particularly indicated when long time-lags exist between the release of carbon and its reuptake, when considering long-life products or when there is a transformation of the carbon to $CH_4$ or other GHG, which are more effective than $CO_2$.

### 5.3.3 Assessing the Impacts of Biogenic Feedstocks

The conversion processes and the production and provision of the feedstock are usually main contributors to the environmental impacts of biochemical products [119]. Apart from the biogenic origin and temporal aspects influencing climate impacts, as outlined in the previous subsection, the use of biogenic feedstock has additional characteristics which must be regarded in sustainability and life cycle assessment and lead to different impact patterns in biogenic feedstock and in comparison to fossil feedstock [32]. The key characteristics of biogenic feedstocks have been outlined in Sect. 4.2, thereby leading to the necessity of a detailed consideration of these in the LCSAs either through suitable data from existing databases or an explicit modelling of the specific feedstock provision chains (c.f., e.g. [62]).

However, despite the importance of accounting for these impacts, methodological issues regarding the environmental life cycle assessment continue to exist. In particular, these encompass one central aim of biogenic feedstock use, the measurement of the climate change mitigation potential, non-$CO_2$ GHG emissions, land use and land use change, as well as biodiversity issues.

With regard to climate change, it is also important to consider non-$CO_2$ GHG emissions from feedstock production and provision. These are influenced by applied machinery and degree of mechanisation [62], handling of by-products [32], and particularly, fertiliser application [120]. Another particularly important factor is the consideration of direct and indirect emissions of nitrous oxide ($N_2O$) in feedstock production. Direct emissions cover the $N_2O$ release of nitrogen added in fertilisation from the soil. Indirect $N_2O$ emissions are caused through two pathways – the volatilisation of ammonia ($NH_3$) and the emission of nitrogen oxides ($NO_x$) from land management, from burning fossil fuels and biomass with a subsequent redeposition of these gases and their products to soil and water and the leaching as well as runoff of nitrogen from managed land [121]. As the GWP of $N_2O$ is 265 times higher than that of $CO_2$ [122], these emissions play an important role in the determination of the climate effects of the agricultural feedstock. Thus, these emissions may lead to a high degree of uncertainties in the impact assessments [120]. As the relevance of these emissions is not questionable, they are regarded in numerous studies on the GHG emissions of biogenic feedstock provision [120]. Common practice and recommendation is to use factors – that is, percentages of applied nitrogen content in the fertiliser. However, there has been a debate on the order of magnitude of these factors. In its 2006 guidelines for the preparation of national GHG inventories [123], the IPCC suggested general default factors and stated uncertainty ranges; these have been criticised (e.g. [124]). Consequently, in the 2019 update of its 2006 guidelines, the IPCC lays out disaggregated factors for different climatic conditions [121].

Agriculture and forestry use fertile land, which is limited on earth. Global land area of the continents, including Antarctica, comprises 14,900 million ha (Mha). In the year 2000, 3,900 Mha thereof was forests, 1,500 Mha arable land, and 3,500 Mha permanent pastures [66, 125]. A growing world population and urbanisation has led to an increasing demand for living space [126]. Together with changing dietary patterns, the supply of biofuels and biomaterials, and land degradation, crop land expands against grasslands, savannahs, and forests. While the net expansion (without compensation for built environment and soil degradation) is estimated to range between 123 and 495 Mha by 2050 in a business-as-usual scenario, the shares for biofuels and biomaterials are estimated to be between 52 and 195 Mha in 2050 [66]. Thus, they have a significant share in this expansion, which becomes even more important through the wide range in the estimate. Thus, sustainability and life cycle assessment for biogenic feedstock must consider the effects of the feedstock use on land use and linked impacts on soil degradation and biodiversity.

These topics have been the focus of methodological development in environmental life cycle assessment. Lindner et al. therefore address these in-depth in this volume [2]. They introduce a method developed by a UNEP-SETAC working group to assess land use by processes through different quality indicators, occupied area (in square metres), and the duration over which the area is used (in years). Thus, the transformation of land from one condition to another can be assessed [127, 128]. Consistent with this approach, they developed the Land Use Indicator Value Calculation Tool (LANCA®) to calculate the impacts of the land use on the ecosystem services of soils. These encompass erosion resistance, mechanical filtration, physicochemical filtration, groundwater regeneration, and biotic production. The indicators can then be used in LCA studies [129, 130]. Methods to assess biodiversity can roughly be divided into two schools of thought – the diversity of species as a key indicator and the more holistic understanding of environmental conditions. However, also other approaches exist. Although the method by Chaudhary et al. [131] following the first school of thought is recommended by a working group of the UNEP-SETAC Life Cycle Initiative [132], thus far, no method has been identified as a standard approach [2].

### 5.3.4 Consideration of Cascade Uses and the End-of-Life (EoL) Phase

Particular attention in sustainability and life cycle assessments studies of IB has to be paid to the end-of-life (EoL) phase. The fact that numerous IB products originate from biogenic raw materials and are produced with IB does not necessarily lead to a problem-free or more sustainable EoL phase. This can be explained rather well with examples from plastics. A bio-based polyethylene terephthalate is the same polymer as a "conventional" one, with the exception that the ethylene glycol (and in the future also the terephthalic acid) components are bio-based [32]. Thus, one can expect the same EoL sustainability impacts for both materials. Different impacts may result from land filling, (bio-)degradation – for example, through composting – and incineration. If a material is securely landfilled, this may not have adverse sustainability impacts apart from resource losses and occupation of landfill space when no decomposition of the material takes place and no harmful substances are emitted into water or air. A landfilling with a degradation of the material (which is rather likely for organic materials other than certain polymers), in turn, could lead to adverse impacts through emissions of $CH_4$, whereas an incineration would avoid these to a large extent and, at least, recover the energy of the material at the cost of $CO_2$ emissions. In addition, there exist several recycling options, depending on the materials. Others, in turn, may not be accompanied with emissions from degradation but lead to microplastic pollution in water bodies. Thus, it is necessary to consider for each material or discarded product the existing EoL options and assess them in order to obtain a comprehensive picture of sustainability impacts.

Consequently, numerous studies deal with these aspects, mostly focused on individual product groups or materials. For example, Hottle et al. [32] compare different biopolymers (i.e. poly-lactic acid (PLA), thermoplastic starch (TPS), Bio-PET, bio-high-density polyethylene (Bio-HDPE), and bio-low-density polyethylene (Bio-LDPE) with PET, and fossil-based HDPE and LDPE for landfilling, recycling, and incineration (PETs, HDPEs, and LDPEs) as well as composting and landfilling (PLA and TPS). They conclude that for TPS and a high-degradation scenario of PLA, higher GHG emissions occur in the landfill scenario due to $CH_4$ emissions than in composting but that composting leads to higher impacts in seven other impact categories (smog, acidification, carcinogenics, non-carcinogenics, respiratory effects, ecotoxicity, and fossil fuel depletion), as composting needs more machinery and water than landfilling. Due to the replacement of virgin material through recycling, recycling reduces environmental impacts of PETs, HDPEs, and LDPEs. This holds true, particularly regarding GWP and fossil fuel depletion. These are further increased by the bio-based polymers since these further replace fossil fuels as raw materials [32]. The large influence of the individual recycling processes on the environmental reliefs was shown in a study by Maga et al. [133] who investigated different emerging recycling options for PLA. They also showed the influence of methodical choices and crucial assumptions in the (uncertain) recycling system. Here, particularly the credits for the recycling products and "correction factors" to account for, for example, quality losses, play an important role.

More generally, circular economy (CE) has become a popular concept with large expectations regarding, for example, resource efficiency, reductions of GHG emissions, and reduced resource criticality (cf. [134] for an overview). In contrast to "technical materials" like metals, for biological materials, circular material flows are rather unrealistic as biological materials are sensitive to degradation processes due to oxidation or mechanical damages like a decreasing fibre length in case of cellulose. Consequently, for biological materials, CE mainly aims at an intensification of use – that is, using less material for the same output (dematerialisation), using materials longer, reusing materials for a similar purpose (mainly for wooden beams), or utilising material cascades. A material cascade is characterised by several subsequent, different types of uses over which the quality of the material decreases, thereby allowing the extension of utilisation time of the resource [135]. In IB, for the increasingly discussed waste biorefineries (e.g. [136, 137]) cascades gain in relevance. To be established, a material cascade requires incentives to form [138] – that is, for input material it has to be attractive to use only materials whose quality is marginally above the quality needed as input and for output material omitting possible further uses must be unattractive. An example of a misguided incentive would be if a combustion facility would pay higher prices than a company reusing the biomaterial. Although prices are set by the market, distributing the environmental impacts arising in the cascade (from biomass supply, steps within the cascade required to pre-process the material or treat it before further use and from end-of-life operations) needs to be done in a manner such that there is an incentive for each participant to join a cascade and to remain in it (cf. [138]). Thus, the allocation problem in cascades can be seen as a particular case of open-loop recycling, bearing in mind the entire cascade ([138] with proposals to use game theory to approach the allocation problem, cf. [139]).

## *5.4 Uncertainties and Risks*

Uncertainties are inherent in any assessment. Thus, it is important to identify sources for uncertainties, estimate their magnitude, and determine their impacts on the results. If reasonable, measures must be taken to reduce uncertainties, for example, via additional data collection. Uncertainties can be differentiated into the following types [140]:

- Aleatory uncertainties arising from natural variability inherent in the system
- Epistemic uncertainties from insufficient knowledge of the system under consideration
- Parameter uncertainties, that is, uncertainties regarding the true value of a parameter in the model
- Model uncertainties, that is, uncertainties regarding the appropriateness of the applied model
- Volitional uncertainties arising from mismatch of what a person announces to do and actually does

For LCA, Huijbregts [141] distinguishes between uncertainties due to variability in the real world (spatial, temporal, between objects) and uncertainties of the product and of the environmental system, each with parameter, model, and choice-induced uncertainties. For LCA, epistemic, parameter, and model uncertainties are likely the types of uncertainties of highest relevance. For the practically working LCA modeller, data uncertainties assume a key role as these uncertainties can be most easily influenced, for example, by more effort in primary data collection and use of appropriate generic data. Equally important are model uncertainties arising from questions regarding the modelling of the actual system in terms of the selected functional unit (e.g. [142]), system boundaries (cf. [143] for an introduction, allocation methods e.g. [144–146]), and life cycle impact assessment (LCIA) model selected (cf. [147] for the effects of different endpoint LCIA models when comparing ethanol with gasoline in Brazil). Therefore, it must be emphasised that uncertainties are unavoidable in any sustainability assessment, and the most severe mistake would be to be non-transparent about them. Therefore, we strongly encourage to estimate the magnitude of any uncertainties and to analyse their effects on the results through sensitivity analysis. For data uncertainties, the pedigree matrix [148] may be helpful to estimate the quality of data. In the pedigree matrix, (generic) data is assessed with respect to five categories (reliability of the data, its completeness, as well as the temporal, geographical, and its technological fit to the problem under analysis) (cf. also [38] for a similar categorisation) at five levels each. Weidema et al. [148] also provide a method (summed variances based on the assumption that the co-variances are zero) to determine the combined effects of all these uncertainty categories as well as default values. In the EU ILCD handbook [100], a relative standard deviation of 7% is considered as very good and of 7–10% as good for single data points, which provides some indication of what can be considered as the least number of uncertainties inherent in LCA data. Due to parameter and model uncertainties as well as error propagation, the total uncertainties are expected to be larger. Uncertainty information is highly relevant to understand LCA study results, even more as results are often presented with spurious accuracy in LCA.

As mentioned above, a variety of methods can be used to address and manage the different types of uncertainties (e.g. [141, 149, 150]), in particular:

- Sensitivity analyses, for example, to analyse the impact of variations of *single* parameters or data values on the results.
- Monte Carlo analysis to analyse the joint effects of parameter and data values uncertainties on the results in more complex settings and without requiring normally distributed data, as in Gaussian error propagation law; density functions are needed as input information (cf. [151] for density functions).
- Scenario analyses, typically used to determine the effects of different modelling choices – for example, end-of-life options (cf. Sect. 5.3.4) – as well as of plausible developments of the actual system, such as the future carbon intensity of the energy system on the results.

Unfortunately, in practice, a large number of LCA studies do not sufficiently address uncertainty using such tools or do not report their efforts in sufficient detail, which puts at least the conclusions drawn from these LCA results into question. An application for uncertainty analysis applied to biofuel systems is provided by [152] with a differentiation into parameter and scenario uncertainties.

Although, to the best of our knowledge, there is no study that particularly analyses the uncertainties in IB assessments, our expectation is that these uncertainties tend to be larger in IB than in numerous other applications, as heterogeneity and variability are typical for biological systems. The variability of the substrate quality alone may result in considerable yield differences of biopharmaceuticals (cf. [153–155]).

Whereas uncertainties refer to ambiguous or vague information [140], a risk is the effect of these uncertainties on objectives which might not be achieved [156]. Although, in principle, risk is neutral, most often, risk is used for negative outcomes only. Risks are typically defined as the severity of an outcome times the probability of the underlying event (e.g. [157]) and are typically presented in a risk matrix with the extent of the damage *E* on the x-axis and probability of occurrence *P* on the y-axis [5, 158, 159]. In principle, the higher the product of both – that is, the risk – is, the lower is the willingness to accept that risk. However, a risk with an excessively severe damage is often considered as not acceptable, even if its probability is minimal and, thus, the product still low [158]. The same applies for risks with a very low extent of damage but a very high probability of occurrence. Benchmarks of what may be considered as acceptable or not vary over time and among cultures, nations, and people, as can be inferred from the debate on nuclear power or use of GMOs. Further challenges arise from the often lacking certainties regarding the true extent of damage, *E*, and/or of the probability of occurrence, *P* (cf. [158, 159] with a risk typology accounting for these knowledge deficits). In LCA, risks are typically neglected, since the LCA approach follows the "best estimate" principle [160] – that is, the most likely situation is focused on ignoring both more positive and more negative outcomes. Consequently, technologies with inherent risks, like nuclear power, are often assessed rather positively in LCA [160]. The same would apply to IB if risks from the use of GMOs are present but not accounted for in LCA. However, since IB mostly uses mild conditions in terms of temperature and pressures and less toxic chemicals, LCA also tends to not sufficiently acknowledge the benefits that IB brings in terms of reduced risks in these areas (cf. also [5] for risks associated with biogas production). In 2017, several authors discussed the example of engineered nanomaterials in the context of whether risk assessment (RA) must be integrated in LCA or merely the results (cf. [161–163]). For a more in-depth-analysis of risks and IB, we refer to [5] (this book).

## 5.5 Grouping, Normalisation, Weighting, and Aggregation

Grouping, normalisation, and weighting are optional steps in the phase of life cycle impact assessment [30, 115]. Grouping is the process of aggregating several impact indicators that share a common characteristic in a group. In normalisation, a reference point is used – for example, a population equivalent – in order to illustrate the *relative* magnitude of the impacts [46]. In weighting, weights are assigned to single impact and damage categories in order to aggregate them subsequently, most often as a weighted sum [164].

In normalisation, impacts are normalised against a reference point, thereby resulting in *relative* magnitudes. These must be interpreted with great care. Suppose a product has a high absolute impact in category A and a low one in category B. In fact, category B does not represent a major environmental problem in that country. Further, suppose that per capita emissions used for normalisation are very high in category A and very low in category B. Then, after normalisation, the normalised value for category A is supposedly considerably lower than that for category B. An inexperienced reader could get the idea that category A is less important, which is not the case. The only information that can be inferred from the relative information is that the share of product A's impacts in the overall national impacts would be higher. This information can be very important when using the results for policy support. Thus, in the normalisation step, the information regarding the absolute magnitude gets lost, and comparisons between and aggregation of normalised indicator values must be avoided.

*Weighting*, the most often applied of the three processes, is typically conducted with the aim of aggregating the weighted categories afterwards. Several weighting techniques are applied in LCA [165]. For environmental indicators in general, Gan et al. [164] provide a comprehensive review on weighting and aggregation techniques. Huppes et al. [165] distinguish value-based weights and preference-based weights, with the latter further differentiated into individual and collective preferences and then further into revealed and stated preferences. With value-based weights, Huppes et al. mean ordinal weights and exclusion criteria like no use of GMOs. *Individual revealed preferences* can be determined by analysing, for example, individual damage prevention costs or individual payments for avoiding damage, and *individual stated preferences* by willingness-to-pay (WTP) for avoided damage or willingness-to-accept extra damage [165]. Further, collective preferences are more important for sustainability and life cycle assessment, and *collective stated preferences* can be determined, for example, by asking representative persons or groups or by distance-to-target approaches. It is obvious that such a weighting is value-laden.

On the other hand, aggregation of environmental impact (midpoint) categories to damage (endpoint) categories – for example, as a *potentially affected fraction* (PAF) of species for damage to ecosystems and DALYs for damage to human health – is largely science-based. However, a number of uncertainties in the impact pathway and its modelling make the results less reliable, and the final weighting and aggregation step at the level of the endpoints remains a value-laden one. There, the *cultural theory of risk*-based approach implemented in several endpoint methods, such as ReCiPe2016 (cf. Sect. 2.3.1), is most widespread in use. To represent an individualistic, a hierarchical, and an egalitarian perspective, weights for the three areas of protection of the natural environment, human health, and resources were derived in public opinion polls. It is obvious that opinions vary over time, between social groups, populations, etc. Ultimately, all weighting techniques have their drawbacks and even inconsistencies for LCA [165], and the statistical methods mentioned in Gan et al. [164] for environmental indicators are uncommon (and less suitable) for LCA applications. However, care must be taken not to conduct a weighting unintentionally when weighting all categories equally during interpretation, for example, when arguing that product A is superior in 10 out of 13 impact categories and, thus, also preferable in general. The criteria for selecting a weighting technique in LCA are reviewed by Johnsen and Løkke [166].

In any case, the large uncertainties induced by weighting and aggregation make a sensitivity analysis (cf. Sect. 5.4) compulsory to achieve higher confidence in the results (cf., e.g. do Carmo et al. [167] with an application for social LCA).

## 5.6 Streamlining and Simplified Methods

Despite the described methodical and application-determined challenges, sustainability and life cycle assessment is a powerful tool to assess product systems and processes comprehensively. However, conducting such studies is a complex, laborious, and time- and resource-consuming process [168]. In addition, the complexity of the toolset is often a barrier for practitioners [169], particularly in small- and medium-sized enterprises (SMEs) [170]. A further aspect hindering the applicability is the challenge of data availability [171]. In particular, in early stage development of products and processes, it is often difficult to fulfil data needs for thorough assessments. However, such assessments are particularly valuable in early stage developments, since unfavourable solutions can be sorted out and developments may be directed towards sustainable solutions as the early development stage determines the sustainability impact of products or services to a large extent (cf. Sect. 4.5).

In order to achieve a broad application, the assessment toolset must be simplified and standardised in numerous cases [172]. Thus, since the early developments of LCA, researchers and practitioners aimed for approaches that reduced the efforts of these studies, while simultaneously delivering sufficient results for the purposes of the study (cf. [173, 174]). Consequently, the idea of streamlining the elaborated toolset emerged rather early. The intention was to make life cycle-based assessment tools more applicable and relevant without compromising the approach as a whole [175]. Over the years, these approaches received increased attention, as demonstrated in the rising number of publications in this field [170].

In the efforts for streamlining, it is essential to ensure that the methodology is consistent with the aims and intended uses so that the results delivered fit the user's purposes [175]. Wittstock et al. [176] distinguish screening LCA, simplified LCA, and full LCA and characterise these in terms of purpose, completeness of assessment and data representativeness, documentation, and communication. Graedel and Allenby [171] state that almost every study is somehow streamlined and, thus, positioned between the two extremes – a "fully comprehensive LCA" and a basic "ecoscreening" approach.

There are a number of approaches to streamline sustainability and life cycle assessment. These approaches are also referred to as simplified or customised approaches that highlight the aspects of simplification or tailoring to the needs of specific purposes or customers. The following are the most important starting points ([cf. [170, 171, 177]):

- Exclusion of products violating given constraints
- Exclusion of or focus on specific life cycle stages, for example, consideration of the biotechnological production steps only
- Streamlining data acquisition, for example, by:
  - Consideration of qualitative information, relying on, for example, expert knowledge

  - Selecting inventory parameters, for example by using threshold values
  - Use of surrogate data such as secondary data from databases
  - Eased data acquisition, for example, by automation

- Streamlining impact analysis

  - Focus on specific impacts only, such as climate change
  - Elimination of the impact analysis step as a whole
  - Easing communication of the results

- Customising: tailoring the approaches to the needs of the users
- Elimination of the interpretation step, for example, merely providing inventory and impact assessment results

Despite the growing number of publications in the field of streamlined assessments, streamlined approaches in general focus mostly on environmental aspects with (purposeful) streamlined approaches for holistic sustainability assessments being rare. Moreover, publications in the IB field are, to the best of our knowledge, rather scarce and so are streamlined approaches for holistic sustainability assessments. The existing streamlined approaches focus mostly on environmental aspects.

Relevant for IB may be publications considering the supply of biogenic resources. These can be found among those for the agri-food industry [169, 170, 178] or biofuel production (e.g. [179]). Other publications address the use of biogenic materials in construction [177]. The focus is mainly on the raw material aspect with the argument that it is there that the major sustainability issues arise. However, this leaves out assessments for the processes themselves and subsequent life cycle stages. The main focus lies – as mentioned above – on environmental aspects (LCA), with few works such as [168, 179] with broader sustainability assessments.

Arzoumanidis et al. [170] compare the results of streamlined approaches with those of LCAs. They state that these approaches help to identify hotspots but that both approaches may lead to different results, particularly if the methods differ (e.g. in terms of different characterisation factors) or if the considered process deviates too much from used surrogate data.

Thus, it can be concluded that definite results require conducting comprehensive assessments. Streamlined methods have proven their worth to identify hotspots and to support *process and product developments*, particularly when data on industrial-scale implementation of these processes is scarce and uncertain. An overview of streamlined methods and tools can be found in Table 1.

## 5.7 *Communication of Sustainability and Environmental Assessment Results*

Apart from the central requirements of comprehensive, reproducible, and purpose-adequate results [180], the results must fit the needs of the decision-makers. These need to be informed in an understandable manner. Jessinghaus states that a good analytical assessment framework and a good set of indicators may enable informed decisions but that this does not necessarily lead to good decisions [181]. On the one hand, non-experts may not be able to understand the thorough and complex results of a sustainability and life cycle assessment. On the other hand, they may not be able to consider the consequences of simplifications, aggregation, and weighting sufficiently. In comparison to economic indicators, where usually a small number of rather intuitive and understandable indicators are used, environmental and social assessments often use larger sets of indicators with different rationales and may confuse the decision-maker. Thus, there is a demand for suitable approaches for the communication of assessment results in an understandable manner that avoids the oversimplification and negligence of important aspects of industrial biotechnologies. Publications in this field are usually closely related to methods for a methodical streamlining (cf. Sect. 5.6) and multicriteria decision support (cf. [182]) and complement these aspects.

On the level of policy evaluation, Jessinghaus [181] proposes to introduce the *Policy Performance Index (PPI)*. For every sustainability pillar, he defines an index comprising aggregated indicators: he distinguishes an Environmental Pressure Index, composed of 6 indicators with 60 components. Analogously, a Social Pressure Index and an Economic Performance Index are constructed. Together, these form the aggregated PPI. For a better communication of the results, these are presented graphically in a pie chart in which the colours represent the performances. The colour scheme advances in seven steps, from red for "crisis" to green for "very good". The pie chart consists of an inner circle, depicting the overall index (PPI); a middle ring, showing the performance regarding the sub-indices (environmental, social, economic); and an outer ring for the sub-sub-indices – that is, the individual indicators or indicator sets (Fig. 4).

The described index is one approach that may help voters in the evaluation of the decisions of their government and help governments to focus on their policies. The approach is referred to by other authors as "Dashboard of Sustainability" [180]. Naturally, it can be criticised for (over-)simplification and normative issues in constructing the indices and used weightings. However, it provides a multi-level investigation within a consistent indicator set. It helps to consider these issues in decision-making, to indicate trade-offs, and thus probably to initiate a further and deeper examination of specific issues. It also avoids neglecting single indicators as total scores or indices do.

For a comparison of product systems, Traverso et al. [180] developed the *Life Cycle Sustainability Assessment Dashboard (LCSD)*. They adopted the approach described above as a basis and substitute the used indicators and sub-indicator sets

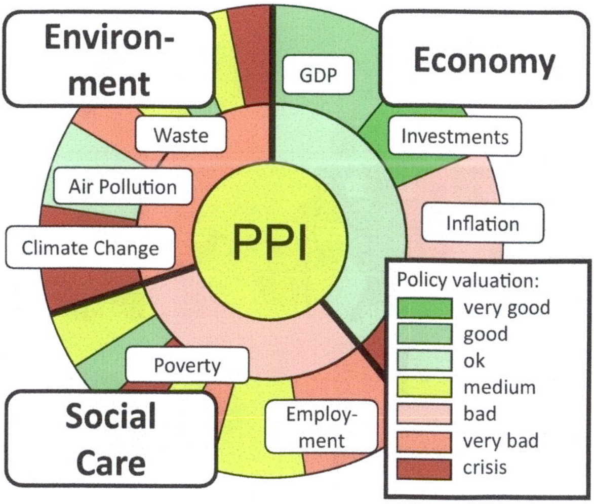


**Fig. 4** The Policy Performance Index modified in accordance with [181]

by the results of LCA, LCC, and sLCA with the indicator sets of these methodologies (cf. [180], Fig. 5). Their approach also allows for different weightings to account for the decision-makers' normative preferences. Again, the communication of assessments on multi-levels is enabled. The approach has the advantage of using well-established consistent and coherent LCSA indicator sets while simultaneously easing the communication of the results.

The *Life Cycle Sustainability Triangle (LCST)* is another approach that displays the results of a sustainability and life cycle assessment. In this approach, assessment criteria in the three sustainability dimensions are aggregated to scores for these dimensions – that is, an economic, a social, and an environmental score. The results are presented in a ternary plot or triangular diagram. Each of the corners depicts one of the dimensions, and the position of a point within the ternary plot describes different weightings of these criteria: the corners represent a 100% weight for one of the dimensions, and the space in between represents combined weightings. The assessment results are presented as colour. In the comparison of alternatives, one can mark areas where one or the other alternative are favourable (cf. Fig. 6). With the knowledge that obtaining weights for criteria is normative and, thus, problematic, decision-makers can use these results in order to take into account the uncertainties regarding the weights when taking decisions. The strengths of this approach lie in the enabling of a better understanding of the influence of criteria weights on the overall assessment results. However, the approach is limited to the level of the results of the three dimensions.

The *Gaia Biorefiner approach* supports sustainability benchmarking of products and processes in areas such as biofuels, biochemicals, and biomaterials. The approach aims at companies in their business development and investors to check for and ensure the realisation of sustainability benefits of products and technologies. These are assessed along their value chains in 10 indicator groups consisting of

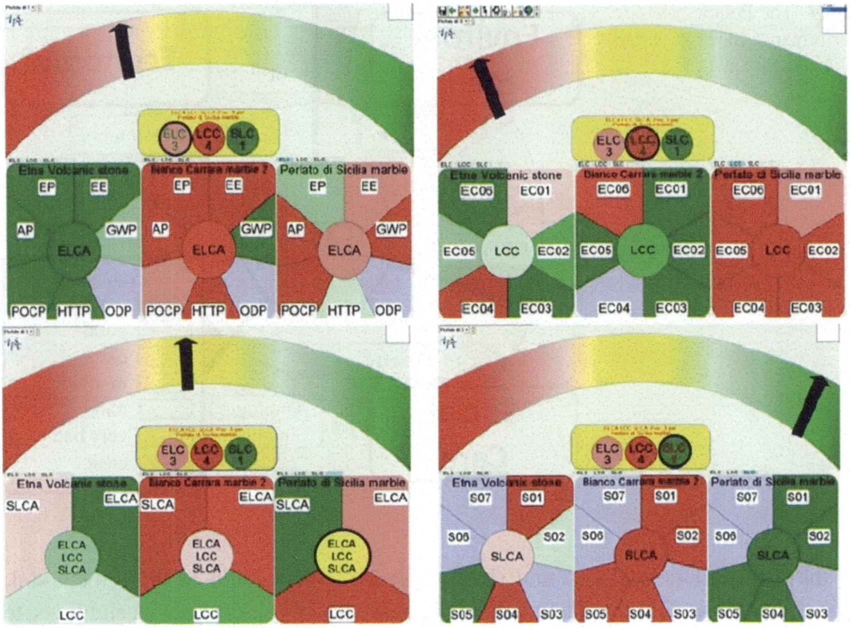


**Fig. 5** Graphical representation of the Life Cycle Sustainability Dashboard (LCSD) displaying results of environmental life cycle assessment (ELCA, upper left), life cycle costing (LCC, upper right), social life cycle assessment (sLCA, lower left), and life cycle sustainability analysis (LCSA, lower right) [183], https://www.mdpi.com/2071-1050/2/10/3309

30 indicators. The indicators can be selected from sustainability and life cycle assessment methods and be adapted for different purposes – that is, apart from applications in bioeconomy [184, 185], for example, in metallurgy as well [185]. The results are presented in a graphical manner (cf., e.g. Fig. 7). Coloured dots represent every indicator. The colour indicates potential competitive advantages (green), indifferences (orange), and potential risks (red) in comparison to a benchmark. The indicators are grouped in 10 environmental sustainability indicator groups that highlight the major impact categories. The approach enables companies and investors to identify the most eco-efficient and advanced solutions and innovations and the most favourable investments and ensure that the benefits of bio-based technologies and products are fully realised.

The approach has been developed for the environmental dimension of sustainability and life cycle assessments only but could be easily adapted to also encompass social and economic aspects within a corresponding framework. Although it was developed specifically for bioeconomy concepts, the methodology is generic and can be easily transferred to other application fields.

As is evident from this brief overview, with the exception of the Gaia Biorefiner approach, there are, to the best of our knowledge, no specific approaches for the presentation and communication of LCSA results in IB. Although adaptable and configurable for this application field, there are only a few studies on the applications

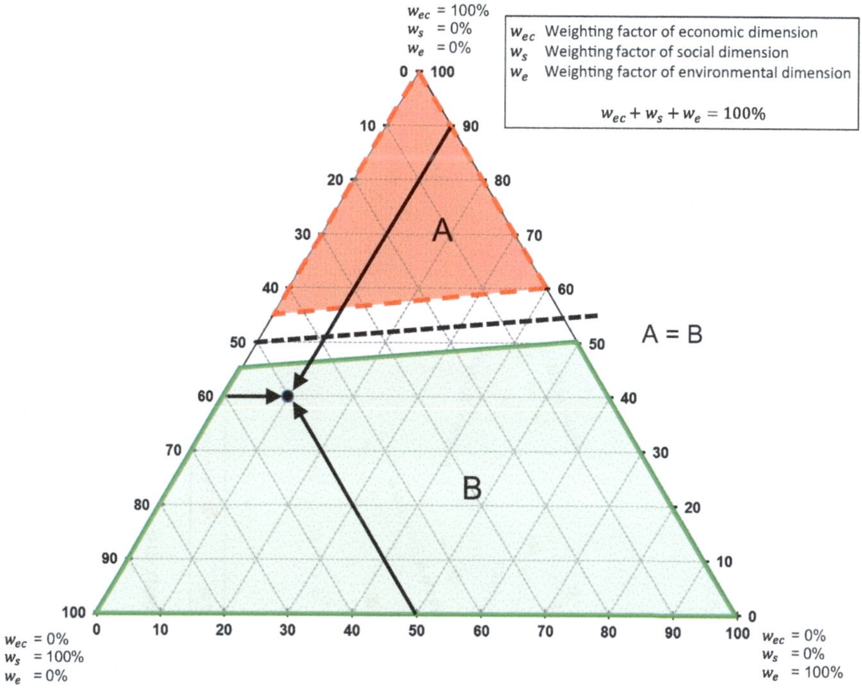


**Fig. 6** Life Cycle Sustainability Triangle for the comparison of alternatives A and B redrawn and slightly modified from [183], https://www.mdpi.com/sustainability/sustainability-02-03309/article_deploy/html/images/sustainability-02-03309-g004.png

in this field, even though there is a huge need for sustainability and life cycle assessment decision support in research, industry, and politics. It can also be stated that the focus of the existing research lies in comparative assessments of products and technologies against benchmarks. Consequential and prospective assessments are not covered and provide an important open research field.

## 6 Conclusions

Sustainability and life cycle assessment can contribute to the further development and implementation of industrial biotechnology (IB) since process alternatives can be assessed, investment decisions and policy development can be supported, and assessment results can be used as a basis for communication with stakeholders. Against the backdrop of a growing and fast-developing field of IB and an increasing number of works in this field, we present an explorative review of methods and applications.

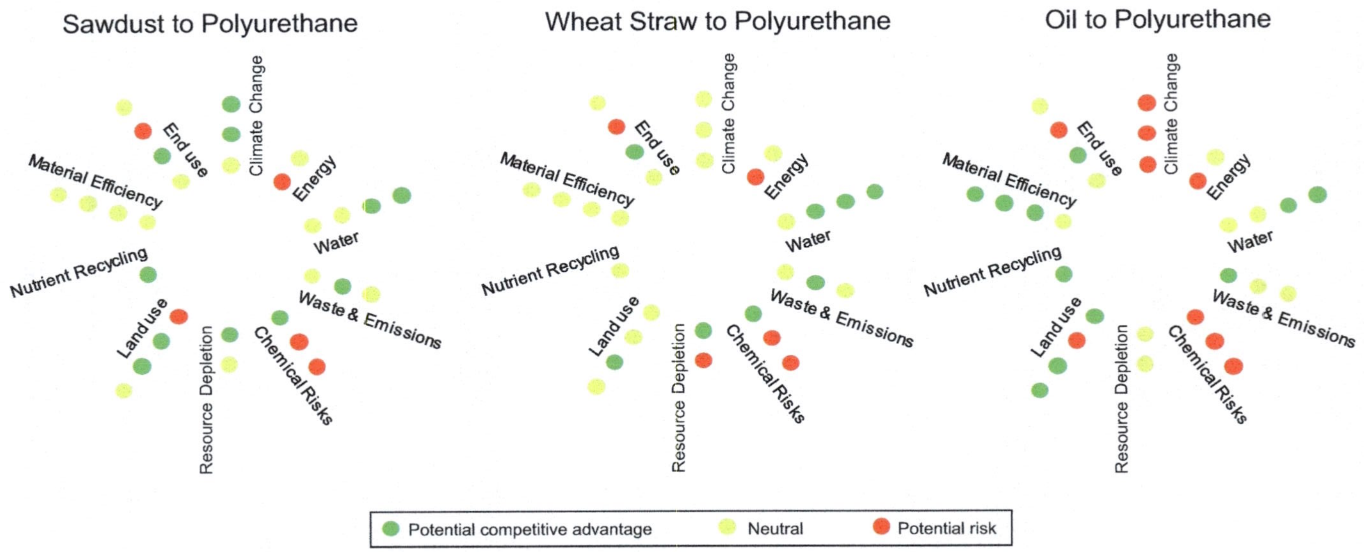


**Fig. 7** Example for the graphical results of the GAIA Biorefiner approach to (environmental) sustainability benchmarking [184]

There exists a comprehensive set of approaches, and tools for sustainability and life cycle assessment of IB. This ranges from simple metrics to elaborate assessment frameworks and combinations of several comprehensive frameworks, partially coupled with further model systems. In the analysed literature, a divide becomes apparent, thereby separating the communities of the (sustainability and life cycle) assessment and "their" journals and the community of researchers in industrial biotechnology who focus more on technical and development aspects. Although an early involvement of assessment in the development process often offers an opportunity to support development and orientate it towards more sustainable solutions, these aspects play only a subordinate role in IB journals. Further, the focus of the existing literature on sustainability and life cycle assessments of IB is on energy use and biofuels, whereas materials and chemical production are rather seldom covered.

The first task in any assessment is the identification and configuration of applicable frameworks, methodologies, methods, and tools. Here, this contribution provides orientation and guidance. In particular, the goal and scope definition of the assessment are decisive, thereby indicating whether rather simple indicators, streamlined approaches, or comprehensive "full-scale" assessments covering all sustainability dimensions are required. In addition, the selected methodologies, methods, and tools must take into consideration the particular characteristics of IB – such as the feedstock originating from forestry or agriculture, associated logistical aspects, the biotechnological conversion processes, and the often low technology readiness level. Due to the biogenic origin of the feedstock, numerous studies regard emissions from the use of biogenic resources as carbon neutral. However, this is only justified as long as no additional GHG emissions are provoked directly – for example, from the conversion to methane emissions – or indirectly from land use changes or fertiliser use. Moreover, temporal shifts of GHG emissions may be relevant, for example, when harvesting timber. Land competition with severe social implications and biodiversity losses are additional aspects that must be considered in sustainability and life cycle assessments since the use of IB can lead here to different impact patterns in comparison to established fossil resource-based technologies and products. IB often utilises genetically modified organisms (GMOs). When reading about LCA studies dealing with GMOs, the reader must keep an eye on which aspects of GMO use were covered and which were not. LCA is not the first choice to assess risks, including those originating from genetic information spills from GMO into the environment. Here, other tools must at least be considered. Apart from the feedstock and production phases, the use of products and materials in cascades and the EoL phase require special attention. Material cycles of biogenic materials are different from those of "technical materials" such as metals – unlike, for example, steel, the degradation processes of biomaterials inhibit recycling without quality losses. Cascades with decreasing material qualities are more relevant, leading to challenges regarding the allocation of impacts. The fate of products and materials of biological origin must also be accounted for when the EoL phase differs – for example, in cases of biodegradability and composting in comparison to recycling, energetic recovery, or landfilling.

In comparison to research in other application fields, sustainability and life cycle assessment for IB also requires special attention with respect to several common methodological aspects. An assessment in IB is often conducted in a prospective manner and must also consider the consequences of, for example, technology diffusion by estimating impacts of a realisation of a process in the future and resultant systemic changes. This poses specific requirements regarding the approach – that is, the data used – and methods and tools applied. This can, for example, also result in explicit endogenous modelling of market behaviour or a coupling with other model types and the treatment of uncertainties and handling of risks. Further, sustainability and life cycle assessment is also a field that requires multicriteria assessment. Approaches of multicriteria decision analysis (MCDA) have proven beneficial in achieving the aims and objectives of the study.

Several approaches have been developed to streamline the comprehensive toolset for sustainability and life cycle assessment in order to reduce the necessary efforts without compromising the general approach and necessary depth of the study. Here, a choice following the purpose of the study is advised, using simplified methods for certain basic estimates, while a full-scale assessment is conducted when detailed and comprehensive results are required.

A further issue is the communication of study results. Again, trade-offs exist between how simply and intuitively the results are presented and communicated to ease explanation and understanding and the danger of oversimplification and negligence of important results. In addition, a few graphical presentation approaches within a multi-level indicator framework are presented. This facilitates the required understandable presentation and communication on the basis of comprehensive full-scale assessments and prohibits or reduces the risk of oversimplification.

In this paper, though we follow a broad approach, ranging from basic to very specific topics, we can only present an overview and selection of those aspects which are – from our rather subjective viewpoint – most relevant. Since entire streams of literature exist for most topics covered in sections and subsections, we tried to refer the interested reader to these further works. However, some of these topics are specific contributions in this book. For others, we reference respective studies.

In principle, we tried to include a broad understanding of sustainability. However, this work mainly focuses on the environmental pillar, leaving out social and economic aspects in the detailed descriptions. We also do not deal with the field of eco- or sustainable design in the field of IB which may – for example, in the combination with synthetic biology – be an interesting field of further research and thus assessment. On the economic side, IB and its characteristics can lead to much smaller scales and lower by-product production as well as lead to new business models with smaller and more decentralised production sites [186]. To the best of our knowledge, this field has barely taken into consideration the larger consequences for the industry structure and, consequently, has different assessment needs and results.

Despite this room for further research, this study provides an overview, orientation, and guidance for own studies of students, researchers, and practitioners on this important and fast-emerging topic and, thus, contributes to the further development and implementation of IB.

## References


1. Venkatesh A, Posen ID, MacLean HL et al (2019) Environmental aspects of biotechnology. Adv Biochem Eng Biotechnol. https://doi.org/10.1007/10_2019_98
2. Lindner JP, Beck T, Bos U et al (2019) Assessing land use and biodiversity impacts of industrial biotechnology. Adv Biochem Eng Biotechnol. https://doi.org/10.1007/10_2019_114
3. Asveld L, Osseweijer P, Posada JA (2019) Societal and ethical issues in industrial biotechnology. Adv Biochem Eng Biotechnol. https://doi.org/10.1007/10_2019_100
4. Macombe C (2019) Social life cycle assessment for industrial biotechnology. Adv Biochem Eng Biotechnol. https://doi.org/10.1007/10_2019_99
5. Chen C, Reniers G (2018) Risk assessment of processes and products in industrial biotechnology. Adv Biochem Eng Biotechnol. https://doi.org/10.1007/10_2018_74
6. Saling P (2019) Assessing industrial biotechnology products with LCA and eco-efficiency. Adv Biochem Eng Biotechnol. https://doi.org/10.1007/10_2019_102
7. Hodgson E, Ruiz-Molina M-E, Marazza D et al (2016) Horizon scanning the European bio-based economy: a novel approach to the identification of barriers and key policy interventions from stakeholders in multiple sectors and regions. Biofuels Bioprod Bioref 10(5):508–522. https://doi.org/10.1002/bbb.1665
8. Festel G (2018) Economic aspects of industrial biotechnology. Adv Biochem Eng Biotechnol. https://doi.org/10.1007/10_2018_70
9. Ness B, Urbel-Piirsalu E, Anderberg S et al (2007) Categorising tools for sustainability assessment. Ecol Econ 60(3):498–508. https://doi.org/10.1016/j.ecolecon.2006.07.023
10. Saurat M, Ritthoff M, Smith L (2016) Overview of existing sustainability assessment methods and tools, and of relevant standards: deliverable 1.1 in the EU project SAMT – Sustainbility assessment methods and tools to support decision-making in the process industries
11. Singh RK, Murty HR, Gupta SK et al (2009) An overview of sustainability assessment methodologies. Ecol Indicators 9(2):189–212. https://doi.org/10.1016/j.ecolind.2008.05.011
12. WCED (1987) Our common future. Oxford University Press, Oxford
13. Purvis B, Mao Y, Robinson D (2019) Three pillars of sustainability: in search of conceptual origins. Sustain Sci 14(3):681–695. https://doi.org/10.1007/s11625-018-0627-5
14. Rockström J, Steffen W, Noone K et al (2009) Planetary boundaries: exploring the safe operating space for humanity. Ecol Soc 14(2). https://doi.org/10.5751/ES-03180-140232
15. Sala S, Ciuffo B, Nijkamp P (2015) A systemic framework for sustainability assessment. Ecol Econ 119:314–325. https://doi.org/10.1016/j.ecolecon.2015.09.015
16. Becker B (1997) Sustainability assessment: a review of values, concepts and methodological approaches. Issues in agriculture, vol 10. CGIAR, Washington, D.C.
17. Sala S, Farioli F, Zamagni A (2013) Progress in sustainability science: lessons learnt from current methodologies for sustainability assessment: Part 1. Int J Life Cycle Assess 18 (9):1653–1672. https://doi.org/10.1007/s11367-012-0508-6
18. Sala S, Farioli F, Zamagni A (2013) Life cycle sustainability assessment in the context of sustainability science progress (part 2). Int J Life Cycle Assess 18(9):1686–1697. https://doi.org/10.1007/s11367-012-0509-5
19. Klöpffer W (2003) Life-cycle based methods for sustainable product development. Int J Life Cycle Assess 8(3):157–159
20. Comes T, Doll C et al (2010) Challenges for national clean air policy: a national integrated assessment model for Germany. Ökologisches Wirtschaften 25(2):34–37
21. Wrisberg N, Udo de Haes HA, Helias A, Triebswetter U et al (eds) (2002) Analytical tools for environmental design and management in a systems perspective: the combined use of analytical tools. Eco-efficiency in industry and science, vol 10. Springer, Dordrecht
22. Finnveden G, Moberg Å (2005) Environmental systems analysis tools – an overview. J Clean Prod 13(12):1165–1173. https://doi.org/10.1016/j.jclepro.2004.06.004
23. Gasparatos A, Scolobig A (2012) Choosing the most appropriate sustainability assessment tool. Ecol Econ 80:1–7. https://doi.org/10.1016/j.ecolecon.2012.05.005

24. Singh RK, Murty HR, Gupta SK et al (2012) An overview of sustainability assessment methodologies. Ecol Indic 15(1):281–299. https://doi.org/10.1016/j.ecolind.2011.01.007
25. de Meester S, van der Vorst G, van Langenhove H et al (2013) Sustainability assessment methods and tools. In: Reniers GLL, Sörensen K, Vrancken K (eds) Management principles of sustainable industrial chemistry, vol 32. Wiley-VCH Verlag GmbH & Co. KGaA, Weinheim, pp 55–88
26. Pihkola H, Pajula T, Federley M et al (2017) Sustainability assessment in the process industries – current practice and paths for future development: conclusions and recommendations from the SAMT project, vol 299. VTT Technology, Espoo
27. Organisation for Economic Co-operation and Development (2010) Guidance on sustainability impact assessment. https://doi.org/10.1787/9789264086913-en
28. Pope J, Annandale D, Morrison-Saunders A (2004) Conceptualising sustainability assessment. Environ Impact Assess Rev 24(6):595–616. https://doi.org/10.1016/j.eiar.2004.03.001
29. Guinée J (2016) Life cycle sustainability assessment: what is it and what are its challenges? In: Clift R, Druckman A (eds) Taking stock of industrial ecology. Springer International Publishing, Cham, pp 45–68
30. DIN EN ISO Environmental management – Life cycle assessment – Principles and framework (ISO 14040:2006). Accessed 20 Mar 2019
31. Baitz M (2017) Attributional life cycle assessment. In: Curran MA (ed) Goal and scope definition in life cycle assessment. Springer, Dordrecht, pp 123–143
32. Hottle TA, Bilec MM, Landis AE (2017) Biopolymer production and end of life comparisons using life cycle assessment. Resour Conserv Recycl 122:295–306. https://doi.org/10.1016/j.resconrec.2017.03.002
33. Maga D, Thonemann N, Hiebel M et al (2019) Comparative life cycle assessment of first- and second-generation ethanol from sugarcane in Brazil. Int J Life Cycle Assess 24(2):266–280. https://doi.org/10.1007/s11367-018-1505-1
34. Haitz F, Radloff S, Rupp S et al (2018) Chemo-enzymatic epoxidation of lallemantia ibericaseed oil: process development and economic-ecological evaluation. Appl Biochem Biotechnol 185(1):13–33. https://doi.org/10.1007/s12010-017-2630-1
35. Laure S, Leschinsky M, Fröhling M et al (2017) Assessment of an organosolv lignocellulose biorefinery concept based on a a material flow analysis of a pilot plant. Cellul Chem Technol 48(9–10):793–798
36. Silalertruksa T, Gheewala SH, Pongpat P et al (2017) Environmental sustainability of oil palm cultivation in different regions of Thailand: greenhouse gases and water use impact. J Clean Prod 167:1009–1019. https://doi.org/10.1016/j.jclepro.2016.11.069
37. Vargas-Ramirez JM, Wiesenborn DP, Ripplinger DG et al (2017) Carbon footprint of industrial-beet sugars stored as raw thick juice for use as a fermentation feedstock. J Clean Prod 162:1418–1429. https://doi.org/10.1016/j.jclepro.2017.06.153
38. DIN EN ISO Environmental management – Life cycle assessment – Requirements and guidelines(ISO 14044:2006 + Amd 1:2017). Accessed 20 Mar 2019
39. Hunkeler DJ (ed) (2008) Environmental life cycle costing. CRC Press, Boca Raton
40. Guinée JB, Heijungs R, Huppes G et al (2011) Life cycle assessment: past, present, and future. Environ Sci Technol 45(1):90–96. https://doi.org/10.1021/es101316v
41. Curran MA (2017) Chapter 1: overview of goal and scope definition in life cycle assessment. In: Curran MA (ed) Goal and scope definition in life cycle assessment. Springer, Dordrecht, pp 1–62
42. IES (2010) ILCD handbook: analysing of existing environmental impact assessment methodologies for use in life cycle assessment. IES, Ispra
43. Huijbregts MAJ, Steinmann ZJN, Elshout PMF et al (2016) ReCiPe 2016: a harmonized life cycle impact assessment method at midpoint and endpoint level report I: characterization. National Institute of Public Health and Environment Ministry of Health, Wellness and Sport, Bilthoven

44. Kühnen M, Hahn R (2017) Indicators in social life cycle assessment: a review of frameworks, theories, and empirical experience. J Indus Ecol 21(6):1547–1565. https://doi.org/10.1111/jiec.12663
45. Benoît C, Mazijn B (2009) Guidelines for social life cycle assessment of products. United Nations Environment Programme, Paris
46. Hauschild MZ (2018) Introduction to LCA methodology. In: Hauschild MZ, Rosenbaum RK, Olsen SI (eds) Life cycle assessment: theory and practice. Springer, Cham
47. Comes T, Hiete M, Schultmann F (2013) An approach to multi-criteria decision problems under severe uncertainty. J Multi-Criteria Decis Anal 20(1–2):29–48. https://doi.org/10.1002/mcda.1487
48. Bjørn A, Laurent A, Owsianiak M et al (2018) Chapter 7: goal definition. In: Hauschild MZ, Rosenbaum RK, Olsen SI (eds) Life cycle assessment: theory and practice. Springer, Cham, pp 67–74
49. Beaussier T, Caurla S, Bellon-Maurel V et al (2019) Coupling economic models and environmental assessment methods to support regional policies: a critical review. J Clean Prod 216:408–421. https://doi.org/10.1016/j.jclepro.2019.01.020
50. Kirchherr J, Reike D, Hekkert M (2017) Conceptualizing the circular economy: an analysis of 114 definitions. Resour Conserv Recycl 127:221–232. https://doi.org/10.1016/j.resconrec.2017.09.005
51. Searchinger T, Heimlich R, Houghton RA et al (2008) Use of U.S. croplands for biofuels increases greenhouse gases through emissions from land-use change. Science 319 (5867):1238–1240. https://doi.org/10.1126/science.1151861
52. Arvidsson R, Tillman A-M, Sandén BA et al (2018) Environmental assessment of emerging technologies: recommendations for prospective LCA. J Indus Ecol 22(6):1286–1294. https://doi.org/10.1111/jiec.12690
53. Olsen SI, Borup M, Andersen PD (2018) Chapter 21: future-oriented LCA. In: Hauschild MZ, Rosenbaum RK, Olsen SI (eds) Life cycle assessment: theory and practice. Springer, Cham, pp 499–518
54. Buyle M, Audenaert A, Billen P et al (2019) The future of Ex-Ante LCA? Lessons learned and practical recommendations. Sustainability 11(19):5456. https://doi.org/10.3390/su11195456
55. Bergesen JD, Suh S (2016) A framework for technological learning in the supply chain: a case study on CdTe photovoltaics. Appl Energy 169:721–728. https://doi.org/10.1016/j.apenergy.2016.02.013
56. Curran MA (ed) (2017) Goal and scope definition in life cycle assessment. Springer, Dordrecht
57. Klöpffer W, Grahl B (2014) Life cycle assessment (LCA): a guide to best practice. Wiley-VCH Verlag GmbH & Co. KGaA, Weinheim
58. Jolliet O, Saadé-Sbeih M, Shaked S et al (2016) Environmental life cycle assessment. CRC Press, Boca Raton
59. Bjørn A, Owsianiak M, Laurent A et al (2018) Scope definition. In: Hauschild MZ, Rosenbaum RK, Olsen SI (eds) Life cycle assessment: theory and practice. Springer, Cham, pp 75–116
60. Stichnothe H (2019) Sustainability evaluation. Adv Biochem Eng Biotechnol 166:519–539. https://doi.org/10.1007/10_2016_71
61. VDI e.V (2016) VDI 6310: Nachhaltigkeitsbewertung und Klassifikation von Bioraffinerien. Beuth, Berlin
62. Fröhling M, Schweinle J, Meyer J-C et al (2011) Logistics of renewable raw materials. In: Ulber R, Sell D, Hirth T (eds) Renewable raw materials: new feedstocks for the chemical industry, vol 47. Wiley, Hoboken, pp 49–94
63. Zimmer T, Rudi A, Müller A-K et al (2017) Modeling the impact of competing utilization paths on biomass-to-liquid (BtL) supply chains. Appl Energy 208:954–971. https://doi.org/10.1016/j.apenergy.2017.09.056

64. Raschka A, Carus M (2012) Stoffliche Nutzung von Biomasse: Basisdaten für Deutschland, Europa und die Welt. Umweltbundesamt Förderkennzeichen 3710 93 109. nova-Institut GmbH, Hürth
65. Jering A, Klatt A, Seven J et al (2013) Sustainable use of global land and biomass resources. Umweltbundesamt, Dessau-Roßlau
66. UNEP (2014) Assessing global land use: balancing consumption with sustainable supply. United Nations Environment Programme, Nairobi
67. Lenk F, Bröring S, Herzog P et al (2007) On the usage of agricultural raw materials–energy or food? An assessment from an economics perspective. Biotechnol J 2(12):1497–1504. https://doi.org/10.1002/biot.200700153
68. Wright MM, Brown RC (2007) Comparative economics of biorefineries based on the biochemical and thermochemical platforms. Biofuels Bioprod Bioref 1(1):49–56. https://doi.org/10.1002/bbb.8
69. Bask A, Rajahonka M (2017) The role of environmental sustainability in the freight transport mode choice. Int J Phys Distrib Logist Manag 47(7):560–602. https://doi.org/10.1108/IJPDLM-03-2017-0127
70. Santos G, Behrendt H, Teytelboym A (2010) Part II: policy instruments for sustainable road transport. Res Transp Econ 28(1):46–91. https://doi.org/10.1016/j.retrec.2010.03.002
71. Kerdoncuff P (2008) Modellierung und Bewertung von Prozessketten zur Herstellung von Biokraftstoffen der zweiten Generation. Karlsruhe, Karlsruhe
72. Schwaderer F (2012) Integrierte Standort-, Kapazitäts- und Technologieplanung von Wertschöpfungsnetzwerken zur stofflichen und energetischen Biomassenutzung. Zugl.: Karlsruhe, Karlsruher Inst. für Technologie (KIT), Diss., 2012, Technische Informationsbibliothek u. Universitätsbibliothek; KIT Scientific-Publ, Hannover, Karlsruhe
73. Hiete M, Ludwig J, Schultmann F (2012) Intercompany energy integration. J Indus Ecol 16 (5):689–698. https://doi.org/10.1111/j.1530-9290.2012.00462.x
74. Jegannathan KR, Nielsen PH (2013) Environmental assessment of enzyme use in industrial production – a literature review. J Clean Prod 42:228–240. https://doi.org/10.1016/j.jclepro.2012.11.005
75. Ma SK, Gruber J, Davis C et al (2010) A green-by-design biocatalytic process for atorvastatin intermediate. Green Chem 12(1):81–86. https://doi.org/10.1039/b919115c
76. Renner I, Klöpffer W (2005) Untersuchung der Anpassung von Ökobilanzen an spezifische Erfordernisse biotechnischer Prozesse und Produkte. UBA-Texte, Berlin
77. Fantke P, Ernstoff A (2018) LCA of chemicals and chemical products. In: Hauschild MZ, Rosenbaum RK, Olsen SI (eds) Life cycle assessment: theory and practice, vol 22. Springer, Cham, pp 783–815
78. Jiménez-González C (2018) Life cycle assessment. In: Constable D, Jiménez-González C, Anastas PT (eds) Green metrics, vol 29. Wiley-VCH, Weinheim, pp 95–124
79. Kralisch D (2008) Application of LCA in process development. In: Lapkin A, Constable DJC (eds) Green chemistry metrics. Wiley, Chichester, pp 248–271
80. Kralisch D, Ott D, Gericke D (2015) Rules and benefits of life cycle assessment in green chemical process and synthesis design: a tutorial review. Green Chem 17(1):123–145. https://doi.org/10.1039/c4gc01153h
81. Sadhukhan J, Ng KS, Hernandez EM (2014) Biorefineries and chemical processes: design, integration and sustainability analysis. Wiley, Chichester
82. Zhang L, Wang Q, Hessel V (2018) Green chemistry metrics and life cycle assessment for microflow continuous processing. In: Constable D, Jiménez-González C, Anastas PT (eds) Green metrics, vol 37. Wiley-VCH, Weinheim, pp 157–206
83. Woodley JM (2018) Benchmarking the sustainability of biocatalytic processes. In: Constable D, Jiménez-González C, Anastas PT (eds) Green metrics, vol 86. Wiley-VCH, Weinheim, pp 207–230
84. Constable D, Jiménez-González C, Anastas PT (eds) (2018) Green metrics. Handbook of green chemistry, vol 11. Wiley-VCH, Weinheim

85. Dicks AP, Hent A (2015) Green chemistry metrics: a guide to determining and evaluating process greenness. SpringerBriefs in molecular science. Springer, Cham
86. Lapkin A, Constable DJC (eds) (2008) Green chemistry metrics. Wiley, Chichester
87. McCarty LS, Borgert CJ (2006) Review of the toxicity of chemical mixtures: theory, policy, and regulatory practice. Regul Toxicol Pharmacol 45(2):119–143. https://doi.org/10.1016/j.yrtph.2006.03.004
88. Smith R (2016) Chemical process design and integration.2nd edn. Wiley, New York
89. Constable DJC (2018) Green chemistry metrics. In: Constable D, Jiménez-González C, Anastas PT (eds) Green metrics. Wiley-VCH, Weinheim, pp 1–27
90. Andraos J (2008) Application of green metrics analysis to chemical reactions and synthesis plans. In: Lapkin A, Constable DJC (eds) Green chemistry metrics. Wiley, Chichester, pp 69–199
91. Sheldon RA, Sanders JPM (2015) Toward concise metrics for the production of chemicals from renewable biomass. Catal Today 239:3–6. https://doi.org/10.1016/j.cattod.2014.03.032
92. Sheldon RA (2017) The E factor 25 years on: the rise of green chemistry and sustainability. Green Chem 19(1):18–43. https://doi.org/10.1039/c6gc02157c
93. Constable DJC, Jimenez-Gonzalez C, Lapkin A (2008) Process metrics. In: Lapkin A, Constable DJC (eds) Green chemistry metrics. Wiley, Chichester, pp 228–247
94. Anastas ND, Leazer J, Gonzalez MA et al (2018) Expanding rational molecular design beyond pharma: metrics to guide safer chemical design. In: Constable D, Jiménez-González C, Anastas PT (eds) Green metrics, vol 37. Wiley-VCH, Weinheim, pp 29–48
95. DOD (2011) Technology readiness assessment guidance. Department of Defense, Wahsington
96. European Commission (2011) High level expert group on key enabling technologies 2011. European Commission, Brussels
97. German Federal Government (2012) Biorefineries roadmap. German Federal Government, Berlin
98. Biwer A, Rudershausen A, Heinzle E et al (2001) Ökologische und ökonomische Evaluation biokatalytischer Prozesse während ihrer Entwicklung. Biokatalyse (Special edition of DBU BioSpektrum) 7:6–8. 15–24
99. Thomassen G, van Dael M, van Passel S et al (2019) How to assess the potential of emerging green technologies? Towards a prospective environmental and techno-economic assessment framework. Green Chem 21(18):4868–4886. https://doi.org/10.1039/C9GC02223F
100. JRC-IEA (2010) ILCD handbook – general guide on LCA – detailed guidance. Publications Office of the European Union, Luxembourg
101. Ekvall T, Azapagic A, Finnveden G et al (2016) Attributional and consequential LCA in the ILCD handbook. Int J Life Cycle Assess 21(3):293–296. https://doi.org/10.1007/s11367-015-1026-0
102. Zamagni A, Guinée J, Heijungs R et al (2012) Lights and shadows in consequential LCA. Int J Life Cycle Assess 17(7):904–918. https://doi.org/10.1007/s11367-012-0423-x
103. Nilsson A, Shabestary K, Brandão M et al (2019) Environmental impacts and limitations of third-generation biobutanol: Life cycle assessment of n-butanol produced by genetically engineered cyanobacteria. J Indus Ecol:6. https://doi.org/10.1111/jiec.12843
104. Parsons S, Chuck CJ, McManus MC (2018) Microbial lipids: progress in life cycle assessment (LCA) and future outlook of heterotrophic algae and yeast-derived oils. J Clean Prod 172:661–672. https://doi.org/10.1016/j.jclepro.2017.10.014
105. Weidema BP, Ekvall T, Heijungs R (2009) Guidelines for application of deepended and broadened LCA. Technical Report of CALCAS project
106. Moni SM, Mahmud R, High K et al (2019) Life cycle assessment of emerging technologies: a review. J Indus Ecol:1. https://doi.org/10.1111/jiec.12965
107. Tsatsakis AM, Nawaz MA, Kouretas D et al (2017) Environmental impacts of genetically modified plants: a review. Environ Res 156:818–833. https://doi.org/10.1016/j.envres.2017.03.011

108. Harder R, Peters GM, Molander S et al (2016) Including pathogen risk in life cycle assessment: the effect of modelling choices in the context of sewage sludge management. Int J Life Cycle Assess 21(1):60–69. https://doi.org/10.1007/s11367-015-0996-2
109. Kodym A, Afza R (2003) Physical and chemical mutagenesis. Methods Mol Biol 236:189–204. https://doi.org/10.1385/1-59259-413-1:189
110. Seager TP, Trump BD, Poinsatte-Jones K et al (2017) Why life cycle assessment does not work for synthetic biology. Environ Sci Technol 51(11):5861–5862. https://doi.org/10.1021/acs.est.7b01604
111. Guinée JB, Heijungs R, van der Voet E (2009) A greenhouse gas indicator for bioenergy: some theoretical issues with practical implications. Int J Life Cycle Assess 14(4):328–339. https://doi.org/10.1007/s11367-009-0080-x
112. Brandão M, Levasseur A, Kirschbaum MUF et al (2013) Key issues and options in accounting for carbon sequestration and temporary storage in life cycle assessment and carbon footprinting. Int J Life Cycle Assess 18(1):230–240. https://doi.org/10.1007/s11367-012-0451-6
113. Garcia R, Freire F (2014) Carbon footprint of particleboard: a comparison between ISO/TS 14067, GHG Protocol, PAS 2050 and climate declaration. J Clean Prod 66:199–209. https://doi.org/10.1016/j.jclepro.2013.11.073
114. Brandão M, Kirschbaum MUF, Cowie AL et al (2019) Quantifying the climate change effects of bioenergy systems: comparison of 15 impact assessment methods. GCB Bioenergy 11 (5):727–743. https://doi.org/10.1111/gcbb.12593
115. Rosenbaum RK, Hauschild MZ, Boulay A-M et al (2018) Life cycle impact assessment. In: Hauschild MZ, Rosenbaum RK, Olsen SI (eds) Life cycle assessment: theory and practice, vol 20. Springer, Cham, pp 167–270
116. Wijnants L, Allacker K, de Troyer F (2019) Life-cycle assessment of timber frame constructions – the case of rooftop extensions. J Clean Prod 216:333–345. https://doi.org/10.1016/j.jclepro.2018.12.278
117. Fenner AE, Kibert CJ, Woo J et al (2018) The carbon footprint of buildings: a review of methodologies and applications. Renew Sustain Energy Rev 94:1142–1152. https://doi.org/10.1016/j.rser.2018.07.012
118. Elshout PMF, van der Velde M, van Zelm R et al (2019) Comparing greenhouse gas footprints and payback times of crop-based biofuel production worldwide. Biofuels:1–7. https://doi.org/10.1080/17597269.2019.1630056
119. Ögmundarson Ó, Sukumara S, Laurent A et al (2020) Environmental hotspots of lactic acid production systems. GCB Bioenergy 12(1):19–38. https://doi.org/10.1111/gcbb.12652
120. Cherubini F, Strømman AH (2011) Life cycle assessment of bioenergy systems: state of the art and future challenges. Bioresour Technol 102:437–451
121. IPCC (2019) 2019 refinement to the 2006 IPCC guidelines for national greenhouse gas inventories. IPCC, Geneva
122. Pachauri RK, Mayer L, IPCC SYR TSU (eds) (2015) Climate change 2014: synthesis report. Intergovernmental Panel on Climate Change, Geneva
123. IPCC (2006) 2006 IPCC Guidelines for National Greenhouse Gas Inventories. IGES, Japan
124. Crutzen PJ, Mosier AR, Smith KA et al (2008) $N_2O$ release from agro-biofuel production negates global warming reduction by replacing fossil fuels. Atmos Chem Phys 8(2):389–395. https://doi.org/10.5194/acp-8-389-2008
125. Bleischwitz R, Bringezu S (2009) Sustainable resource management: global trends, visions and policies. Greenleaf Publication, Sheffield
126. UNESDA (2019) World urbanization prospects the 2018 revision. United Nations, Department of Economic and Social Affairs, Population Division, Bruxelles
127. Koellner T, de Baan L, Beck T et al (2013) UNEP-SETAC guideline on global land use impact assessment on biodiversity and ecosystem services in LCA. Int J Life Cycle Assess 18 (6):1188–1202. https://doi.org/10.1007/s11367-013-0579-z

128. Milà i Canals L, Bauer C, Depestele J et al (2007) Key elements in a framework for land use impact assessment within LCA (11 pp). Int J Life Cycle Assess 12(1):5–15. https://doi.org/10.1065/lca2006.05.250
129. Beck T, Bos U, Wittstock B et al (2016) LANCA®: land use indicator value calculation in life cycle assessment. Fraunhofer-Verlag, Stuttgart
130. Bos U, Horn R, Beck T et al (2010) LANCA® – characterization factors for life cycle impact assessment: Version 2.0. Fraunhofer-Verlag, Stuttgart
131. Chaudhary A, Verones F, de Baan L et al (2015) Quantifying land use impacts on biodiversity: combining species-area models and vulnerability indicators. Environ Sci Technol 49 (16):9987–9995. https://doi.org/10.1021/acs.est.5b02507
132. Teixeira RFM, Maia de Souza D, Curran MP et al (2016) Towards consensus on land use impacts on biodiversity in LCA: UNEP/SETAC life cycle initiative preliminary recommendations based on expert contributions. J Clean Prod 112:4283–4287. https://doi.org/10.1016/j.jclepro.2015.07.118
133. Maga D, Hiebel M, Thonemann N (2019) Life cycle assessment of recycling options for polylactic acid. Resour Conserv Recycl 149:86–96. https://doi.org/10.1016/j.resconrec.2019.05.018
134. Butterworth J, Morlet A, Nguyen HP et al (2014) Towards the circular economy: accelerating the scale-up across global supply chains. World Economic Forum, Geneva
135. Sirkin T, Houten M (1994) The cascade chain: a theory and tool for achieving resource sustainability with applications for product design. Resour Conserv Recycl 10(3):213–276. https://doi.org/10.1016/0921-3449(94)90016-7
136. OECD (2018) Realising the circular bioeconomy: DSTI/STP/BNCT(2017)7/FINAL. OECD Science Technology and Industry Policy Paper
137. Venkata Mohan S, Nikhil GN, Chiranjeevi P et al (2016) Waste biorefinery models towards sustainable circular bioeconomy: critical review and future perspectives. Bioresour Technol 215:2–12. https://doi.org/10.1016/j.biortech.2016.03.130
138. Rehberger M (2019) Essays on decision support for allocation problems in the life cycle management of innovative technologies. Dissertation, Ulm University
139. Frischknecht R (2000) Allocation in life cycle inventory analysis for joint production. Int J Life Cycle Assess 5(2):85–95. https://doi.org/10.1007/BF02979729
140. Thunnissen DP (2003) Uncertainty classification for the design and development of complex systems. In: 3rd annual predictive methods conference. Newport Beach, California, pp 1–16
141. Huijbregts MAJ (1998) Part I: a general framework for the analysis of uncertainty and variability in life cycle assessment. Int J Life Cycle Assess 3:273–280
142. Matheys J, van Autenboer W, Timmermans J-M et al (2007) Influence of functional unit on the life cycle assessment of traction batteries. Int J Life Cycle Assess 12(3):191–196. https://doi.org/10.1065/lca2007.04.322
143. Tillman A-M, Ekvall T, Baumann H et al (1994) Choice of system boundaries in life cycle assessment. J Clean Prod 2(1):21–29. https://doi.org/10.1016/0959-6526(94)90021-3
144. Cottle DJ, Cowie AL (2016) Allocation of greenhouse gas production between wool and meat in the life cycle assessment of Australian sheep production. Int J Life Cycle Assess 21(6):820–830. https://doi.org/10.1007/s11367-016-1054-4
145. Gac A, Lapasin C, Tribot Laspière P et al Co-products from meat processing: the allocation issue. In: Schenck R, Huizenga D (eds) LCA Food 2014. 9th international conference on life-cycle assessment in the agri-food sector, 8–10 October 2014, San Francisco, USA. American Centre for Life Cycle Assessment, San Francisco, pp 438–442
146. Rehberger M, Hiete M (2019) Allocation procedures for generic cascade use cases – an evaluation using monte carlo analysis. Mater Sci Forum 959:32–45. https://doi.org/10.4028/www.scientific.net/MSF.959.32
147. Cavalett O, Chagas MF, Seabra JEA et al (2013) Comparative LCA of ethanol versus gasoline in Brazil using different LCIA methods. Int J Life Cycle Assess 18(3):647–658. https://doi.org/10.1007/s11367-012-0465-0

148. Weidema BP, Bauer C, Hischier R et al (2013) Overview and methodology. Data quality guideline for the ecoinvent database version 3. ecoinvent report. The Ecoinvent Centre, St. Gallen
149. Lloyd SM, Ries R (2007) Characterizing, propagating, and analyzing uncertainty in life-cycle assessment: a survey of quantitative approaches. J Indus Ecol 11(1):161–179. https://doi.org/10.1162/jiec.2007.1136
150. Rosenbaum RK, Georgiadis S, Fantke P (2018) Uncertainty management and sensitivity analysis. In: Hauschild MZ, Rosenbaum RK, Olsen SI (eds) Life cycle assessment: theory and practice, vol 7. Springer, Cham, pp 271–321
151. Frey C, Penmanm J, Hanle L et al (2006) Uncertainties. In: Eggleston S, Buendia L, Miwa K et al (eds) 2006 IPCC guidelines for national greenhouse gas inventories. Institute for Global Environmental Strategies, Hayama
152. Malça J, Freire F (2010) Uncertainty analysis in biofuel systems. J Indus Ecol 14(2):322–334. https://doi.org/10.1111/j.1530-9290.2010.00227.x
153. Gilbert A, Huang Y-M, Ryll T (2014) Identifying and eliminating cell culture process variability. Pharm Bioprocess 2(6):519–534. https://doi.org/10.4155/pbp.14.35
154. Jose GE, Folque F, Menezes JC et al (2011) Predicting Mab product yields from cultivation media components, using near-infrared and 2D-fluorescence spectroscopies. Biotechnol Prog 27(5):1339–1346. https://doi.org/10.1002/btpr.638
155. McGillicuddy N, Floris P, Albrecht S et al (2018) Examining the sources of variability in cell culture media used for biopharmaceutical production. Biotechnol Lett 40(1):5–21. https://doi.org/10.1007/s10529-017-2437-8
156. DIN ISO 31000 (2009) Risk management – Principles and guidelines(DIN ISO 31000:2009)
157. Meyer T, Reniers GLL (2016) Engineering risk management. De Gruyter, Berlin
158. German Advisory Council on Global Change (WBGU) (2000) Strategies for managing global environmental risks: annual report 1998. World in transition, vol 1998. Springer, Berlin
159. Renn O, Klinke A (2004) Systemic risks: a new challenge for risk management. EMBO Rep 5 Spec No: S41-6. https://doi.org/10.1038/sj.embor.7400227
160. Bjørn A, Owsianiak M, Molin C et al (2018) Main characteristics of LCA. In: Hauschild MZ, Rosenbaum RK, Olsen SI (eds) Life cycle assessment: theory and practice, vol 319. Springer, Cham, pp 9–16
161. Guinée JB, Heijungs R, Vijver MG et al (2017) Setting the stage for debating the roles of risk assessment and life-cycle assessment of engineered nanomaterials. Nat Nanotechnol 12 (8):727–733. https://doi.org/10.1038/nnano.2017.135
162. Linkov I, Trump BD, Wender BA et al (2017) Integrate life-cycle assessment and risk analysis results, not methods. Nat Nanotechnol 12(8):740–743. https://doi.org/10.1038/nnano.2017.152
163. Tsang MP, Kikuchi-Uehara E, Sonnemann GW et al (2017) Evaluating nanotechnology opportunities and risks through integration of life-cycle and risk assessment. Nat Nanotechnol 12(8):734–739. https://doi.org/10.1038/nnano.2017.132
164. Gan X, Fernandez IC, Guo J et al (2017) When to use what: methods for weighting and aggregating sustainability indicators. Ecol Indic 81:491–502. https://doi.org/10.1016/j.ecolind.2017.05.068
165. Huppes G, van Oers L, Pretato U et al (2012) Weighting environmental effects: analytic survey with operational evaluation methods and a meta-method. Int J Life Cycle Assess 17(7):876–891. https://doi.org/10.1007/s11367-012-0415-x
166. Johnsen FM, Løkke S (2013) Review of criteria for evaluating LCA weighting methods. Int J Life Cycle Assess 18(4):840–849. https://doi.org/10.1007/s11367-012-0491-y
167. do Carmo BBT, Margni M, Baptiste P (2017) Addressing uncertain scoring and weighting factors in social life cycle assessment. Int J Life Cycle Assess 22(10):1609–1617. https://doi.org/10.1007/s11367-017-1275-1

168. Pesonen H-L, Horn S (2013) Evaluating the sustainability SWOT as a streamlined tool for life cycle sustainability assessment. Int J Life Cycle Assess 18(9):1780–1792. https://doi.org/10.1007/s11367-012-0456-1
169. Renouf MA, Renaud-Gentié C, Perrin A et al (2018) Effectiveness criteria for customised agricultural life cycle assessment tools. J Clean Prod 179:246–254. https://doi.org/10.1016/j.jclepro.2017.12.170
170. Arzoumanidis I, Salomone R, Petti L et al (2017) Is there a simplified LCA tool suitable for the agri-food industry? An assessment of selected tools. J Clean Prod 149:406–425. https://doi.org/10.1016/j.jclepro.2017.02.059
171. Graedel T, Allenby BR (2010) Industrial ecology and sustainable engineering. Prentice Hall, Boston
172. Hellweg S, Milà i Canals L (2014) Emerging approaches, challenges and opportunities in life cycle assessment. Science 344(6188):1109–1113. https://doi.org/10.1126/science.1248361
173. Todd JA (1996) Streamlining LCA concepts and thoughts. In: Curran MA (ed) Life cycle assessment. McGraw-Hill, New York
174. Weitz KA, Todd JA, Curran MA et al (1996) Streamlining life cycle: assessment considerations and a report on the state of practice. Int J Life Cycle Assess:79–85
175. Todd JA, Curran MA (eds) (1999) Streamlined life-cycle assessment: a final report from the SETAC North America Streamlined LCA Workgroup
176. Wittstock B, Gantner J, Lenz K, Saunders T, Anderson J, Carter J, Gyetvai Z, Kreißig J., Braune A, Lasvaux S, Bosdevigie B, Bazzana M, Schiopu N, Jayr E, Nibel S, Chevalier J, Hans J, Pullana-i-Palmer P, Gazulla C, Mundy J, Barrow-Williams T, Sjöström C (2012) EeBGuide Guidance Document Part B: Buildings – Operational guidance for life cycle assessment studies of the Energy Efficient Buildings Initiative. https://www.eebguide.eu/. Accessed January 2020
177. Heidari M, Mathis D, Blanchet P et al (2019) Streamlined life cycle assessment of an innovative bio-based material in construction: a case study of a phase change material panel. Forests 10(2):160. https://doi.org/10.3390/f10020160
178. Bellon-Maurel V, Short MD, Roux P, Schulz M, Peters GM (2014) Streamlining life cycle inventory data generation in agriculture using traceability data and information and communication technologies – part I: concepts and technical basis. J Clean Prod 69:60–66. https://doi.org/10.1016/j.jclepro.2014.01.079
179. Zah R, Faist M, Reinhard J et al (2009) Standardized and simplified life-cycle assessment (LCA) as a driver for more sustainable biofuels. J Clean Prod 17:S102–S105. https://doi.org/10.1016/j.jclepro.2009.04.004
180. Traverso M, Finkbeiner M, Jørgensen A et al (2012) Life cycle sustainability dashboard. J Indus Ecol 16(5):680–688. https://doi.org/10.1111/j.1530-9290.2012.00497.x
181. Jessinghaus J (2000) On the art of aggregating apples & oranges. Nota di Lavoro, Milano
182. Dias LC, Freire F, Geldermann J (2019) Perspectives on multi-criteria decision analysis and life-cycle assessment. In: Doumpos M (ed) New perspectives in multiple criteria decision making: innovative applications and case studies, vol 1. Springer, Cham, pp 315–329
183. Finkbeiner M, Schau EM, Lehmann A et al (2010) Towards life cycle sustainability assessment. Sustainability 2(10):3309–3322. https://doi.org/10.3390/su2103309
184. Pursula T, Aho M, Rönnlund I et al (2018) Environmental sustainability indicators for the bioeconomy. In: Filho WL, Pociovalisteanu DM, de Brito B, Roberto P et al (eds) Towards a sustainable bioeconomy: principles, challenges and perspectives. Springer, Cham, pp 43–61
185. Rönnlund I, Reuter M, Horn S et al (2016) Eco-efficiency indicator framework implemented in the metallurgical industry: part 2 – a case study from the copper industry. Int J Life Cycle Assess 21(12):1719–1748. https://doi.org/10.1007/s11367-016-1123-8
186. Falcone PM, Hiete M (2019) Exploring green and sustainable chemistry in the context of sustainability transition: the role of visions and policy. Curr Opinion Green Sustain Chem 19:66–75. https://doi.org/10.1016/j.cogsc.2019.08.002

Adv Biochem Eng Biotechnol (2020) 173: 205–230
DOI: 10.1007/10_2019_99

Published online: 6 June 2019

# Social Life Cycle Assessment for Industrial Biotechnology

**Catherine Macombe**

## Contents



**Abstract** This chapter focuses on social assessment methods (known as SLCA) applied to industrial biotechnology (IB), which are part of the "life cycle" approach. IB is a heading for a set of different technologies. The first section presents a review of the literature to provide an analysis of the results and limitations of IB SLCA studies, with the main focus on the biofuel industries. Often conducted via a social performance analysis based on CSR (corporate social responsibility) criteria, most studies provide little new information. Nevertheless, there are some studies on the change caused by the emergence of an IB. These studies use national accounts input-output tables, which allow us to predict impacts. The second section suggests rules to follow in order to achieve a "good" SLCA in the field of IB, in other words, to be

---

C. Macombe (✉)
ITAP, University of Montpellier, IRSTEA, Montpellier SupAgro, Montpellier, France
e-mail: Catherine.macombe@irstea.fr

able to anticipate the main known impacts or at least to carry out the assessment in a rigorous and transparent fashion. The conclusion focuses on the prospects and challenges of IB SLCA, in a world which is experiencing immense upheaval.

**Graphical Abstract**

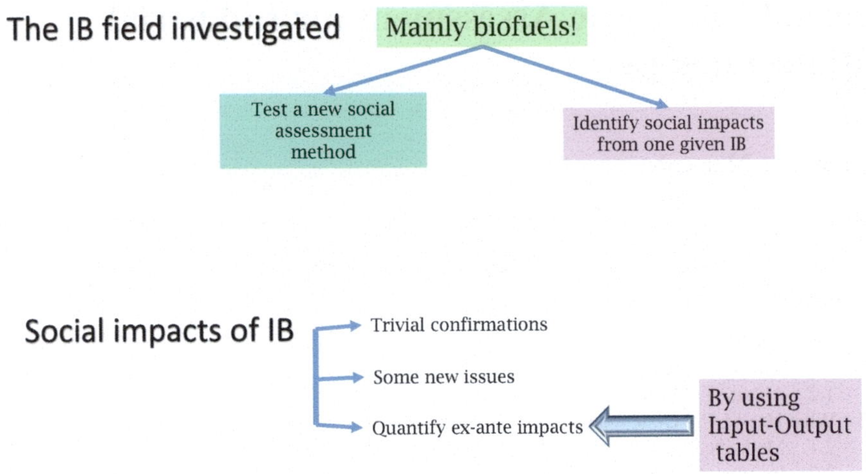


**Keywords** Biofuels, Life cycle, Methods, Review, Social criteria

## 1 Introduction

Matos and Hall [1] reported the case of a company seeking to market genetically modified organisms in Brazil but had neglected the social aspects (gene invasiveness, allergies, and other unexpected effects like habit of the parasites) on the grounds that they were inconvenient or not very well known ([1], p. 1093). As you can easily imagine, it experienced setbacks in getting established! This anecdote highlights the importance of assessing the social effects of industrial biotechnology (IB).

### *1.1 Aim and Scope*

IB systems raise the question: *how can we build IB with the most favorable social impacts possible?* The answer is not a simple one, which helps to explain why social impacts tend to be neglected in public policy, as shown by[1] German and Schoneveld [2] for the European Commission-approved schemes of biofuel feedstock production, processing, and trade. Many researchers start out with laudable "social" ambitions [3] and publish accurate and remarkable results of a purely technical-economic nature only [4]. So, is there some "social assessment" of IB's impacts?

The increasing focus on IB by policies and research projects (see [5]) has gone hand in hand with the emergence of the environmental life cycle assessment (hereafter ELCA) as the main method of assessing environmental impacts. It therefore makes sense to question the social life cycle assessment (hereafter SLCA) – with regard to social effects – as with its precursor. This chapter will focus on the methods of social assessment (grouped under the term SLCA). In principle, SLCA assesses social impacts in advance at all stages of the chain in order to show the transmedia effects. Indeed, when moving from one scenario to another, three kinds of transmedia effects can occur: (1) between different natures of impacts (e.g., effects upon inequalities versus effects upon health), (2) between different steps of the life cycle (e.g., production step versus transportation step), and (3) between actors (e.g., workers versus users). In practice, however, we will see that things are not so simple.

SLCA deals with industries created by the implementation of IB applications and not the biotechnological processes themselves. The industries tested by SLCA are at various stages of development, from the prototype stage to factories in commercial operation. The social investigations mainly deal with biofuel industries. Indeed, worrying social impacts have been found [6] but also positive effects [7], and these are summarized in the report of the International Union for Conservation of Nature [8]. A special issue of the *Ecology and Society* journal [9] is dedicated to them.

Section 1.2 is devoted to presenting the SLCA methods, depicting the two main approaches and highlighting synergies between IB and SLCA fields. The following section (cf. Sect. 2) analyzes the results and limitations of IB SLCA, based on a review of the literature. The section after that (cf. Sect. 3) examines what would constitute a "good" SLCA in the field of IB and suggests rules to follow. We will conclude with the challenges and prospects of SLCA for IB.

## 1.2 About SLCA and IB

This paragraph presents the available methods to perform a social evaluation in IB field (cf. Sect. 1.2.1), then the two types of SLCA (cf. Sect. 1.2.2), and finally the reciprocal merits of IB and SLCA to go together (cf. Sect. 1.2.3).

---

[1]The authors note the absence of any food security impact assessment or mitigation requirements despite the impacts of recent food crises on rural livelihoods and political stability in the Southern countries. Only one scheme goes far beyond mitigating negative impacts. The others place the main emphasis on mitigating negative socioeconomic impacts [2].

### 1.2.1 Available Methods

What do we hope to get from a social assessment method?: (1) the quantification of impacts and (2) the identification of transmedia effects.

"Quantitation is key since it enables rigorous comparison and assessments [and] also brings the advantages of rapid computational methods. Such simulations allow alternatives to be quickly evaluated" ([10], p. 1012). Identifying transmedia effects is the characteristic of the methods that are in line with life cycle thinking. The aim is to avoid "false good ideas," which remove a negative impact from the first scenario but generate an undesirable impact of another nature, at another stage of the life cycle or for another group of people [11].

Mattila et al. [12] recently presented the main social assessment methodologies adapted for bioeconomics. We have reproduced a table (Table 1) from this paper as it will help to classify the social assessment methods in this chapter among the four

**Table 1** Features of the categories of methods and tools for value chain social assessment

| Name | Method or tool | Goal | Object | Focuses on | Comments |
|---|---|---|---|---|---|
| 1. Performance accounting | Methods using tools like the SHDB | Calculation of social performances | Industry, company, product | Comparison Dissemination | Allows evolution to be tracked over time and benchmarking |
| 2. Modeling consequence of change | Methods using cause-effect relationships | Calculation of social impacts of one change | Industry, company | Designing the system boundary Comparison | Fast, few data needed, but few relationships available to date |
| 3. Participatory methods | Methods using different tools for participation | Getting knowledge from laymen and key stakeholders about social performances or impacts regarding different scenarios | One local company, one small territory | Problem definition Designing the system boundary Dissemination | Time-consuming, expensive, sensitive. To be used when appropriation by stakeholders is the purpose |
| 4. Multi-criteria decision analysis | Methods like AHP or ELECTRE | Decision-making support in choosing between alternatives, by classifying the importance of impacts | Any objects | Classifying | Can include other criteria than social |

Source: From Mattila et al. [12]

*SHDB* social hotspots database [13], *AHP* analytic hierarchy process [14], *ELECTRE* a specific multi-criteria decision method [15]

categories presented. Group 1 includes performance assessments (observing the state of certain criteria at a given time). Group 2 employs a variety of methods (input-output (IO) tables, scenario generation, pathway calculations based on past data, etc.) to *simulate the effects of a change*. Group 3 includes participatory methods: it involves capturing the representations of the players involved with regard to the categories of performance or significant impacts or even at the indicator level. Group 4 covers the multi-criteria methods, the aim of which is not to assess but to aggregate assessments. It should be noted that these four groups are not exclusive since there are many studies that combine two, or even three, of them to answer the same research question (e.g., see [16]).

There is no widely agreed method of social assessment that would apply to IB. Starting from 2006, von Geibler et al. [17] analyzed interviews with six biotechnology experts and 33 other stakeholders to extract a list of eight key "social aspects." For liquid biofuels, Markevičius et al. [18] selected 15 social criteria from the literature, but no consensus was reached among the 46 international experts from academia, business, and NGOs who were responsible for choosing the most important ones. Scott et al. [19] suggested several classic performance criteria. For the emergence of bioeconomy industries, Hasenheit et al. [20] defined a system of relationships between potential social impacts and criteria lists. Martin et al. [21] conducted a review of the sustainability assessments of bio-based value chains, published between 2010 and 2015 (limited to items produced or used in Sweden), and organized an Open Space Workshop composed of Swedish industry researchers and practitioners. The workshop would only consider "working conditions" among the criteria submitted. Falcone et al. [22] used the same method (not limited in terms of area) to identify the main social impact categories, interviewing three experts and a number of stakeholders and concluding that their different visions must be taken into account. Rafiaani et al. [23] collated 59 studies (published between 1990 and 2016) and identified three main methods: *impact assessment*, *socioeconomic impact assessment*, and *social life cycle analysis* (SLCA). They favor the latter. Unfortunately, their view obliterates the profound differences that have shaped the SLCA approaches. There are in fact multiple SLCA that belong in any one of the four methodologies identified in Table 1!

### 1.2.2 Two Types of SLCA

SLCA has its roots in the history of the environmental life cycle assessment (see [24, 25], this publication). O'Brien et al. [26] questioned the socioeconomic impacts of a change, the environmental effects of which they had just been studying. Ever since it was created, the method has been facing a dilemma which is markedly illustrated by Norris' proposals [27]. In the same paper, Norris intuited that there are two SLCA. They are later named type 1 SLCA and type 2 SLCA [28]. Type 1 SLCA, also known as "life cycle CSR" ([29], p. 28), involves collecting static "social performance" information (often derived from criteria collected as part of corporate social responsibility (CSR)) from the organizations that comprise the life cycle and then pooling them to establish a "social score" of sorts for the life cycle. This is category 1 in Table 1. Conversely, in line with O'Brien et al. [26] and Weidema [30], type 2 SLCA aims to assess *the social effects of a change*. It is based on the methods described in row 2 (or 3) of Table 1. Norris [27] predicts that an improvement in national income per capita (GDP/capita) will have a positive impact on the life expectancy of the poor. He drafted the idea of a "Preston pathway," which was later developed by Feschet et al. [31].

Type 1 SLCA provides static information which has little or nothing to do with the social impacts of a transformation. It is interesting to take stock of the situation and track the evolution of the CSR indicators over time. Type 2 SLCA gives an indication of the likely social effects of a change. It is based both on calculations that quantify the magnitude of the impacts (from known pathways, row 2 of Table 1) and on scenarios (row 3 of Table 1). Because only a small number of pathways are known [32], and the foresight of the interviewees is limited, it will only provide guidance on a few social impacts.

### 1.2.3 Industrial Biotechnology for SLCA

In the theory, the merits of SLCA to achieve social evaluation of IB are identical to those favoring ELCA (see [33]): LCAs apply to value chains made up of multiple companies, and they are designed to highlight transmedia effects and report on large-scale standardized industrial processes while being applied to organic processes such as those found in agriculture or agribusiness. At the first glance, SLCA shows promise when it comes to assessing the social impacts of IB. Conversely, IB is a fertile field for SLCA as its three peculiarities [10] raise some original social issues: (1) there are often several biotechnological routes that can be taken to obtain a chemical or fuel, and, conversely, the same product (e.g., bioethanol) can be used for different purposes, e.g., as fuel or as a platform chemical. (2) There is a challenge of replacing at least some existing platform chemicals by others of organic origin. This challenge will radically alter the social impacts as the life cycles of the new molecules (e.g., glucose) will involve an agricultural and conversion phase not involved in the life cycles of the standard molecules. (3) Due to the large quantities of freshwater used, IB raise questions about water use and reuse. In some regions, these impacts will need to be anticipated in order to avoid any new conflicts over water usage.

## 2 What Has SLCA Done for Industrial Biotechnology?

We seek to illustrate the various contributions of SLCA to IB through the variety of the cases presented. We prefer analyzing methods rather than highlighting results as an unsuitable method will give irrelevant results.

## 2.1 Methods

The review brings together documents that deal with both IB and social assessment, either in the form of discussion or case studies, or a sustainability assessment, provided that it considers social criteria for all or part of the life cycle. The documents can be published papers, conference minutes, or reports. Indeed, any sort of literature may be relevant when it comes to prescience, which this field of knowledge falls under. We did not select documents that only used cost-benefit or socioeconomic studies or financial studies for a single stage of the life cycle, nor did we include those which tried to assess a single impact such as "acceptability," "safety," or "number of jobs created" or even "health,"[2] unless they were explicitly part of a broader SLCA. We retained social effect studies which covered several stages of the life cycle, regardless of the method of data collection and analysis.

We examined the databases stated in Table 2 for the period 2008–2018, when it was possible to filter it.

We added the relevant papers presented in the "International Seminars in Social Life Cycle Assessment"[3] 2–6 to this first document search. We then added a few documents that were already known about or were identified by snowballing. After eliminating double counting, the corpus brings together 58 articles, representing fewer than 15 separate case studies.

In order to analyze the literature review, we look at which IB the documents are discussing and what the scope is (the whole value chain, a part of it). Is there another topic of interest (e.g., agricultural raw materials, waste, etc. which may be precursors of the IB)? What are the author's main motives: is their motive really to find out the effects of a particular IB application? Or is it to use IB as the basis for a social assessment method – which is in fact their main concern? Finally, what results are obtained in terms of performance or social impacts?

## 2.2 Results of the Literature Review

Here we present the IB fields investigated and the different motivations of the authors.

[2]On the subject of health, it should be noted that the software used to perform an ELCA routinely offers the calculation of a "human health" impact which does not require data other than the data used to carry out the rest of the ELCA, but fails to address the social determinants of health.

[3]This seminar cycle was launched in Montpellier in 2011 by Cirad and Irstea, who published the presentations of seminars 4 and 6 in the FruiTrop Thema collection of works (Cirad, Montpellier).

**Table 2** Databases searched and hits

| Databases searched | Query | Number of hits | Documents selected |
|---|---|---|---|
| Web of Science core | Industrial biotechnology∗ social life cycle analysis | 0 | |
| Web of Science all databases | Industrial biotechnology∗ social life cycle analysis | 29 | 5 |
| Google Scholar (all dates) | Industrial biotechnology∗ social life cycle analysis | Indefinite | 3 |
| Scopus (2008–2018) | Industrial biotechnology∗ social life cycle analysis | 5 | 1 |
| Atoz | Industrial biotechnology∗ social life cycle analysis | 0 | |
| Atoz | Industrial biotechnology∗ | 30 | 0 |
| Web of Science core | Biofuel∗ social life cycle | 128 | 11 |
| Web of Science all databases | Biofuel∗ social life cycle | Too many | |
| Google Scholar (all dates) | Biofuel∗ social life cycle | Indefinite | 9 |
| Scopus | Biofuel∗ social life cycle, key-word life cycle | 89 | 7 |
| Atoz | Biofuel∗ social life cycle | 0 | |
| Web of Science core | Biodiesel∗ social life cycle | 76 | 5 |
| Web of Science core | Social life cycle enzyme∗ (in fact, we get some biofuels) | 24 | 1 |
| Web of science all databases | Social life cycle enzyme∗ | 226 too many | |
| Scopus (2008–2018) Restricted to biochemistry, genetics and molecular biology and agricultural and biological sciences | Social life cycle enzyme∗ | 34 | 2 (on biofuels only) |
| Atoz | Social life cycle enzyme∗ | 0 | |

### 2.2.1 The IB Fields Investigated

The main IB investigated by social life cycle is biofuels, usually on its own. In some cases, also by-products were investigated [3, 34]. The area under scrutiny may be the assessment of feedstocks and products as well as processes including energy and mass integration [10], use phase (e.g., in vehicle), or end-of-life. Exceptions to biofuel studies are rare, to the point that anybody wanting to carry out a review of the social impacts of the bio-based plastic industries [35] will only find one study of this type.

Three case studies are cited. The consulting firm Evea [36] carried out a type 1 SLCA study for a cosmetics company on a new bio-based thermoplastic, derived from sugarcane and manufactured in Brazil. Alvarez-Chavez et al. [37] investigated the industry to identify health and safety impacts during the life cycle of the bioplastics (e.g., “use of GMOs and hazardous pesticides to grow the feedstock to produce the bio-based plastic; use of hazardous chemicals or petroleum-based co-polymers during plastic production and processing [-]") ([37], p. 49). Perez-Lopez et al. [38][4] examined the sustainability of the production of the *red carotenoid astaxanthin* by the green microalga *Haematococcus pluvialis*, by data collection in the laboratory and on a pilot and semi-pilot scale. They combine a type 1 SLCA study with a cost-benefit approach in order to identify "the main strengths of the process from a holistic perspective" ([38], p. 3).

### 2.2.2 What Are the Authors' Motivations?

Many papers use the case of biofuels as a test for the method they want to promote. For others, the priority is to identify the social impacts caused by the existence or creation of a biofuel industry. The scenarios are thus much more documented. Finally, a third category of papers covers a material which could be turned into biofuels (cereals, wood, waste, etc.) but which has a number of intended destinations.

#### Biofuel Testing Methods

The type 1 approach (row 1, Table 1) based on the recommendations of UNEP-SETAC [39] is the most common. Bailey et al. [40] used a derived method. They tested it on a small anaerobic digestion plant (breakdown of organic wastes by bacteria in the absence of oxygen) producing biogas, in order to identify the process changes needed for it to be applied to agriculture. Chingono and Mbohwa [41] looked at biofuels produced in the KwaZulu-Natal and Western Cape Regions of South Africa as a testing ground for a performance SLCA. Do Carmo et al. [42] mentioned that comparing different options of biodiesel supply was an "illustrative case study," to apply their method of performance assessment. Ekener et al. [43] sought to "introduce the positive impacts" in a type 1 SLCA, and chose the case of vehicle fuels, comparing fossil fuels and biofuels. Other authors consider the Dutch biodiesel industry as a case study to test their own methodology, mixing the recommendations of UNEP-SETAC and monetarization [44].

Henke and Theuvsen [45] compiled their own list of indicators [46]. As an example, they used the biogas industry, and the electricity production based on woodchips derived from coppicing, in different regions of Germany.

Other methods combine performance and participatory approaches. Halog and Manik [16] suggest a methodology which includes type 1 LCAs, stakeholder analysis supported by multi-criteria decision analysis, and dynamic system modeling. They test this framework on the development of biofuel supply chain networks.

[4]It should be noted that the first author mainly published on the assessment of microalga as a source of biofuels.

Also based on a participatory assessment, the method of Ren et al. [47] (the originality of which lies in translating linguistic terms into numbers through fuzzy theory) was tested on three alternative bioethanol production scenarios in China.

### A Real Investigation into the Social Impacts of Biofuels

This category includes participatory assessments (row 3, Table 1), with or without a performance assessment (row 1, Table 1), and assessments of the impacts of changes (row 2, Table 1) introduced by biofuels.

Renn [48] assessed the change in the current energy landscape (including the social effects) by stakeholders for four energy policy scenarios, some of which use waste to produce gas. Mbohwa and Myaka [49] gave an overview of the impact of biofuels (manufactured in small plants from used soybean oil) in South Africa, combining field investigations and desk screening of literature. Manik et al. [50] researched the social implications of current practices in palm oil biodiesel in the case study conducted in the Jambi province of Indonesia. They combined a type 1 SLCA with an expert survey supported by a literature review. The challenge posed by Somé and Revéret [51] is to assess the social impacts of biofuels in Africa. They suggest using a combination of methods (capability assessment, performance SLCA, validation by experts in the field). Valente et al. [34] combined a type 1 approach with a bottom-up survey of workers in a bioethanol and biochemical industry (chemical, rubber, and plastic products) in Norway. In order to assess the social effects of implementing a glycerol biorefining project in the Netherlands, Cadena et al. [52] explored the sustainability reports of companies representing the life cycle stages, analyzed by type 1 SLCA and with the help of three experts. Kaltenegger and Schwarzinger [53] compared the traditional production of steel to the new process where waste wood is used as a feedstock converted to bio-coal, itself replacing fossil coal in a steel mill furnace. Moreover, the carbon monoxide in exhaust fumes is microbially fermented to yield bioethanol. The type 1 assessment will be conducted at a district and national level with the help of the steel mill's CSR department. Through experts, Peñaloza and Keller [54] obtained 18 scenarios for the creation of a microalgae (*Dunaliella salina*) biorefinery and used the social hotspots database [13] to assess the social impacts created by each one, mainly according to the country where the plant was based.

Siegel Moecke et al. [55] studied the local social effects of the operation of a cooperative biodiesel plant which salvages waste cooking oils from restaurants in the State of Santa Catarina (Southern Brazil) in order to power the local fishing fleet's boat engines. They thoroughly questioned the people involved, from the restaurant owners to the fishermen, about the changes in their lives. What are the social impacts caused by the sudden emergence of a *Jatropha*-based biofuel sector? This is the question asked by Boonkum et al. [56] in Thailand, who used eight criteria of social changes[5] developed by the Global Bioenergy Partnership Sustainability Indicators for Bioenergy [57]. Similarly, Muñoz Mayorga et al. [58] questioned the social effects (at the international, national, and local level) of replacing petroleum with *Jatropha* oil to generate electricity on the island of Floreana in Ecuador. The authors identified the most vulnerable players using a huge range of sources.

On a larger scale, the assessments of the effects of a change are *based on input-output (IO) tables*. A project by Malik et al. [59] assessed the impact of establishing a future cellulose-refining industry located in the south of Australia. In order to assess job losses/creation and loss/creation of economic activity, they used a multi-region (IO) analysis, which takes into account the entire refinery supply chain. This is the same method that this team uses to compare the bio-crude algae industry with an equivalent conventional crude oil industry in Western Australia [60]. Using a similar IO method, Cambero and Sowlati [61] sought to optimize job creation (taking into account the quality of these jobs), the net present value, and greenhouse gas emissions from a Canadian forest-based biorefinery industry located in British Columbia, by testing several scenarios. Souza et al. [62] compared the social effects on workers of three implementation scenarios for a sugarcane-based bioethanol industry in Brazil. They selected the criteria of job creation, occupational accidents, wage profile, education profile, and gender profile. Their method combined an inventory with an IO method, aiming to expand the social effects of a given scenario over its supply chain (simulating the impacts of ethanol production on the Brazilian economy). Using the same approach, Wang et al. [63] assessed the social impacts (employment, GDP development, and trade balance) of four biofuel use scenarios for aviation in Brazil.

#### What About the IB Source Materials Industry?

The raw material studied can be as follows:

1. Waste oil. Vinyes et al. [64] compared the social effects of three existing *waste oil* collection systems (through schools, through door to door, and through urban collection centers) near Barcelona, Spain. They used several UNEP-SETAC [39] indicators.
2. Guar gum (to be used as a bio-based thickening agent in personal care products). The social effects of this industry were studied before and after a "Sustainable Guar Initiative" program is implemented in Rajasthan, India [65].
3. Cereals. Delcour et al. [66] defined four usage scenarios for Walloon *cereals* (business-as-usual, environmental, social, and economic improvement; new recycling plants in Wallonia; focus on cereals for biorefining and bio-based chemicals) to compare their socioeconomic impacts (added value and work-time distribution) in 2030.

[5]For example, allocation and tenure of land for new bioenergy production, effects of bioenergy use, and domestic production on the price and supply of a food basket, change in income, etc.

4. Mostly wood. Siebert and Bezama [67] looked at the future of the *wood-based bioeconomy* in rural regions in Germany. The authors adapted a type 1 method [68] to examine three levels: national, regional, and sectoral. Stakeholders were identified at all three levels and asked about the socioeconomic issues. The wood-based bioeconomy was also looked at by Mattila et al. [12], who justified the combination of a participatory approach and an IO method to assess the social effects of its development in Finland.

## 2.3 What Are the Impacts of IB?

Studies geared toward learning about the effects of IB often provide confirmations, but they sometimes raise new issues. There are a few outstanding studies quantifying the effects according to the magnitude of the change and therefore comparing scenarios. These are the only ones that can claim to provide forecasts.

### 2.3.1 Confirmations. . .

An SLCA carried out by Henke and Theuvsen [45] reports that electricity produced by woodchips derived from coppicing sometimes takes up large amounts of arable land. The most illustrative example is provided by a Master's thesis [69] from Luleå University in Sweden, which compares three "theoretical" biofuels (Brazilian ethanol, Swedish biogas from waste, and Swedish biodiesel from rapeseed oil) using a type 1 SLCA. Ethanol receives a very poor score due to child labor and forced labor. Moreover, cane is denounced for occupying land that was previously intended for food production. This is the only criticism directed at biogas, which could encourage the production of dedicated crops. The biogas industry would create jobs and act as a catalyst for new technologies. With regard to the growing demand for biodiesel, it could indirectly increase imports of palm oil, which is known to cause deforestation and excessive carbon emissions in Indonesia. Blom and Solmar [69] suggested that the biodiesel industry does not seem to have any effect on the job market. Conversely, according to Boonkum et al. [56], increased income and job creation in the local community are the major social effects of using jatropha as a source of biofuel in Thailand. According to Manik et al. [50], the Indonesian palm oil case study indicated that working conditions and cultural heritage are problems. Vinyes et al. [64] found that the door-to-door waste oil collection system employs more staff than other ones do, including more disabled staff.

### 2.3.2 Interesting Findings. . .

Mbohwa and Myaka [49] disclosed nothing new in observing low wages and the employment of illegal migrants in agriculture in South Africa. But they also suggested educating workers on AIDS prevention alongside the development of new soybean production facilities for biodiesel. This is a critical issue in rural areas. Vuaillat et al. [36] have identified several sensitive issues (hotspots) in the life cycle of the new Brazilian bio-based thermoplastic. Competition for land leads to killings in areas where there are large indigenous populations who do not benefit from jobs created by the industry. As for bioplastics, Alvarez-Chavez et al. [37] stated that none of them are currently "sustainable" in terms of environmental, health, and safety aspects. Delcour et al. [70] noted that employees in the Walloon biofuel industry have the best benefits and wages and that their training is the most expensive; there is no part-time work and the retention rate is as good as in the food industry. The workers are more likely to be men on permanent contracts and slightly older than the average across all industries. According to Siegel Moecke et al. [55], the creation of a Brazilian biofuel cooperative unit using waste cooking oil has generated better income for the cooperators and fishermen; more surprisingly, the empowerment of around 50 people; and solid education in helping protect the local environment.

### 2.3.3 Forecasting Becomes a Possibility...

The assessment by Malik et al. [59] of gains/losses in jobs and economic activity (brought about by the diversion of forest biomass from the paper industry to the biofuel industry) showed that the loss only represents 10% of new jobs and new activities created. As for biofuels produced from algae in Western Australia, they stimulate economic activity a lot more, and, while they employ fewer staff on-site than the conventional crude oil industry, they actually employ more when one considers the new industry as a whole [60]. The Pareto optimal solutions for producing forest-based biofuels in Canada show that there is a correlation between the potential to generate high-impact jobs in the region and its potential to generate greenhouse gas emission savings ([61], p. 721). According to Souza et al. [62], the simulation of a growth in demand of one billion of Brazilian Reais in the three biofuel scenarios was able to reveal the job creation (differentiated by sector of activity), the number of likely occupational accidents, the future wage profiles, the rate of female employment, and the future education-level profiles. The majority of the jobs created by the development of biofuels for aviation using biomass came from the agriculture, forestry, chemical, and transportation sectors. For example, the scenario involving "Fischer-Tropsch with eucalyptus" tended to create 15% more jobs than "alcohol to jet with sugarcane" [63].

## 3 What Should SLCA Do for IB?

SLCA should be used "to keep social and societal aspects under control" [36], to prevent a situation from deteriorating, or even to improve it as soon as possible. The studies carried out using SLCA could contribute to this, as long as they follow certain rules (cf. Sect. 3.4). However, SLCA has its own limitations (cf. Sect. 3.1), which diminish its relevance in certain cases (cf. Sect. 3.2). The economic and social contexts in which IB have emerged are also impediments (cf. Sect. 3.3.).

## *3.1 Discussion and Limitations of the Results*

Authors genuinely concerned with the social impacts of a case combine several investigation methods. This confirms that no one method is currently polished enough to give satisfactory results on its own. In the interest of clarity, but especially because this is what makes the difference, we will analyze the results according to the method used.

### 3.1.1 Type 1 SLCA Provides Little New Information

In practice, there are very few complete assessments that use performance lists. Many papers remain at the method discussion stage. Some studies describe interesting existing social aspects of production (e.g., [36, 70]).

More often than not, the results are disappointing and comparable to what one would achieve if they were to conduct an in-depth Internet investigation [71]. This is because the criteria for type 1 SLCA are often borrowed from biofuel certification schemes, which are not effective in social terms ([6], p. XV). Their criteria were chosen to speak to buyers' emotions or even to provide a distraction from real problems [72]. SLCA conducted on this basis also adopted these flaws. In the case of biofuels, "critical social factors – such as participatory processes, common management of resources, health implications and other aspects of poverty reduction or smallholder inclusiveness - are not typically addressed as primary concerns of existing certification schemes" ([6], p. 9). Real social problems are rarely taken into account by this type of SLCA. Moreover, studies that use a type 1 SLCA get interesting results when they go beyond this scope (e.g., [49, 58]) to see "what has changed" or "what can change."

### 3.1.2 Participatory Methods Are Always Right

In order to find out the social impact of the establishment (dating back 70 years) of a certain sugarcane bioethanol family mill in Sao Paulo (Brazil), Grigoletto Duarte et al. [73] interviewed 36 people who were mainly locals or factory workers. The results were remarkably informative (e.g., they showed how well the factory had stabilized the local society and created common goods). Without a doubt, the best method of assessing social impacts is to go and visit the people who are experiencing these effects. However, this approach involves waiting for a change to take place and intelligently questioning the lucid (or even omniscient) players about their consensual representation. This very sentence demonstrates the main limitations of participatory methods. First of all, SLCA is supposed to be able to anticipate impacts. The players therefore have to be interviewed before the change has taken place. Secondly, they all have to be located and interviewed correctly, which requires real skills [74], something which tends to be overlooked by LCA surveyors. The players are expected to know everything (and say everything). However, they may forget important aspects. In the Brazilian study by Siegel Moecke et al. [55], the effects selected only reflect what the researcher perceived from local representations. What about the procurement of new oil, for example? Moreover, two players in the same category of stakeholders are supposed to have the same opinion. However, as noted by Ekener et al. [75], the classification of various fuels in terms of social performance depends on the classifier's preferences. All of the conceivable manipulations in human groups [76] highlight the limitations of participatory studies when used alone to carry out a social assessment.

### 3.1.3 Forecasting Is Made Possible with Type 2 SLCA, but…

Forecasting potential impacts becomes possible when authors use a type 2 SLCA because it offers an approximate quantification of the magnitude of impacts according to the magnitude of the change. Furthermore, several impacts are considered. Manik et al. [50] pointed to the respective contributions of the different stages of the industry in terms of jobs and economic activity. The correlation highlighted by Cambero and Sowlati [61] for forest-based Canadian biofuels suggests that there is a relationship to be explored between greenhouse gas emission savings and high-impact jobs. Souza et al. [62] suggest a "socio-design" quasi-tool for Brazilian biofuel growth scenarios.

These studies using type 2 SLCA outline a social assessment of IB that can predict impacts and potential transmedia effects. However, the question of the nature of the impacts to be assessed remains unresolved. Indeed, are we not forgetting some significant social impacts (e.g., widening inequality)? On this basis, we will discuss what constitutes a "good" IB SLCA study.

## *3.2 When Should an IB SLCA Be Conducted?*

When all of the processes are located in the same country (or if the sponsors have no interest in what happens in other countries!), and if we have the time and opportunity to invest in participation properly, we can do a much more precise and ad hoc job by surveying people rather than conducting an SLCA. A remarkable example is given by Renn [48] who spent 3 years regularly calling upon the industry, unions, public utilities, operators of small power plants, churches, scientists, philosophers, and environmentalist representatives for the assessment of five waste energy systems.

These works led to the consensual classification of scenarios in order of preference, a valuable insight for decision-makers.

In contrast, SLCA seems to be justified in the following circumstances: (1) when it is not possible to carry out a quality participatory assessment, (2) when it is believed that participation alone will not give complete results, and (3) when the assessment criteria are intended to be generic.

### *3.3 Context of the Emergence of IB*

IB have been developed to cause less pollution and, more importantly, to prepare for the replacement of declining fossil resources.[6] The effects of IB on the natural environment nor the risks will not be discussed at this point (see [24, 77], this publication). By contrast, the decline of fossil resources places IB in an unprecedented social context. With an energy returned on energy invested ratio[7] varying between 1:1 and 10:1 [78] compared to 10:1 and 20:1 for conventional oil [79], biofuels are relevant, especially in a context where *oil is very expensive or even unavailable*. Certainly, very high oil prices will also drive up biomass prices [80], but at least biomass will remain available in the long run, while oil will no longer be available. On the other hand, it is doubtful that the ecological and social merits of the bioeconomy alone are enough to warrant its widespread adoption, despite the opinion of Falcone et al. [22]. However, in a world where oil is becoming incredibly expensive and rare, the problems that arise will be of a severity and magnitude unseen in the West for 200 years. It is not clear whether SLCA will be called upon in this new context.

### *3.4 Proposal of Rules for Assessing Industrial Biotechnology Issues with SLCA*

We suggest a few rules below to help conduct an SLCA as soundly and honestly as possible. The purpose of the study is (1) either to fine-tune a project from a social perspective (by adapting certain aspects that were not taken into account in previous technical feasibility, economic, and environmental studies) (2) or to provide reliable data to encourage ownership of a project (provided that active participation phases are included).

[6]"In a world with limited (or very expensive) oil it is less clear where the chemical of the future will originate." ([10], p. 1012).

[7]In short, the quantity of energy obtained in relation to the amount invested.

### 3.4.1 Choosing a "Simple" Problem

Matos and Hall [1] believe that an LCA is appropriate for a situation of low complexity or when the simplification introduced by the method does not distort the problem. However, when there is a large number of interactions, LCA is too simplistic. In the same way, SLCA is contraindicated in situations of high complexity with a large number of interactions. Here are two extreme examples: when there are significant and unpredictable rebound effects (e.g., the spread of biofuel use across the globe), SLCA is unable to predict the rebound effects; at the opposite, some local cases which mobilize a whole set of "small" interactions in the social fabric without any particular salient element emerging (e.g., the manufacture of an enzyme in a small industry, limited to a developed country) are difficult to handle by SLCA. The typical IB case in which SLCA is applicable is the introduction of a new globalized[8] industry to a given region and under several scenarios (biofuels, fine chemicals, etc.). In this case, we can anticipate discrepancies between the scenarios, for example, in terms of employees' health, widening inequality, and even public health in the developing countries in question.

### 3.4.2 Being Explicit About the Nature of the Impacts Deemed to Be Important

Whichever impact one chooses to study, the choice will always be questioned. It is therefore necessary to be very clear about exactly "what matters" in the social world. For example, in all of our work, we suggest that the fairest criterion is public health so that, wherever possible, the social effects identified are translated into health criteria. As with the ELCA, an evaluation using SLCA frequently includes indicators expressed in different units. For a study at the national level, this will be the increase or decrease in infant mortality and life expectancy. For a study at the regional level, this will be the increase or decrease in self-reported health and the regional accident rate. For a local study, this will be variations in work accident rates or local morbidity for a particular disease, for example. In addition, there are classifications[9] that prioritize the social determinants of health between themselves. Comparison of different scenarios, placed in various contexts, then becomes possible. However, any other hierarchy is acceptable, provided that it is clear.

---

[8]Biofuels offer an ideal field of study in the future, because of "the expected large gap between future demand and potential domestic supply in the North" the production of biofuels for the North will increase in countries in the South ([6], p. xiii).

[9]Such as the classification of social determinants for health developed by the WHO's dedicated commission [81].

### 3.4.3 Describing the Baseline

SLCA assesses social progress or regressions in relation to a reference situation (previous situation, business-as-usual, another scenario). Of course, the criteria chosen for the description depend on the nature of the impacts to be assessed. If public health is the focus, then one would research the mortality rate, average life expectancy, etc. For the same actual system, the description used to conduct a social assessment is not the one that would be used to conduct an environmental assessment. The actual system is the same, but the two models differ in their chosen descriptors.

### 3.4.4 Expanding the Scope of the Study to the Strategic Arena

The scope of a scenario is threefold: spatial, temporal, and effects (main players impacted). The most difficult thing is defining the spatial scope. In ELCA, there is a category of so-called "consequential" studies "exploring [-] the wider changes to the overall system that may arise from using that product" [33]. As with the consequential ELCA, in all studies conducted using SLCA, the boundaries of the system studied need to be set broadly enough so that they *include all of the organizations that are heavily affected* by the change in the study, as in the agricultural example given by Lagarde and Macombe [82]. Thus, to assess the social impacts of a future cellulose-refining industry located in South Australia, Malik et al. [59] take into account "the loss in economic activity and employment in the paper, pulp and paperboard industry resulting from the diversion of forestry biomass to biofuel production" ([59], p. 96). When Wang et al. [63] study job creation in the biofuel industry for aviation, they account for the displacement effects in the fossil sector ([63], p. 254). As with an ELCA, parts of the life cycle that are not affected by the change are not recorded in the spatial scope [33]. The setting of the spatial and temporal boundaries determines the identification of the key players affected, with the level of fineness adapted to the case (the employees of company G and company F, the farmers in the P region, village R, but not village Q, etc.). Therefore it is not possible to designate categories of "stakeholders" until the spatial scope has been determined [82].

### 3.4.5 Assessing the Phenomenon on Multiple Scales

Since the responses to change in social phenomena can be reversed when the scale is changed,[10] and because each level of decision-making raises new social problems, it

[10]Rostila et al. [83] demonstrate the reversal of the link between income inequality and self-reported health, depending on whether the study is at the Stockholm municipality level or at the neighborhood level.

must be made clear which scale is being investigated. The social effects of a biofuel policy at the national level are not the same as those at the district level where a factory is to be set up, nor are they the same as those experienced by the company building the factories. Similarly, the tools and criteria differ according to the scale [84]. At the state level (or even the regional level when the data is available), we can use (1) national accounts IO tables to predict effects in terms of the flow of workers and economic activity (like [60]) and (2) the macroeconomic pathways ("Preston" from [31]; "Wilkinson" from [85, 86]). At the company level, we can use the "Siegrist" pathway (from [87]) and the "Wesseling" pathway (from [88]), or we can adopt the method by Arvidsson et al. [89]. The different quoted pathways are summarized in Table 3.

It should be noted that, regardless of the scale, interviews can be used to gather opinions on the potential impacts. In cases where the change concerns several spatial scales, ideally SLCA would be carried out for the same change at different scales.

#### 3.4.6 Taking Threshold Effects into Account

SLCA assesses the main significant social effects of a marginal change represented by scenario A compared to the effects of scenario B, with both being placed in the same context. The impacts caused by the occurrence of scenario C (which would produce X times the service given by scenario A) are not automatically X times the impacts caused by the occurrence of A! In other words, the relationship between the amount of change and the amount of impact is not linear [11]. The *effects of thresholds must be taken into account* individually for each impact.

### 3.5 Toward Better Assessing IB with SLCA

The methodological developments needed depend on the purpose of the IB evaluation. Is it to monitor the principal features of existing IB facilities along time, or is it to anticipate the social impacts before setting up new facilities?

If the aim is to monitor existing IB facilities, SLCA of type I can help. By providing the follow-up of the same indicators along the time, this approach supports the efforts of companies toward social and environmental progress. One practical improvement should be reducing the number of criteria (the usual case studies involves more than 150 indicators) to focus on the few relevant ones for IB. When benchmark and comparisons are at stake between different scenarios (e.g., between different sources of energy), it should be of the utmost importance to define a strict protocol to be sure that the compared *services* rendered by the two scenarios are the same (e.g., 1,000 KWh is not the same depending on whether they are storable or not).

If the aim is trying to anticipate important social impacts, it is still a long way to go. From our knowledge, despite the topic has shaken the ELCA field for 10 years,

**Table 3** The different pathways available to assess a change by SLCA of type II

| Scale | Main data | Pathway name | Principles | References |
|---|---|---|---|---|
| State level (or the regional scale if data are available) | National accounts Input-output tables | Input-output | Trace the flows of economic activity and job creation/loss across the economic sectors | Malik et al. [60] |
| State level | Added value generated by the activity, GDP, life expectancy | Preston | Calculates the gain in life expectancy caused by the new added value generated by the activity | Feschet et al. [31] |
| State level | Reference Gini coefficient, infant mortality, input-output tables, income per job category | Wilkinson | Calculates the new Gini caused by the activity and the results in terms of infant mortality | Bocoum et al. [85] |
| State level | Input-output tables | Social footprint | Calculates income redistribution and loss of productivity because of missing governance | Weidema [86] |
| Company level | Survey about the psychosocial resources and stress factors in the company | Siegrist | Accounting for psychosocial resources at work to balance stress factors, identification of the stressful cases, and calculation of the workers' average morbidity | Silveri [87] |
| Company level | Survey regarding the technical agricultural itineraries to be assessed | Wesseling | Accounting for real practices of exposition to pesticides, calculation of the "human cost of pesticides" for workers regarding different ITK (technical agricultural itineraries) | di Cesare et al. [88] |
| Company level | Health impacts from ELCA of the product, number of human lives saved by the product | / | Comparing in DALY the health impacts caused by manufacturing one product with the human lives saved by the product (e.g., an airbag) | Arvidsson et al. [89] |

*DALY* disability-adjusted life years (unit of the measure of overall disease burden)

there is no work in SLCA about the social impacts of water scarcity nor of reuse. To cope with fair evaluation of IB, it is important to include the social impacts regarding water. More generally, as highlighted by the literature review, there are too few case studies of IB performed by SLCA of type 2 at the local or facility scale. Performing several case study surveys would teach us where are the important "missing pathways" (to assess the IB impacts at the local scale) and which are general enough to deserve long scientific work in order to obtain a well-established pathway. Finally,

SLCA is disarmed in front of "rebound effects," because it misses methodical developments in their anticipation. If SLCA researchers do not grasp this question, they will soon be overwhelmed by those who will grasp it, especially in the field of IB. Indeed, we have the intuition that the emergence of IB will accompany or entail huge social transformation. Thus, the rebound effects will make the difference between good and bad solutions.

# 4 Conclusions

This chapter discusses SLCA's contribution to the social assessment of IB in three steps: by analyzing the work carried out in social life cycle analysis in this field, by discussing the interests and limitations of the methods employed, and by systematizing the characteristics of what constitutes a "good" SLCA in the IB field. Not only is there no widely accepted approach to SLCA – which means that the field is not fully developed – but the authors who really want to learn about the social impacts of IB use a combination of several methods. When it comes to IB, the biofuel industries are the preferred area for analyzing social impacts. Often conducted using a social performance analysis based on CSR (corporate social responsibility) criteria, most studies provide little new information. Conversely, there are a few studies on the change caused by the appearance of an IB. They are based on the use of input-output tables of the national accounts, and these allow us to simulate and predict the impacts of various scenarios. The rules suggested for conducting an IB SLCA ensure that the studies can be useful for decision-making and/or can improve interactions between players (particularly when the work involves meaningful player participation).

Life cycle thinking also helps to question the future. For the time being, refineries are built thanks to oil; their structures are manufactured in rolling mills powered by fossil fuels; and materials transformed by IB are transported thanks to the use of oil. In the future, it is likely that the scarcity and/or the increased cost of fossil fuels will shape other types of refineries. It is conceivable that they will be located in close proximity to the resource and will be smaller as their dimensions will depend on the local availability of biomass. These refineries of tomorrow will look more like those advocated in Southern countries ([6], p. 10) than the large plants found in industrialized countries. Some of them are likely to be mobile, as described in the European project MOBILE FLIP [90] in order to seek out the resource in response to seasonal availability. The social impacts brought about by the creation of the three types of refinery are clearly different.

More generally, energy returned on energy invested rates of primary energy sources have been declining since the 1980s [91], while those expected by developed societies are much too high to be provided by new renewable energies (including biofuels) [92]. This gap is leading us toward social transformations, far from marginal changes. When we assess today what the future social impacts of IB will be, we are implicitly in the same context (background) as today. Yet it is conceivable that the effects of climate change could quickly become so severe that they restrict access to resources, to the point that they completely disrupt the balance of power between states and social relations within a state. The new rich countries would be those who have a large quantity of biomass resource per capita [93]. If the old rich countries let it happen, the distribution of power and development centers will completely change. For this reason, forecasts made in the field of IB must humbly acknowledge that they are referring to the current socioeconomic context, which will be turned upside down in the next 30 or even 10 years.

## References

1. Matos S, Hall J (2007) Integrating sustainable development in the supply chain: the case of life cycle assessment in oil and gas and agricultural biotechnology. J Oper Manag 25(6):1083–1102
2. German L, Schoneveld GC (2012) A review of social sustainability considerations among EU-approved voluntary schemes for biofuel, with implications for rural livelihood. Energy Policy 51:765–778. ISSN 0301-4215
3. Nitzsche R et al (2016a) Process simulation and sustainability assessment during conceptual design of new bioeconomy value chains. In: European biomass conference and exhibition proceedings 2016 (24th EUBCE), pp 1723–1726
4. Nitzsche R, Budzinski M, Gröngröft A (2016b) Techno-economic assessment of a wood-based biorefinery concept for the production of polymer-grade ethylene, organosolv lignin and fuel. Bioresour Technol 200:928–939
5. Schürrle K (2018) History, current state and emerging applications of industrial biotechnology. In: Fröhling M, Hiete M (eds) Sustainability and life cycle assessment in industrial biotechnology, Advances in biochemical engineering/biotechnology. Springer, Heidelberg
6. Elbehri A, Segerstedt A, Liu P (2013) Biofuels and the sustainability challenge: a global assessment of sustainability issues, trends and policies for biofuels and related feedstocks. FAO, Rome, p xvi + 174. ISBN: 9789251074145
7. Ejigu M (2008) Toward energy and livelihoods security in Africa: smallholder production and processing of bioenergy as a strategy. Nat Resour Forum 32:152–162
8. IUCN (2014) A global assessment of the environmental and social impacts caused by the production and use of biofuels. IUCN, Gland. 42 pp
9. German L, Schoneveld GC, Pacheco P (2011) The social and environmental impacts of biofuel feedstock cultivation: evidence from multi-site research in the forest frontier. Ecol Soc 16(3):24. Special issue: special feature on local, social, and environmental impacts of biofuels
10. Jiménez-González C, Woodley JM (2010) Bioprocesses: modeling needs for process evaluation and sustainability assessment. Comp Chem Eng 34(7):1009–1017
11. Macombe C, Loeillet D (2013) Social life cycle assessment, for who and why? In: Macombe C (coord) LCAs socio-economic effects in value chains, Nov 2013. FruitTrop Théma, Montpellier, pp 35–51
12. Mattila TJ, Judl J, Macombe C, Leskinen P (2018) Evaluating social sustainability of bioeconomy value chains through integrated use of local and global methods. Biomass Bioenergy 109:276–283
13. SHDB (2013) Social hotspots database. http://www.socialhotspot.org/. Accessed Nov 2017
14. Saaty TL (1980) The analytic hierarchy process. McGraw-Hill, New York
15. Roy B (1991) The outranking approach and the foundations of electre methods. Theor Decis 31(1):49–73. https://doi.org/10.1007/BF00134132
16. Halog A, Manik Y (2011) Advancing integrated systems modelling framework for life cycle sustainability assessment. Sustainability 3:469–499

17. von Geibler J, Liedtke C, Wallbaum H, Schaller S (2006) Accounting for the social dimension of sustainability: experiences from the biotechnology industry. Bus Strateg Environ 15:334–346
18. Markevičius A, Katinas V, Perednis E, Tamašauskienė M (2010) Trends and sustainability criteria of the production and use of liquid biofuels. Renew Sustain Energy Rev 14 (9):3226–3231
19. Scott F, Quintero J, Morales M, Conejeros R, Cardon C, Aroca G (2013) Process design and sustainability in the production of bioethanol from lignocellulosic materials. Electron J Biotechnol 16(3):13
20. Hasenheit M, Gerdes H, Kiresiewa Z et al (2016) Summary report on the social, economic and environmental impacts of the bioeconomy. http://bio-step.eu/fileadmin/BioSTEP/Bio_documents/BioSTEP_D2.2_Impacts_of_the_bioeconomy.pdf
21. Martin M, Røyne F, Ekvall T et al (2018) Life cycle sustainability evaluations of bio-based value chains: reviewing the indicators from a Swedish perspective. Sustainability 10(2):547
22. Falcone PM, Imbert E, Tani A, Tartiu VE, Morone PG (2018) Transitioning towards bioeconomy: assessing the social dimension through the lenses of the stakeholders. In: Cirad (ed) People and places for partnership, 6th international seminar in social LCA, Pescara, 10–12 Sept 2018. Thema, FruiTrop, Cirad, Montpellier, pp 189–193
23. Rafiaani P, Kuppens T, Van Dael M, Azadi H, Lebailly P, Van Passel S (2018) Social sustainability assessments in the biobased economy: towards a systemic approach. Renew Sustain Energy Rev 82(Part 2):1839–1853
24. Venkatesh A, Posen ID, MacLean HL, Chu PL, Griffin WM, Saville BA (2018) Environmental aspects of biotechnology. In: Fröhling M, Hiete M (eds) Sustainability and life cycle assessment in industrial biotechnology, Advances in biochemical engineering/biotechnology. Springer, Heidelberg
25. Fröhling M, Hiete M (2018) Sustainability and life cycle assessments in industrial biotechnology: a review of current approaches and future needs. In: Fröhling M, Hiete M (eds) Sustainability and life cycle assessment in industrial biotechnology, Advances in biochemical engineering/biotechnology. Springer, Heidelberg
26. O'Brien M, Doig A, Clift R (1996) Social and environmental life cycle assessment (SELCA) approach and methodological development. Int J Life Cycle Assess 1(4):231–237
27. Norris G (2006) Social impacts in product life cycles: towards life cycle attribute assessment. Int J Life Cycle Assess 1:97–104
28. Parent J, Cucuzzella C, Revéret J-P (2010) Impact assessment in SLCA: sorting the sLCIA methods according to their outcomes. Int J Life Cycle Assess 15:164–117
29. Macombe C, Falque A (2013) An alternative to life cycle CSR. In: Macombe C (coord) LCAs socio-economic effects in value chains, Nov 2013. FruitTrop Théma, Montpellier, pp 21–33
30. Weidema B (2006) The integration of economic and social aspects in life cycle impact assessment. Int J Life Cycle Assess 11(1):89–96
31. Feschet P, Macombe C, Garrabé M, Loeillet D, Benhmad F, Rolo Saez A (2012) Social impact assessment in LCA using the Preston pathway – the case of banana industry in Cameroon. Int J Life Cycle Assess 18:490–503
32. Macombe C (ed) (2017) Social LCA researcher school book: social evaluation of the life cycle, application to the agriculture and agri-food sectors. FruiTrop Thema, Cirad, Montpellier
33. McManus MC, Taylor CM (2015) The changing nature of life cycle assessment. Biomass Bioenergy 82:13–26
34. Valente C, Brekke A, Saur Modahl I (2018) Testing environmental and social indicators for biorefineries: bioethanol and biochemical production. Int J Life Cycle Assess 23:581–596
35. Spierling S, Knüpffer E, Behnsen H, Mudersbac M et al (2018) Bio-based plastics – a review of environmental, social and economic impact assessments. J Clean Prod 185:476–491
36. Vuaillat M, Jouanne G, Le Pochat S, Yousnadj D (2013) Environmental and social life cycle assessment of a new biobased material in Brazil. In: 3rd international seminar in social LCA, Montreal, 6–7 May 2013

37. Alvarez-Chavez CR, Edwards S, Moure-Eraso R, Geiser K (2012) Sustainability of biobased plastics. General comparative analysis and recommendations for improvement. J Clean Prod 23(1):47–56
38. Perez-Lopez P, Feijoo G, Moreira M (2017) Sustainability assessment of blue biotechnology processes: addressing environmental, social and economic dimensions. In: LCM 2017 (8th international conference on life cycle management), Luxembourg, Sept 2017
39. UNEP-SETAC (2009) Guidelines for social life cycle analysis of products. UNEP, Geneva
40. Bailey JA, Amyotte P, Khan FI (2010) Agricultural application of life cycle iNdeX (LInX) for effective decision making. J Clean Prod 18(16–17):1703–1713
41. Chingono TT, Mbohwa C (2015) Social impacts of biofuels production in the Kwa-Zulu Natal and Western Cape regions of South Africa. In: Ao SI, Douglas C, Grundfest WS, Burgstone J (eds) World congress on engineering and computer science, WCECS 2015. Lecture notes in engineering and computer science, vol 2, pp 956–960
42. do Carmo BBT, Margni M, Baptiste P (2017) Addressing uncertain scoring and weighting factors in social life cycle assessment. Int J Life Cycle Assess 22(10):1609–1617
43. Ekener E, Hansson J, Gustavsson M (2018a) Addressing positive impacts in social LCA-discussing current and new approaches exemplified by the case of vehicle fuels. Int J Life Cycle Assess 23(3):556–568
44. Sajid Z, Lynch N (2018) Financial modelling strategies for social life cycle assessment: a project appraisal of biodiesel production and sustainability in Newfoundland and Labrador, Canada. Sustainability 10:3289
45. Henke S, Theuvsen L (2014) SLCA: regionally differentiated evaluation of biogas plants and short rotation coppices. J Austrian Soc Agric Econ 23:81–90
46. Henke S, Theuvsen L (2013) Social life cycle assessment: socioeconomic evaluation of renewable energy. In: Koroneos C, Rovas D, Dompros A (eds) Proceedings of ELCAS 2013, pp 1035–1047
47. Ren J, Manzardo A, Mazzi A, Zuliani F, Scipioni A (2015) Prioritization of bioethanol production pathways in China based on life cycle sustainability assessment and multicriteria decision-making. Int J Life Cycle Assess 20:842–853
48. Renn O (2004) Social assessment of waste energy utilization scenarios: fuel and energy abstracts. Energy 45(2):150
49. Mbohwa C, Myaka N (2010) Social life cycle assessment of biodiesel in South Africa: an initial assessment. In: Proceedings of the 9th international conference on ecobalance, Tokyo, 9–12 Nov 2010, 4 pp
50. Manik Y, Leahy J, Halog A (2013) Social life cycle assessment of palm oil biodiesel: a case study in Jambi Province of Indonesia. Int J Life Cycle Assess 18:1386
51. Somé AKZ, Revéret J-P (2014) Evaluation de la durabilité sociale de la filière des agrocarburants en Afrique suivant la perspective cycle de vie et à l'aide d'une utilisation combinée de méthodes. In: Macombe C, Loeillet D (eds) Social LCA in progress, 4th SocSem, pre-proceedings, Montpellier, 19–21 Nov 2014, pp 188–195
52. Cadena E, Rocca FC, Gutierrez JA, Barona A, Carvalho A (2018) Social life cycle assessment for a biorefinery project. In: Cirad (ed) People and places for partnership, 6th international seminar in social LCA, Pescara, 10–12 Sept 2018. Thema, FruiTrop, Cirad, Montpellier, pp 141–146
53. Kaltenegger I, Schwarzinger S (2018) The challenge of quantification: Social Life Cycle Assessment (s-LCA) for advanced biofuel from waste wood integrated in the steel industry. In: Cirad (ed) People and places for partnership, 6th international seminar in social LCA, Pescara, 10–12 Sept 2018. Thema, FruiTrop, Cirad, Montpellier, pp 102–104
54. Peñaloza D, Keller H (2018) Using the social hotspots database to assess the social risks of prospective value chains: the case of D-Factory. In: Cirad (ed) People and places for partnership, 6th international seminar in social LCA, Pescara, 10–12 Sept 2018. Thema, FruiTrop, Cirad, Montpellier, pp 197–202
55. Siegel Moecke EH, Feller R, dos Santos HA et al (2016) Biodiesel production from waste cooking oil for use as fuel in artisanal fishing boats: integrating environmental, economic and social aspects. J Clean Prod 135:679–688

56. Boonkum P, Nohtomi M, Mungkalasiri J, Thanangkano W, Nagata K, Onoda H (2014) Environmental and social impacts of Jatropha-based biodiesel: a case study in Thailand. In: IEEE, proceedings of the 2014 international conference and utility exhibition of green energy for sustainable development (ICUE)
57. GBEP (Global Bioenergy Partnership Sustainability Indicator for Bioenergy) (2011) 1st edn, FAO, Rome. ISBN 978-92-5-107249-3
58. Muñoz Mayorga MA et al (2018) Social life cycle assessment through the framework Multi-Level Social Life Cycle Assessment (ML-SLCA) of the bioelectricity generation in Floreana Island. In: Cirad (ed) People and places for partnership, 6th international seminar in social LCA, Pescara, 10–12 Sept 2018. Thema, FruiTrop, Cirad, Montpellier, pp 171–176
59. Malik A, Lenzen M, Geschke A (2016) Triple bottom line study of a lignocellulosic biofuel industry. GCB Bioenergy 8:96–110
60. Malik A, Lenzen M, Ralph PJ, Tamburic B (2015) Hybrid life-cycle assessment of algal biofuel production. Bioresour Technol 184:436–443
61. Cambero C, Sowlati T (2016) Incorporating social benefits in multi-objective optimization of forest-based bioenergy and biofuel supply chains. Appl Energy 178:721–735
62. Souza A, Watanabe MDB, Cavalett O, Ugaya CML, Bonomi A (2018) Social life cycle assessment of first and second generation ethanol production technologies in Brazil. Int J Life Cycle Assess 23:617–628
63. Wang Z, Kamali FP, Osseweijer P, Posada Duque J (2018) The role of social sustainability in aviation biofuel supply chains. In: Cirad (ed) People and places for partnership, 6th international seminar in social LCA, Pescara, 10–12 Sept 2018. Thema, FruiTrop, Cirad, Montpellier, pp 252–254
64. Vinyes E, Oliver-Solà J, Ugaya CML, Rieradevall J, Gasol CM (2013) Application of LCSA to used cooking oil waste management. Int J Life Cycle Assess 18:445–455
65. Vuaillat M, Wathelet A, Viot JF, Flore L et al (2018) Sustainable Guar Initiative (SGI) – social impact characterization of an integrated sustainable project. In: Cirad (ed) People and places for partnership, 6th international seminar in social LCA, Pescara, 10–12 Sept 2018. Thema, FruiTrop, Cirad, Montpellier, pp 63–65
66. Delcour A et al (2012) Contribution of social life cycle assessment to the evaluation of more sustainable scenarios of cereal uses in Wallonia (Belgium). In: Proceedings 2nd LCA conference, Lille, 6–7 Nov 2012
67. Siebert A, Bezama A (2014) Regional social life cycle assessment of wood-based products. In: Macombe C, Loeillet D (eds) Social LCA in progress, 4th SocSem, pre-proceedings, Montpellier, 19–21 Nov 2014, pp 167–171
68. Siebert A, Sinéad O'Keeffe AB, Thrän D (2018) Social life cycle assessment: in pursuit of a framework for assessing wood-based products from bioeconomy regions in Germany. Int J Life Cycle Assess 23:651–662
69. Blom M, Solmar C (2009) How to socially assess biofuels? A case study of the UNEP/SETAC code of practice for social-economical LCA. Master thesis, Lulea University of Technology
70. Delcour A et al (2014) ASCV comparative des filières céréalières en Wallonie (Belgique). In: Macombe C, Loeillet D (eds) Social LCA in progress, 4th SocSem, pre-proceedings, Montpellier, 19–21 Nov 2014, pp 93–96
71. Russo-Garrido S, Ménard-Chicoine M, Beaulieu L (2014) Identification des points chauds potentiels: SHDB versus collecte de bureau. In: Macombe C, Loeillet D (eds) Social LCA in progress, proceedings, 4th SocSem, Montpellier, 19–21 Nov 2014, p 123
72. Laufer WS (2003) Social accountability and corporate greenwashing. J Bus Ethics 43 (3):253–261
73. Grigoletto Duarte C, Wegg T, Dibo AP, Gallardo A (2014) Distribution of impacts in biofuels industry: applying social life-cycle assessment. In: IAIA14 conference proceedings. Impacts assessment for social and economic development. 34th annual conference of the International Association for Impact Assessment, Chile

74. Mucchielli A (2009) Dictionnaire des méthodes qualitatives en sciences humaines, 3rd edn. Armand Colin, Paris
75. Ekener E, Hansson J, Larsson A, Peck P (2018b) Developing life cycle sustainability assessment methodology by applying values-based sustainability weighting – tested on biomass based and fossil transportation fuels. J Clean Prod 181:337–351
76. Macombe C, Loeillet D, Gillet C (2018) Extended community of peers and robustness of social LCA. Int J Life Cycle Assess 23(3):492–506
77. Chen C, Reniers G (2018) Risk assessment of processes and products in industrial biotechnology. In: Fröhling M, Hiete M (eds) Sustainability and life cycle assessment in industrial biotechnology, Advances in biochemical engineering/biotechnology. Springer, Heidelberg
78. Murphy DJ, Hall CAS (2010) Year in review – EROI or energy return on (energy) invested. Ann N Y Acad Sci 1185:102–118. Issue: ecological economics reviews
79. Gagnon N, Hall CAS, Brinker L (2009) A preliminary investigation of energy return on energy investment for global oil and gas production. Energies 2(3):490–503
80. Nordhoff S, Höcker H, Gebhardt H (2007) Renewable resources in the chemical industry – breaking away from oil? Biotechnol J 2(12):1505–1513
81. CSDH (2009) Combler le fossé en une génération : instaurer l'équité en santé en agissant sur les déterminants sociaux de la santé, rapport final de la Commission des Déterminants sociaux de la santé. Organisation Mondiale de la Santé, Geneva. ISBN 978 92 4 256370 2
82. Lagarde V, Macombe C (2012) Designing the social life cycle of products from the systematic competitive model. Int J Life Cycle Assess 18:172–184
83. Rostila M, Kölegård ML, Fritzell J (2012) Income inequality and self-rated health in Stockholm, Sweden: a test of the 'income inequality hypothesis' on two levels of aggregation. Soc Sci Med 74(7):1091–1098
84. Macombe C, Leskinen P, Feschet P, Antikainen R (2013) Social life cycle assessment of biodiesel production at three levels: a literature review and development needs. J Clean Prod 52:205–216
85. Bocoum I, Macombe C, Revéret J-P (2015) Anticipating impacts on health based on changes in income inequality caused by life cycles. Int J Life Cycle Assess 20:405–417
86. Weidema B (2018) The social footprint – a pratical approach to comprehensive and consistent social LCA. Int J Life Cycle Assess 23(3):700–709
87. Silveri F (2017) The Siegrist pathway: work conditions and health. Anticipating psychosocial factors effects in the agro-food sector. In Macombe C (ed) Social LCA researcher school book. FruiTrop Thema, Cirad, Montpellier, pp 148–162
88. di Cesare S, Loeillet D, Macombe C (2017) The Wesseling pathway: the assessment of farmworkers exposure to pesticides. In: Macombe C (ed) Social LCA researcher school book. FruiTrop Thema, Cirad, Montpellier, pp 164–173
89. Arvidsson R, Hildenbrand J, Baumann H, Islam KMN (2018) A method for human health impact assessment in social LCA : lessons from three case studies. Int J Life Cycle Assess 23(3):690–699. in Special issue: social LCA in progress
90. Brunklaus B, Molnar S, Sandin G, Torén J, Mangold M (2018) Choice of social indicators within technology development – the case of mobile biorefineries in Europe. In: Cirad (ed) People and places for partnership, 6th international seminar in social LCA, Pescara, 10–12 Sept 2018. Thema, FruiTrop, Cirad, Montpellier, pp 162–166
91. Gagnon L (2008) Civilisation and energy payback. Energy Policy 36(9):3317–3322
92. Lambert JG, Hall CAS, Balogh S, Gupta A, Arnold M (2014) Energy, EROI and quality of life. Energy Policy 64:153–167
93. Friedrichs J (2010) Global energy crunch: how different parts of the world would react to a peak oil scenario. Energy Policy 38:4562–4569

# Part IV
# Specific Methods and Applications

Adv Biochem Eng Biotechnol (2020) 173: 233–254
DOI: 10.1007/10_2019_114

Published online: 9 December 2019

# Assessing Land Use and Biodiversity Impacts of Industrial Biotechnology

**Jan Paul Lindner, Tabea Beck, Ulrike Bos, and Stefan Albrecht**

## Contents



**Abstract** One of the many promises of biotechnology is that it allows societies to move away from a fossil-based industry toward a bio-based industry, with positive implications for anthropogenic climate change and resource dependency. The provision of biomass from agriculture or forestry is, however, linked to specific environmental implications that cannot be disregarded in an informed discussion about the role of biotechnology in the twenty-first century. In this chapter, we discuss landuse-related

J. P. Lindner (✉)
Bochum University of Applied Sciences, Bochum, Germany

Department of Life Cycle Engineering, Fraunhofer IBP, Stuttgart, Germany
e-mail: jan-paul.lindner@hs-bochum.de

T. Beck
International Aid Services Germany, Kirchheim am Neckar, Germany

U. Bos
Thinkstep AG, Leinfelden-Echterdingen, Germany

S. Albrecht
Department of Life Cycle Engineering, Fraunhofer IBP, Stuttgart, Germany

effects of biomass provision such as landscape homogenization, eutrophication, erosion, biodiversity, and others. We also discuss how these effects are represented in Life Cycle Assessment, which is a powerful tool for product sustainability evaluation.

**Graphical Abstract**

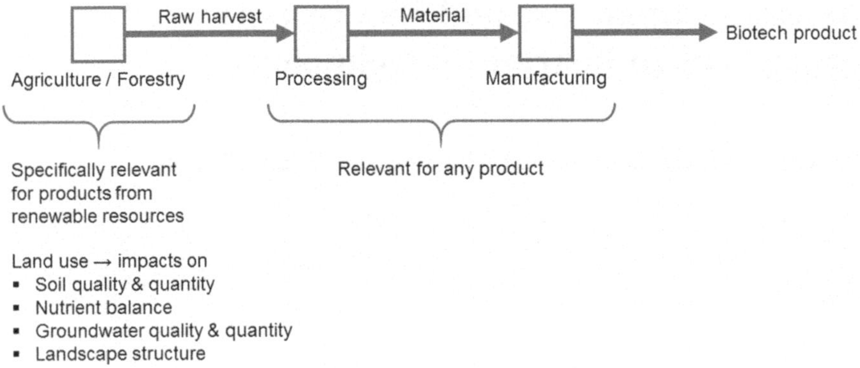


**Keywords** Biotechnology, Land use, Life Cycle Assessment

## 1 Background

Biotechnology is defined in the UN Convention on Biological Diversity [1] as "any technological application that uses biological systems, living organisms, or derivatives thereof, to make or modify products or processes for specific use." Closely related is the more recent concept of the bioeconomy. According to the European Commission, bioeconomy aims to "reduce the dependence on natural resources, transform manufacturing, promote sustainable production of renewable resources from land, fisheries and aquaculture and their conversion into food, feed, fibre, bio-based products and bio-energy, while growing new jobs and industries" [2]. The concept of bioeconomy is deeply integrated in Europe's Framework Programme for Research and Innovation as an integral part of the strategic development of the European economy. There are corresponding national strategies of the EU members. The German Government, for example, aims to establish an internationally competitive bioeconomy through the "National Research Strategy Bioeconomy 2030," in which five main fields of action are specified: securing global nutrition, sustainable agricultural production, producing safe and healthy foods, industrial use of bio-based resources, and energy on the basis of biomass [3].

Industrial biotechnology as part of the bioeconomy as a whole endeavors to meet the increasing need for both conventional mass market products and new innovative products from biomass like starch, sugar, cellulose, fat, oil, proteins, fibers, etc. for several applications (see also Venkatesh et al. [4], as well as Nova [5]).

- Starch and sugar: starch is typically converted to sugar. Sugar is the basis for further modification, for example, to lactic acid, the monomer of the most common bio-based plastic polylactic acid (PLA). C5 sugars can be used as building blocks, e.g., for polyethylene terephthalate (PET) alternatives.
- Cellulose can be 7used directly (e.g., as the main ingredient for paper) or modified, for example, to carboxymethyl cellulose (CMC) or cellulose acetate carboxylates (such as cellulose acetate propionate (CAP) and cellulose acetate butyrate (CAB)).
- Fats are typically cleaved into fatty acids and glycerol. Glycerol is a base chemical in itself but is also a high-energy substrate for anaerobic digestion (with the goal of producing biogas). Oils and fatty acids can be used as food additives, as building blocks for plasticizers, for example, and as substrates for anaerobic digestion.
- Proteins can be used in pharmaceutic products, as food supplements, or as precursors of specialized plastics.
- Natural fibers can be used, for example, directly in yarns or in fiber-reinforced composites.
- Specialty products of biogenic origin are, for example, vitamins, fragrances, enzymes, and more.

Scarlat et al. [6] give an overview of the "role of biomass and bioenergy in a future bioeconomy." They show the current and future demand and needs for a sustainable bioeconomy in Europe and in the world. A special focus is given to the current biomass supply worldwide, in Europe and in the European Union in 2011 regarding agriculture, food and beverage, agro-industrial products, fisheries and aquaculture, forestry logging, food-based industry, biochemicals, bioplastics, biolubricants, biosolvents, biosurfactants, enzymes, biopharmaceuticals, biofuels, and bioenergy. It is completed with an outlook to the increasing need and demand for biomass in the respective applications. One of the major findings is that a "significant increase in the demand for biomass for bio-based materials, together with the predictable increase in biomass demand for bioenergy will increase the competition for natural resources, in particular for land and water resources with potential negative impact on the land use patterns, biodiversity and environment" [6]. Venkatesh et al. [4] also refer to the environmental implications of biotechnology. They stress that "environmental impacts of industrial biotechnology may be significant across a number of categories that include, but may not be limited to, non-renewable resource depletion, water withdrawals and consumption, climate change and natural land transformation/occupation."

Thus a transition is needed toward optimal use of renewable biological resources, i.e., toward a use pattern that maximizes biomass output while minimizing environmental impacts. To face all efforts for the transformation toward a sustainable bioeconomy in Europe and worldwide in a responsible way, it is mandatory to expand the view on less environmental impacts and reduced greenhouse gas emissions of available and future products with a distinct view on land use effects, soil quality changes, and biodiversity in global supply chains. See also Fröhling and Hiete [7] for a review of current approaches and future needs in sustainability assessment related to industrial biotechnology.

## 2 The Link Between Biotechnology, Land Use, and Biodiversity

In most cases biotechnology demands for biogenic feedstock, be it for substrates or for precursor material. When looking at biotechnology on an industrial scale, it is quite likely that this feedstock is produced by large-scale agriculture or intensive forestry. For example, starch and sugar crops (e.g., wheat, corn/maize, sugar beets, sugar cane) provide feedstock for digestion products such as lactic acid, the precursor of the prevalent bioplastic PLA. Both agriculture and forestry demand space in terms of land surface as a production factor.

Land that can be used for biomass production is a scarce good. Table 1 shows that already about 75% of the Earth's land surface is anthropogenically used. The remaining area is nonproductive, remote, or wilderness area (as classified, e.g., by Erb et al. [8]). Regarding the world population growth rate of currently 1.1% per year [9], it is evident that the land demand for food production will likely continue to rise; UNEP [10] estimates an expansion between 710,000 and 3,000,000 $km^2$ until 2050. Land demands for industrial biotechnology, estimated to expand to between 40,000 and 1,115,000 $km^2$ until 2050 [10], aggravate the competition for productive land.

Besides land being regarded as a scarce resource or a resource producer itself, it also provides a number of essential services to humans and ecosystems.

The Millennium Ecosystem Assessment (MA), an extensive study conducted from 2001 to 2005 under the auspices of the United Nations, has made the importance of functioning ecosystems for human well-being on Earth indisputably clear [11]. The aim of the MA was to assess the consequences of anthropogenic changes in ecosystems on human well-being and to provide the scientific basis for measures needed for a sustainable use of ecosystems. One core message of the MA is that humanity has invaded the surrounding ecosystems more quickly and more extensively than ever before and it underscores the global dependency of mankind on nature and ecosystem services as the basis for a healthy and safe life [12]. About 50% of Earth's land area is strongly affected by mankind [13], and this land use does have enormous effects on ecosystem services.

**Table 1** Global land cover and use, source: Erb et al. [8], altered

| Compartment | Area [$km^2$] | Area [%] | |
|---|---|---|---|
| Earth | 510,100,000 | 100 | |
| Seawater | 362,171,000 | 71 | Of Earth |
| Land | 130,406,000 | 26 | Of Earth |
| Settlement and infrastructure | 1,360,000 | 1 | Of land |
| Cropland | 15,225,000 | 12 | Of land |
| Forestry | 34,958,000 | 28 | Of land |
| Total forest area | 41,126,000 | 32 | Of land |
| Grazing land | 46,911,000 | 36 | Of land |
| Nonproductive area | 16,163,000 | 12 | Of land |
| Remote area, wilderness | 15,788,000 | 12 | Of land |

Following the Millennium Ecosystem Assessment [12], the services provided by ecosystems include, for example, climate regulation by incorporation and mid- to long-term storage of atmospheric carbon, as well as nutrient regulation by maintaining the cycle of biomass production and degradation. Water levels are regulated by surface vegetation and soils controlling floods and infiltration to groundwater. Water is purified by physical and chemical soil processes. Diseases or pests are regulated by species or biocoenoses. Cultural aspects of land use are also to be considered, such as the provision of spiritual and religious services, recreation, and tourism. To maintain these services provided by ecosystems, in many cases manifestations of specialized biodiversity are needed that are deeply rooted on the land they are located on.

Most anthropogenic land using systems, including conventional agriculture and intensive forestry, modify land and ecosystems in a way that is detrimental to soil functions and the diversity of life on the land. The following sections show examples, as well as approaches to assess these impacts within a Life Cycle Assessment framework.

In addition to the relation between biotechnology and land use described above, there may be cases where industrial biotechnology does not lead to an increased demand or use of land. If such systems replace conventional systems which have a high demand for land or a high impact on the used land, comparison studies may even show significant improvements. For example, algae cultivated in seawater environments demand very little area on dry land (e.g., [14]). Techniques such as biological heap leaching involve sprinkling low-grade ore (tailings) with a microbial suspension to extract metals from a stock that would otherwise be unsuitable for more expensive procedures. The organisms and biochemical mechanisms involved in bioleaching are the subjects of ongoing research; see, for example, Cárdenas et al. [15] and Glombitza et al. [16].

## 3 Effects of Agriculture

Biotechnology feedstock is typically biomass, which is produced in either agricultural or silvicultural systems (see previous section above). This section deals with the environmental impacts of agriculture as a production system for biotechnology feedstock.

Conventional, high-intensity agriculture uses huge inputs in terms of technology, energy, and resources to produce highest possible yields of standardized crops: Fields are processed using big machinery several times per year. Energy and material resources are consumed for fertilizer and pesticide production; and the application of both on arable land has the potential to highly influence underlying ecosystems in a negative way. As regulations for nonfood agricultural goods are lower than for food, critics assume that agriculture-related environmental impacts will be exacerbated by the growing importance of biomass for bioenergy and in industrial processes (e.g., [17]).

Nitrogen and phosphorous fertilizers are used in high amounts in conventional agriculture and are produced with a high-energy input [18]. Plants need nitrogen compounds for growth and convert it during their metabolic processes: Atmospheric $N_2$ is incorporated, fixed, and assimilated by soil organisms. Mineralized compounds ($NH_4^+$, $NO_3^-$) can then be taken up by plants. Ammonium can also be fixed to clay minerals in the soil, which act as a buffer for inorganic nitrogen compounds. Processes bringing nitrogen compounds back to the atmosphere are evaporation and denitrification; eluviation processes transport nitrogen to lower soil layers and to aquifers, where they have the potential to pollute groundwaters. Nitrogen inputs into the system from agriculture that cannot be assimilated by plants therefore cause both gaseous emissions ($N_2O$, $NH_3$) to the atmosphere that potentially increase global warming and acidification and nitrate pollution of groundwater bodies. Similar processes occur in the phosphorous cycle, with less importance of atmospheric emissions but high eutrophication impacts of phosphorous eluviation to waterbodies: High loads of nutrients lead to an increased phytoplankton production. This again leads to a decrease of oxygen in the water. In both aquatic and terrestrial ecosystems, high nutrient levels lead to an extinction of species that are adapted to low-nutrient conditions. For further reading on nutrient balances of natural or agricultural production systems, we recommend, for example, Odum [19] or Schubert [20]. Especially emissions of nitrogen compounds (e.g., $NH_3$ and $N_2O$ to air, $NO_3^-$ to ground- and surface water) can be considered in LCA by using specific emission factors for cultivation processes.

Soil loss by erosion is another issue that influences ecosystem services provided by land or soil to a large extent and that can largely be influenced by agriculture: Besides rainfall intensities, topography, and basic soil erodibility based on grain size, soil erosion is determined by the vegetation cover, organic matter content of the soil, and agricultural practices such as tillage or contour farming [21, 22].

Soil erosion has been identified as a major threat for global fertile lands by the United Nations Food and Agricultural Organisation (FAO): "Erosion carries away 25 to 40 billion tonnes of topsoil every year, significantly reducing crop yields and the soil's ability to store and cycle carbon, nutrients and water." If action is not taken to reduce erosion, total crop yield losses projected by the year 2050 would be equivalent to removing 1.5 million $km^2$ of land from crop production – or roughly all the arable land in India [23]. Eroded soils cause problems both in the eroded places, where the affected soils loose productivity, and at the alluviation locations (where the soil is deposited) where the eroded soil may, for example, cause problems with infrastructure [24].

Besides polluting the groundwater with nutrients or pesticides, due to high volumes of water withdrawn for irrigation, agriculture in many cases is responsible for groundwater level changes. These again can have severe impacts on surrounding ecosystems, as ecosystems always are very well adapted to the surrounding conditions, and may change significantly when confronted with altered conditions, and also on humans being deprived of clean drinking water [25].

Pesticide application, as is common in conventional agriculture, aims at a reduction of harmful species but very often leads to a drastic reduction of species in

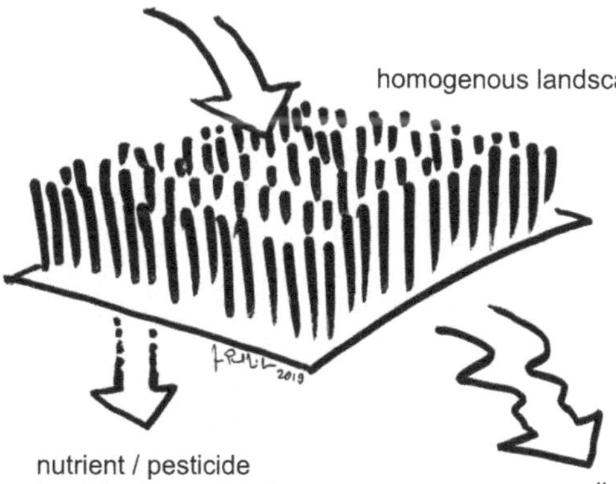


**Fig. 1** Illustrated summary of the environmental effects of agriculture (illustration: J.P. Lindner)

general on agricultural land. Sometimes pesticide use also leads to accumulation of toxic substances in organisms, food, soil, or water. Health effects generated by this for humans or ecosystems, called ecotoxicity or human toxicity, can be addressed in advanced Life Cycle Assessment (LCA) studies that go beyond using standard databases [26].

Another issue closely related to this is the homogenization of the landscape: As only the species supported by agriculture survive, monocultures arise. The destruction of small landscape structuring and biodiversity enhancing elements such as hedges or swamplands leads to marginalization of ecotopes with only some species or biocoenoses surviving [27].

Many of the agriculture-related environmental effects mentioned in this section (Fig. 1) come together in the accelerated loss of biodiversity in agricultural landscapes. The latest Living Planet Report [28] lists the threats to biodiversity: habitat degradation, habitat loss, overexploitation, climate change, pollution (mostly nutrients, but also toxins), and invasive species. Agriculture significantly contributes to at least four of these, i.e., habitat degradation/loss, climate change, and pollution. Consequently, biotechnology proponents need to be aware of the environmental implications of increased agricultural output. More on biodiversity is found in the respective section below.

## 4 Effects of Forestry

Biotechnology feedstock is typically biomass, which is produced in either agricultural or silvicultural systems (see respective section above). This section deals with the environmental impacts of forestry as a production system for biotechnology feedstock.

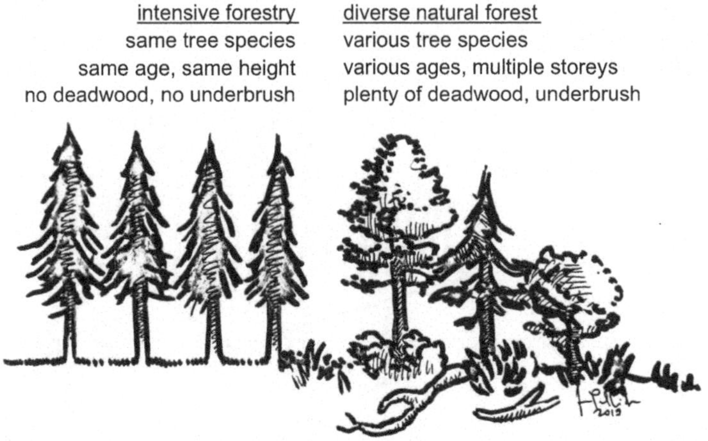


**Fig. 2** Key differences between intensive forestry and natural forests (illustration: J.P. Lindner)

Forestry yields primarily wood as a base product for value chains. The main consumer of wood is the pulp and paper industry, but the wood is actually fractioned and the various fractions are used for different purposes. Wood has already been discovered as an interesting feedstock for industrial biotechnology (see, e.g., [5]). This section deals with the environmental implications of forestry for wood provision.

Forestry is generally considered more environmentally benign than agriculture. It usually doesn't need any fertilizer and pesticides, and productive forests resemble natural forests more than, say, fields resemble bogs. There are, however, a number of environmental implications from forestry.

The main impact of forestry on the forest ecosystem is the homogenization of the landscape (Fig. 2). Intensive forestry promotes monocultures (most prevalent in temperate regions are various spruce species) and thus reduces the suitability of the forest as habitat for a broader variety of species. This is changing as some forestry companies gradually introduce more benign management practices (see, e.g., [29, 30]). Many productive forests, though not all, have a unified age structure (all trees are the same age), and they especially lack old trees. Trees, throughout their individual life cycle, provide habitat and nutrition for a variety of species. Deadwood is often cleared from managed forests. Same as with living trees, deadwood also provides habitat and nutrition for yet another cornucopia of species. Homogenization of landscapes is also connected to the lack of ecotones. Natural forest edges are gradients between the neighboring open landscapes and the full-height closed forest. This gradient, again, provides a broad variety of habitat types and foraging opportunities for a corresponding variety of species. If it is gone, so are the species that depend on it.

In some areas of the world, forestry relies on introduced exotic species. The most prevalent example is eucalyptus in tropical and subtropical regions. The high and straight eucalyptus visibly stands out from, for example, the low and crooked Cerrado vegetation in Brazil. These trees are not suitable as habitat or food for many species that inhabit the remaining areas of natural vegetation in the region (see, e.g., [31]).

Natural forests experience fires in irregular intervals. In managed forests, fire is suppressed. What may seem like a benefit for the forest ecosystem at first sight is actually the loss of a major disturbance factor that the ecosystem needs for rejuvenation and to instigate succession. When a space is cleared in a forest, it is subsequently colonized by pioneer plants that specialize in the rapid colonization of clearances but is overcrowded by other plants in the long run. Different pioneer plants specialize in different stages of the forest regrowth, which leads to the typical succession sequence that can be observed in natural forest ecosystems. Many pioneer plants thrive only in this short timeframe and thus depend on regular disturbance of the forest. But it is not only pioneer plants that rely on disturbances. In North America, the seeds of the giant sequoia tree *Sequoiadendron giganteum* typically sprout after a fire has cleared the area of competitors [32]. Even though the tree grows old and very tall, it still maximizes its survival chances with this behavior. In Scandinavia, the black fire beetle *Melanophila acuminata* lays its eggs only in freshly burnt wood [33].

Forests in the vicinity of rivers may experience semi-regular flooding rather than fire. While the specific cause-effect relationships differ, the overall relevance of disturbance dynamics is similar for flooding as for fire. Flooding appears more often and more regularly than fires (depending on the dynamics of the river, up to yearly flooding may occur) [34, 35].

Secondary effects are a consequence of the infrastructure development that comes with a forestry operation. Roads are built into pristine areas to make them accessible. Apart from the noise pollution and the fragmentation effects of roads, they may also attract nuclei of informal settlements, leading to further development and encroachment into pristine areas [36, 37].

## 5 Land Competition

Land, or surface area, fulfills many purposes for human society, the most straightforward of which is space for dwellings and for the cultivation of food. With fossil energy resources declining, more surface area of Earth will be used to catch solar energy. Traffic infrastructure covers a substantial share of the area of the more densely populated countries in the world. Wildlife conservation needs large undisturbed areas. All these purposes compete for the given surface area of Earth, and most of them for dry surface area. This conundrum is called land competition.

The total dry surface area of Earth is limited and a lot of the available area is under management (see Table 1 above). In world regions with old civilizations, the most suitable areas have been cultivated long ago. While there is potential for increasing yields (e.g., [38–40]), growing demand for biogenic products is often met by new cultivation of pristine areas. Even if new cultivation is deliberately avoided, switching from one crop to a different crop on existing farmland means that the demand for the phased-out crop is likely to be satisfied elsewhere. Even supplying the same product for another application (e.g., switch from food to energy) may trigger this so-called indirect land use change (ILUC).

Considering both the many competing purposes of land and its limited availability, there are two broad strategies for guiding land use change globally: intensification and extensification. Intensification means to maximize the productivity of land that is already in use, thus sparing pristine areas. The environmental impact on the used areas would very likely be high, but the spared areas would remain free from human intervention. Extensification means to spread out the anthropogenic interventions so that the damage is small. This would mean more total area under management and less total pristine area, but the managed area would retain more ecosystem dynamics and more biodiversity than intensively managed areas. This debate is also known as "land sparing vs. land sharing." Arguments for either strategy have been brought forth [41, 42], and it is unclear whether there actually is a winning strategy in the face of constant population growth, economic growth, and the resulting consumption increase. The WWF promotes agroecology as a form of ecological intensification [43]. The German biodiversity strategy demands both more organic agriculture, which can be interpreted as an extensification approach, and extension of nature reserves [44], which would imply intensification on the used area outside the reserves.

## 6 Land Use in Life Cycle Assessment

Life Cycle Assessment is an internationally standardized method established in research and industry for identifying the potential environmental impacts that occur during the entire life cycle of a product or process. The objective is to aggregate all resources and emissions along the value chain of a product or process and calculate the resulting effects on the environment in impact categories such as the contribution to climate change, acidification, or eutrophication. The LCA community is aware of the links between land use and human well-being. In order to cover all relevant environmental impacts of a product or process, land use aspects impacting ecosystem services have to be integrated into methods like LCA. In recent years a framework as well as methods for the consideration of impacts of land use on ecosystem services have successfully been developed and applied in LCA. These methods address, for example, soil quality indicators like erosion, soil organic matter, desertification, soil compaction, physicochemical filtration, mechanical filtration, groundwater regeneration, and salinization.

For more than 20 years, international working groups have been developing, and still are developing, a framework on how land use issues shall be integrated into Life Cycle Assessment [45, 46]. It is known as the UNEP-SETAC framework since it was for the most part developed under the auspices of the Life Cycle Initiative, hosted by both the United Nations Environment Programme (UNEP) and the Society for Environmental Toxicology and Chemistry (SETAC). According to the framework, the impact of a land using process on the land on which it is anchored depends on the quality of a certain indicator, the affected area (given in $m^2$), and the time period the land is occupied (given in years). Quality indicators for use in this framework can be as specific as biotic production or mechanical filtration and as broad as "naturalness" (see reference to the hemeroby concept in biodiversity section below). In general, any desirable property of planetary surface can be interpreted as a quality in the sense of the UNEP-SETAC framework, but LCIA methods exist only for a selection of properties.

One essential concept to understand the integration of effects on ecosystem services and biodiversity in LCA is the concept of transformation and occupation of land: Occupation of land means the condition of a patch of land while it is used. It is assumed that there is no change of land quality throughout the entire time period of usage (e.g., 20 years for a poplar plantation). Occupation is expressed as level of ecosystem quality during use, compared to a specific reference quality, for example, potential natural vegetation. Transformation defines a change in the ecosystem quality of the studied patch between the initial quality of the ecosystem and the quality after the use phase has ended and the land is regenerated. Transformation is separated into a "transformation from" and "transformation to" phase as stated by Koellner et al. [45].

The transformation as well as occupation values present the ecosystem quality difference ($\Delta Q$) between the reference situation ($Q_{\text{ref}}$) and the respective chosen land use ($Q_{\text{LU}}$). The reference situation in life cycle impact assessment describes a reference in a region in relation to the current intervention. This might be a situation without any anthropogenic influence (i.e., without land use at all). The reference situation is used as common basis to calculate the quality differences.

"Transformation from" means the transformation from a previous land use type and respective ecosystem quality $Q_{\text{previous}}$ (e.g., from grassland) to the reference situation $Q_{\text{ref}}$. As depicted in Fig. 3, positive values regarding the ecosystem quality mean an improvement of the ecosystem quality, negative values a decline.

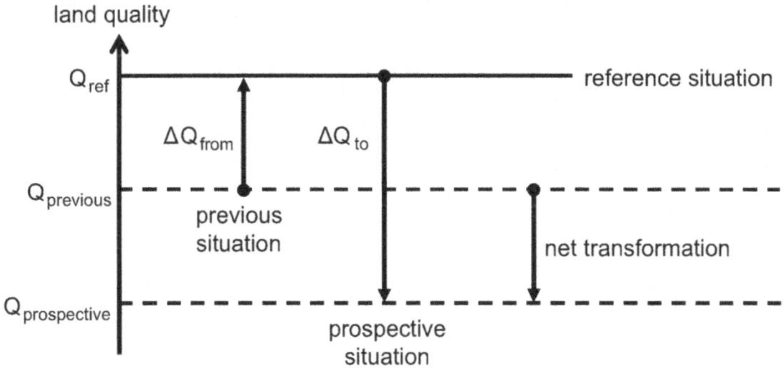


**Fig. 3** Schematic representation of land transformation along the quality of a land use indicator

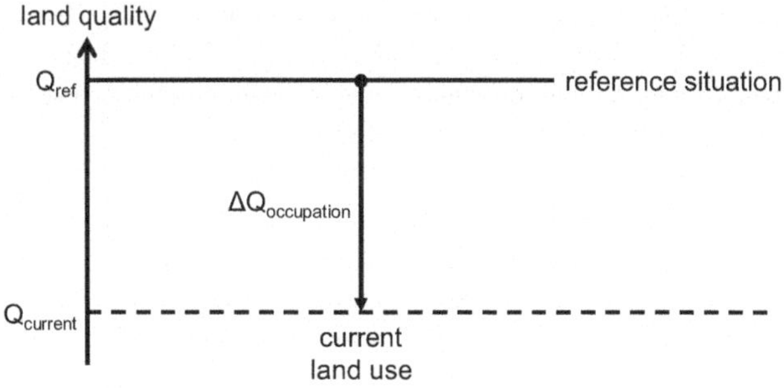


**Fig. 4** Schematic representation of the land occupation along the quality of a land use indicator

"Transformation to" means the transformation from a reference situation $Q_{\mathrm{ref}}$ to a prospective land use type and ecosystem quality $Q_{\mathrm{prospective}}$ (e.g., artificial areas). Positive values imply an improvement of the ecosystem quality, whereas negative values stand for a decline of the ecosystem quality.

The characterization factors for transformation impacts (from/to) are defined as the respective quality differences, as formulated in Eqs. (1) and (2):

$$\mathrm{CF}_{\text{transformation from}} = -\Delta Q_{\mathrm{from}} = -\left(Q_{\mathrm{ref}} - Q_{\mathrm{previous}}\right) \tag{1}$$

$$\mathrm{CF}_{\text{transformation to}} = -\Delta Q_{\mathrm{to}} = -\left(Q_{\mathrm{prospective}} - Q_{\mathrm{ref}}\right) \tag{2}$$

Using the characterization factors as well as the life cycle inventory information given in $m^2$ used land, the impacts are defined as the sum of the characterization factors multiplied by the area per functional unit $A_{\mathrm{FU}}$:

$$\text{Transformation impact} = \left(\mathrm{CF}_{\text{transformation from}} + \mathrm{CF}_{\text{transformation to}}\right) \times A_{\mathrm{FU}} \tag{3}$$

"Occupation" represents the quality difference between the reference situation and the current land use type and respective ecosystem quality $Q_{\mathrm{LU,\ current}}$ (e.g., the excavation area for occupation and mineral extraction). As shown in Fig. 2, positive quality levels mean a degradation, negative values an improvement of the ecosystem quality (Fig. 4).

The characterization factors for occupation impacts are defined as the quality difference between the quality in the occupied state and the reference quality, as given in Eq. (4):

$$\mathrm{CF}_{\mathrm{occupation}} = -\Delta Q_{\mathrm{occupation}} = -\left(Q_{\mathrm{LU,current}} - Q_{\mathrm{ref}}\right) \tag{4}$$

Using the characterization factors as well as the life cycle inventory information given in $m^2a$ land occupied, the impacts are calculated as the product of the characterization factor and the area time per functional unit $(A \times t)_{\mathrm{FU}}$:

$$\text{Occupation impact} = \text{CF}_{\text{occupation}} \times (A \times t)_{\text{FU}} \tag{5}$$

It is a core principle of LCA to multiply an inventory quantity (usually the mass of an emission) with its respective characterization factor. The framework described above introduces area and area time as new inventory quantities and quality differences as characterization factors. It thus allows the integration of impacts from structural changes to land surfaces into the LCA framework which has historically focused on mass and energy balances.

## 7 Soil Quality in Life Cycle Assessment

In the last 10 years, several approaches were developed and published to address soil quality indicators in Life Cycle Assessment: soil organic matter [47–49], desertification [50], compaction [51], erosion [52–54], mechanical and physicochemical filtration [52–54], groundwater regeneration, and biotic production [52–54].

The latter five indicators – erosion, mechanical and physicochemical filtration, groundwater regeneration, and biotic production – are known under the overarching label of LANCA® (Land Use Indicator Value Calculation Tool). This method is in line with the UNEP-SETAC framework for addressing land use in LCA ([45, 46], described above) and recommended by the European Commission to be used for the Product Environmental Footprint (PEF) [55].

LANCA® was developed at the University of Stuttgart, Department Life Cycle Engineering, on the basis of Baitz [52], and has been applied in many projects. With LANCA®, indicator values are calculated that describe the environmental impacts of land using processes on various ecosystem services. These indicators can then be integrated into LCA studies. The following environmental impact categories are calculated on the basis of methods from ecology and pedology:

*Erosion resistance* describes the detachment and transport of soil material [56]. Two types are distinguished: water erosion and wind erosion. Water erosion can be estimated through the Universal Soil Loss Equation (USLE) [57]. The indicator of the impact category is given in [kg soil/a] for transformation impacts and [kg soil] for occupation impacts.

*Mechanical filtration* describes the ability of the soil to mechanically clarify a suspension [58] and thus represents the filter capacity. According to Blume et al. [56], the filtration capacity of a soil indicates the amount of water that can pass through the soil per unit of time. The indicator of the impact category is given in [$m^3$ water/a] for transformation and [$m^3$ water] for occupation impacts.

*Physicochemical filtration* is the ability of the soil to absorb dissolved substances from the soil solution [52, 56]. It strongly depends on the effective cation exchange capacity of the soil and is given in [mol] for transformation and [mol∗a] for occupation impacts.

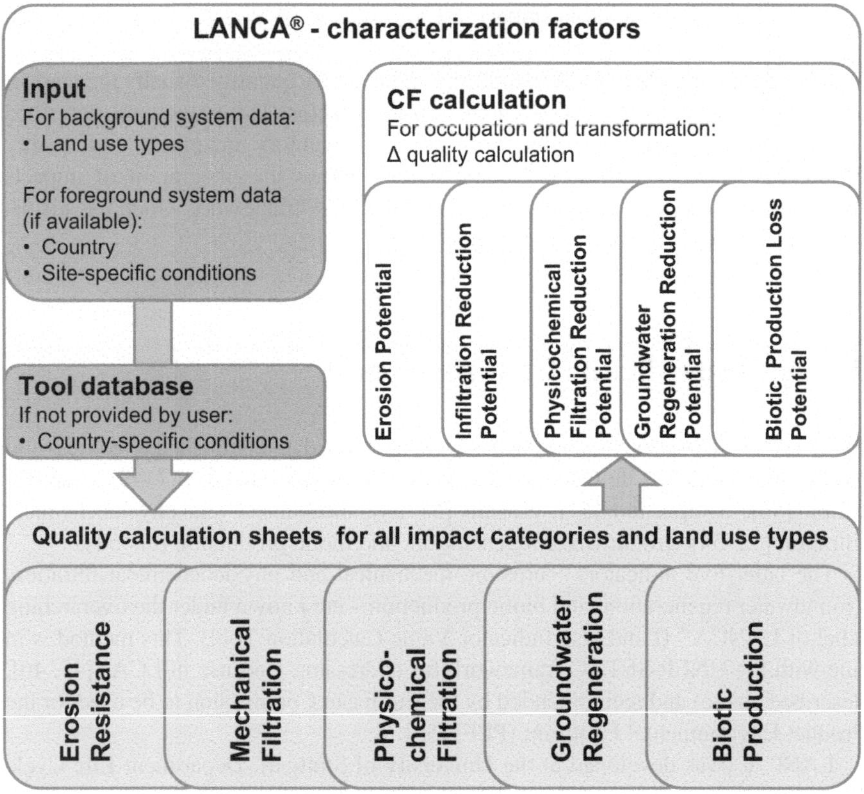


**Fig. 5** LANCA® calculation framework [53, 54]

*Groundwater regeneration* means the capacity of the soil to generate groundwater. It depends on the precipitation, evapotranspiration, and surface discharge. The parameter is measured in [$m^3$ water/a] for transformation and [$m^3$ water] for occupation impacts.

*Biotic production* describes the ability of an ecosystem to produce biomass at a given time. As an approach for biotic production, net primary production (NPP) is used in LANCA®. NPP is the net buildup of biomass in an ecosystem. This is achieved by primary producers (i.e., plants) through photosynthetic fixation of atmospheric carbon and subsequent transformation into organic carbon. The part that is lost through autotrophic breathing is subtracted. The parameter is measured in [kg carbon/a] for transformation and [kg carbon] for occupation impacts.

Figure 5 shows the calculation principle behind LANCA®. First of all, the quality indicators are calculated using site-specific information as well as input data from a database. The quality indicators are then used to obtain the characterization factors by calculating the delta between the quality of the reference situation and the quality of the land under the specific land use.

Site-specific input data can be used in order to calculate data for the foreground system. If these data are not available, country specific representative average data are used. For example, where a LANCA® calculation demands the annual precipitation at the specific site, the country average precipitation can be used instead if the site-specific value is unknown. For the background system, country specific characterization factors can be calculated using the respective average country specific input data from the database.

Both site-specific results and generic, country average results are calculated with LANCA® and can be integrated as characterization factors into any LCA study, adding five new impact categories. The Product Environmental Footprint (PEF) methodology developed by the European Commission distinguishes 75 land use flows such as arable, forest, extensive, forest intensive, etc. For each of these flows, LANCA®-based characterization factors are provided in the flow list of the International reference Life Cycle Data system (ILCD), which provides guidance and data corresponding to the PEF methodology [59]. The characterization factors can be found in Bos et al. [54] and are updated regularly (most recently by [60]). For more information about the PEF methodology, see European Commission [61], Lehmann et al. [62], and Manfredi et al. [63].

## 8 Biodiversity in Life Cycle Assessment

Biodiversity is defined by the United Nations Convention on Biological Diversity as the diversity within species, diversity between species, and the diversity of ecosystems [1]. This broad definition has later been elaborated, for example, by the Millennium Ecosystem Assessment [64] and The Economics of Ecosystems and Biodiversity [65]. The details are beyond the scope of this article. For the interested reader, the Living Planet Report [28, 43] offers a quite digestible synopsis in addition to the Millennium Ecosystem Assessment [11] and TEEB [65].

In LCA, biodiversity is understood as a quality of an area, i.e., as one definition of the quality axis in the UNEP-SETAC framework (see above). This does not necessarily align with the view of conservationists – because quality, time, and space are linearly interchangeable in the framework, among other reasons – but it does allow biodiversity impacts to be addressed consistently with other impacts from land use. There are two main schools of thought in the LCA community about biodiversity, as well as some singular methods.

The first school understands species diversity as the key indicator for biodiversity. Methods from this school of thought tend to be strictly empiric and data-driven. Species diversity is given as number of species per area (species density), backed up by field data about occurrences of species in different land use classes (e.g., forest, pasture, agriculture, urban). If, for example, the reference situation in a given ecoregion is primary forest with 100 species per $km^2$ and the land use assessed is intensively managed arable land with 10 species per $km^2$, then the quality difference $\Delta Q$ is $-90$ sp/$km^2$.

Example methods of the first school are, among others, Koellner [66], Koellner et al. [67], Itsubo and Inaba [68], de Baan et al. [69], Chaudhary et al. [70], and Yamaguchi et al. [71]. Many of these methods are based on the species-area relationship that was originally formulated by Arrhenius [72]. In essence, it states that more species are found in larger areas than in smaller areas. The strength of the link between species richness and area size depends on the land cover (e.g., grassland, forest, urban), as well as the world region specified as, i.e., the biome or ecoregion [73] in which a land using process is situated. Itsubo and Inaba [68] and Yamaguchi et al. [71] take only threatened species into account. They go beyond occurrence data in that they calculate the change in extinction risk for a list of threatened species due to land use change. A "threatened" species in this context is signified by being listed in the IUCN Red List of Threatened Species provided by the International Union for the Conservation of Nature [74]. While the list does distinguish various threat levels, this distinction is lost in most LCIA methods referring to it.

The second school describes biodiversity indirectly via environmental conditions. Methods from this school of thought aim for more holistic coverage of the various aspects of biodiversity (rather than only species diversity). The empirical basis of these methods tends to be weaker than for species diversity alone, though, and the results are abstract indicators rather than concrete species numbers. If, for example, the reference situation is primary forest in pristine condition, it could be assigned a biodiversity value of 100%. Intensively managed arable land would – taking into account the homogenized landscape structure, the nutrient overload, and the pesticide input – be assigned a lower value, say 10%. The quality difference $\Delta Q$ would then be $-90$%.

Example methods of the second school are, among others, Michelsen [75], Coelho and Michelsen [76], Lindner [77], and Fehrenbach [78]. Michelsen describes the biodiversity of a patch of land (on which a land using process is situated) as a function of parameters which relate to ecosystem quality, one of which is deadwood availability in forests. Lindner refines Michelsen's methodology with a more defined mathematical framework. Coelho and Michelsen [76] and Fehrenbach et al. [78] use hemeroby as an indicator for human interference. It is not a direct representation of biodiversity or ecosystem quality, but it can serve as a respective proxy in many world regions.

Apart from these major schools of thought, other methods have been proposed that cannot be associated with either school. For example, Maia de Souza et al. [79] propose functional diversity as a more meaningful indicator than species diversity, focusing on relations between species in ecosystems. Curran et al. [80] offer an overview of existing methods with and without explicit LCA relation. So far, no method has been adapted as a formal or de facto standard in LCA. The Chaudhary [70] method has been recommended by a recent UNEP-SETAC Life Cycle Initiative working group [81]. However, the group also stressed that this is not the ultimate method to address biodiversity in LCA.

## 9 Land Competition in Life Cycle Assessment

There is no denial in the LCA community that land use change is progressing (more area is being converted from its pristine state to a managed state) and that it has to do with the growth of both the global economy and population. Also uncontested is the understanding of progressing land conversion as an environmental problem. The plain use of area can be addressed in LCA. Respective software and database tools include information about the used area and the time over which the area is used as part of the inventory information of many land using processes. This can be seen as a first degree proxy to address land scarcity and land competition.

Beyond plain area use, no methodology has been universally accepted to address market-driven land use change (or indirect land use change, ILUC). The LCA community does not have a unified stance regarding ILUC. Some argue that ILUC cannot be meaningfully calculated [82]. Others claim to be able to do exactly that; for example, Schmidt et al. [83] calculate ILUC based on the assumption of perfect elasticity in the markets for commodities dependent on land use.

Indirect, market-driven effects are generally hard to address in LCA. Most importantly, the attribution of a specific effect to a specific product is very difficult. It is a matter of model accuracy, but even more so a philosophical matter. LCA typically models physical cause-effect relationships. Market-driven effects are based on the free decisions of participants in the global economy (to buy or sell commodities at a certain price). While it is true that many participants of any free market economy make their decisions under severe constraints, treating them like cog-wheels in a cause-effect chain would do them no justice as morally capable beings either.

## 10 Conclusions

The broad use of biotechnology has been a reality of organized human societies for millennia, for example, in the form of yogurt and beer production and more recently in the form of biogas generation. Beyond these traditional applications, the broad introduction of biotechnology into the global economy has the potential for more or less ecological damage, depending on how supply chains and product systems are set up. Biotechnology, if applied carefully and consciously, offers a way to reduce GHG emissions (see, e.g., [84]) and other ecological effects. It also has the potential to do more harm than good if applied thoughtlessly [85].

LCA can be one tool in the box to set up the supply chains of the future. This chapter describes the connections from bio-based products to land use and further to the ecological effects of land use. Methods that allow the effects of land use on soil quality and biodiversity to be quantified within the LCA framework are presented. They deviate to some degree from the traditional LCA modeling principles that focus on mass and energy exchanges between product systems and their environments.

However, LCA is but one tool in the box. LCA results are efficiency indicators, i.e., they convey information about the environmental impact of a product per unit of that product. The vast majority of LCA studies do not consider the effectiveness perspective, which would be about the absolute amounts emissions and resources, as well as related environmental impacts. For example, the Safe Operating Space approach [86, 87] addresses effectiveness. Both perspectives are valuable and necessary for a complete picture, and LCA typically provides about half of that picture.

The links between bio-based products and land use are known. Many of the effects propagated through these links can be quantified and attributed to products. It is imperative to take these links and effects into account when evaluating the sustainability of biotechnological processes and products – even more so when designing future products and entire future value chains.

## References

1. UN (1992) Convention on biological diversity. United Nations, New York. https://www.cbd.int/convention/text/. Accessed 26 Sep 2017
2. EC (2017) Bioeconomy. European Commission, Brussels. http://ec.europa.eu/programmes/horizon2020/en/h2020-section/bioeconomy. Accessed 26 Sep 2017
3. BMBF (2011) National Research Strategy BioEconomy 2030 – Our Route towards a biobased economy. Federal Ministry of Education and Research (BMBF), Berlin
4. Venkatesh A, Posen ID, MacLean HL, Chu PL, Griffin WM, Saville BA (2018) Environmental aspects of biotechnology. In: Fröhling M, Hiete M (eds) Sustainability and life cycle assessment in industrial biotechnology. Advances in biochemical engineering/biotechnology. Springer, Berlin
5. Nova Institute (2014) Industrial material use of biomass in Europe 2013. Poster. Nova Institute, Hürth
6. Scarlat N, Dallemand J-F, Monforti-Ferrario F, Nita V (2015) The role of biomass and bioenergy in a future bioeconomy: policies and facts. Environ Dev 15:3–34
7. Fröhling M, Hiete M (2019) Sustainability and life cycle assessments in industrial biotechnology: a review of current approaches and future needs. In: Fröhling M, Hiete M (eds) Sustainability and life cycle assessment in industrial biotechnology. Advances in biochemical engineering/biotechnology. Springer, Berlin
8. Erb KH, Gaube V, Krausmann F, Plutzar C, Bondeau A, Haberl H (2007) A comprehensive 5 min resolution land-use data set for the year 2000 consistent with national consensus data. J Land Use Sci 2(3):191–224
9. UN (2017) World population prospects: the 2017 revision, key findings and advance tables. Working paper no. ESA/P/WP/248. United Nations, Department of Economic and Social Affairs, Population Division (DESA), New York
10. UNEP (2014) Assessing global land use: balancing consumption with sustainable supply. A report of the Working Group on Land and Soils of the International Resource Panel. Bringezu S, Schütz H, Pengue W, O'Brien M, Garcia F, Sims R, Howarth R, Kauppi L, Swilling M, Herrick J. ISBN 978-92-807-3330-3
11. Millennium Ecosystem Assessment (2005) Ecosystems and human well-being: current state and trends. Island Press, Washington
12. MA (2005) Ecosystems and human well-being: synthesis. A report of the millennium ecosystem assessment. Island Press, Washington

13. Hooke RL, Martín-Duque JF, Pedraza J (2012) Land transformation by humans: a review. GSA Today 22(12):4–10
14. Pei H, Jiang L (2018) Mixing seawater with a little wastewater to produce bioenergy from limnetic algae. Trends Biotechnol 36(5):480–483
15. Cárdenas JP, Quatrini R, Holmes DS (2016) Genomic and metagenomic challenges and opportunities for bioleaching: a mini-review. Res Microbiol 167(7):529–538
16. Glombitza F, Kermer R, Reiche S (2019) Application of geobiotechnology in mining, mineral processing and metal recycling. In: Fröhling M, Hiete M (eds) Sustainability and life cycle assessment in industrial biotechnology. Advances in biochemical engineering/biotechnology. Springer, Berlin
17. Zah R, Böni H, Gauch M, Hischier R, Lehmann M, Wäger P (2007) Ökobilanz von Energieprodukten: Ökologische Bewertung von Biotreibstoffen. Bundesamt für Energie, Bundesamt für Umwelt, Bundesamt für Landwirtschaft, Bern. http://www.news.admin.ch/NSBSubscriber/message/attachments/8514.pdf. Accessed 26 Sep 2017
18. Böcker N, Grahl M, Tota A, Häussinger P, Leitgeb P, Schmücker B (2013) Nitrogen. In: Ullmann's Encyclopedia of Industrial Chemistry. Wiley, published online. http://onlinelibrary.wiley.com/doi/10.1002/14356007.a17_457.pub2/full. Accessed 26 Sep 2017
19. Odum E (1991) Prinzipien der Ökologie: Lebensräume, Stoffkreisläufe, Wachstumsgrenzen. Spektrum Akademischer Verlag, Heidelberg
20. Schubert S (2006) Pflanzenernährung. Grundwissen Bachelor. Ulmer, Stuttgart
21. BGR (2005) Bodenkundliche Kartieranleitung. Bundesanstalt für Geowissenschaften und Rohstoffe, Hannover
22. FAO (2017) Voluntary guidelines for sustainable soil management. Food and Agriculture Organization of the United Nations (FAO), Rome. http://www.fao.org/3/a-bl813e.pdf. Accessed 6 Jul 2018
23. FAO (2015) Status of the world's soil resources. Food and Agriculture Organization of the United Nations (FAO), Rome. http://www.fao.org/3/a-i5228e.pdf. Accessed 26 Sep 2017
24. Dasgupte P, Mäler K-G (1997) The environment and emerging development issues, vol 2. Oxford University Press, New York
25. Masdramootoo C (2012) Sustainable groundwater use in agriculture. Irrig Drain 61:26–33
26. Margni M, Rossier D, Crettaz P, Jolliet O (2002) Life cycle impact assessment of pesticides on human health and ecosystems. Agric Ecosyst Environ 93:379–392
27. Geiger F, Bengtsson J, Berendse F, Weisser W, Emmerson M, Morales M, Ceryngier P, Liira J, Tscharntke T, Winqvist C, Eggers S, Bommarco R, Pärt T, Bretagnolle V, Plantegenest M, Clement L, Dennis C, Palmer C (2010) Persistent negative effects of pesticides on biodiversity and biological control potential on European farmland. Basic Appl Ecol 11(2):97–105
28. WWF (2018) Living planet report 2018: aiming higher. WWF International, Gland
29. SCA (2017) Sustainability report 2016. Svenska Cellulosa Aktiebolaget, Stockholm
30. UPM (2017) Aiming higher with biofore. Annual report 2016. UPM-Kymmene Oy, Helsinki
31. Millan CH, Develey PF, Verdade LM (2015) Stand-level management practices increase occupancy by birds in exotic Eucalyptus plantations. For Ecol Manag 336:174–182
32. Quammen D (2012) Forest giant. A tree-climbing scientist and his team have learned surprising new facts about giant sequoias by measuring them inch by inch. National Geographic Magazine, US Edition, December 2012
33. Graham SA (1939) Principles of forest entomology, 2nd edn. McGraw-Hill, New York
34. Kozlowski TT (2002) Physiological-ecological impacts of flooding on riparian forest ecosystems. Wetlands 22(3):550–561
35. Naiman RJ, Decamps H (1997) The ecology of interfaces: riparian zones. Annu Rev Ecol Syst 28(1):621–658
36. Coffin AW (2007) From roadkill to road ecology: a review of the ecological effects of roads. J Transp Geogr 15(5):396–406

37. Forman RTT, Sperling D, Bissonette JA, Clevenger AP, Cutshall CD, Dale VH, Fahrig L, France R, Goldman CR, Heanue K, Jones JA, Swanson FJ, Turrentine T, Winter TC (2003) Road ecology: science and solutions. Island Press, Washington
38. Bortesi L, Fischer R (2015) The CRISPR/Cas9 system for plant genome editing and beyond. Biotechnol Adv 33:41–52
39. Hawkesford MJ, Araus JL, Park R, Calderini D, Miralles D, Shen T, Zhang J, Parry MA (2013) Prospects of doubling global wheat yields. Food Energy Secur 2(1):34–48
40. Yuan Y, Scheben A, Batley J, Edwards D (2019) Using genomics to adapt crops to climate change. In: Sarkar A, Sensarma SR, van Loon GW (eds) Sustainable solutions for food security. Springer, Cham
41. Grau R, Kuemmerle T, Macchi L (2013) Beyond 'land sparing versus land sharing': environmental heterogeneity, globalization and the balance between agricultural production and nature conservation. Curr Opin Environ Sustain 5(5):477–483
42. Phalan B, Onial M, Balmford A, Green RE (2011) Reconciling food production and biodiversity conservation: land sharing and land sparing compared. Science 333(6047):1289–1291
43. WWF (2016) Living planet report 2018: risk and resilience in a new era. WWF International, Gland
44. BMUB (2007) Nationale Strategie zur biologischen Vielfalt. Federal Ministry for the Environment Nature Conservation, Building and Nuclear Safety (BMUB), Berlin
45. Koellner T, De Baan L, Beck T, Brandão M, Civit B, Margni M, Canals LMI, Saad R, De Souza DM, Müller-Wenk R (2013) UNEP-SETAC guideline on global land use impact assessment on biodiversity and ecosystem services in LCA. Int J Life Cycle Assess 18(6):1188–1202
46. Milà I, Canals L, Bauer C, Depestele J, Dubreuil A, Freiermuth Knuchel R, Gaillard G, Michelsen O, Müller-Wenk R, Rydgren B (2007) Key elements in a framework for land use impact assessment within LCA. Int J Life Cycle Assess 12(1):5–15
47. Brandão M, Milà I, Canals L (2013) Global characterisation factors to assess land use impacts on biotic production. Int J Life Cycle Assess 18(6):1243–1252
48. Brandão M, Milà I, Canals L, Clift R (2011) Soil organic carbon changes in the cultivation of energy crops: implications for GHG balances and soil quality for use in LCA. Biomass Bioenergy 35(6):2323–2336
49. Morais TG, Domingos T, Teixeira RFM (2016) A spatially explicit life cycle assessment midpoint indicator for soil quality in the European Union using soil organic carbon. Int J Life Cycle Assess 21(8):1076–1091
50. Núñez M, Civit B, Muñoz P, Arena AP, Rieradevall J, Antón A (2010) Assessing potential desertification environmental impact in life cycle assessment: part 1: methodological aspects. Int J Life Cycle Assess 15(1):67–78
51. Garrigues E, Corson MS, Angers DA, van der Werf, Hayo MG, Walter C (2013) Development of a soil compaction indicator in life cycle assessment. Int J Life Cycle Assess 18(7):1316–1324
52. Baitz M (2002) Die Bedeutung der funktionsbasierten Charakterisierung von Flächen-Inanspruchnahmen in industriellen Prozesskettenanalysen: Ein Beitrag zur Ganzheitlichen Bilanzierung. Doctoral thesis, University of Stuttgart, Germany
53. Beck T, Bos U, Wittstock B, Baitz M, Fischer M, Sedlbauer K (2010) LANCA®: land use indicator value calculation in life cycle assessment. Fraunhofer-Verlag, Stuttgart
54. Bos U, Horn R, Beck T, Lindner JP, Fischer M (2016) LANCA® – characterization factors for life cycle impact assessment. Version 2.0. Fraunhofer Verlag, Stuttgart
55. European Commission (2017) PEFCR Guidance document [online]. Guidance for the development of Product Environmental Footprint Category Rules (PEFCRs). Version 6.3
56. Blume H-P, Brümmer GW, Horn R, Kandeler E, Kögel-Knabner I, Kretzschmar R, Stahr K, Thiele-Bruhn S, Welp G, Wilke B-M, Scheffer F, Schachtschabel P (2010) Scheffer/Schachtschabel: Lehrbuch der Bodenkunde, 16 Auflage. Spektrum Akademischer Verlag, Heidelberg
57. Wischmeier WH, Smith DD (1978) Predicting rainfall erosion losses. A guide to conservation planning. U.S. Department of Agriculture, Agriculture Handbook No. 537. Gov. Print. Off.

Agriculture Handbook, Prep. by Science and Education Administration, U.S. Department of Agriculture, Washington, DC

58. Marks R, Müller MJ, Leser H, Klink HJ (eds) (1989) Anleitung zur Bewertung des Leistungsvermögens des Landschaftshauhaltes. Forschungen zur Deutschen Landeskunde, vol 229. Zentralausschuss für deutsche Landeskunde, Trier
59. EC (2015) EPLCA – European reference Life-Cycle Database. ELCD 3.2. European Commission – Joint Research Centre – Institute for Environment and Sustainability – Unit H08 Sustainability Assessment. European Commission, Brussels. http://eplca.jrc.ec.europa.eu/ELCD3/index.xhtml?stock=default. Accessed 30 Nov 2015
60. Horn R, Maier S (2018) LANCA® characterization factors for life cycle impact assessment: version 2.5. In: Bos U, Horn R, Beck T, Lindner JP, Fischer M (eds) LANCA® – characterization factors for life cycle impact assessment. Version 2.0. Fraunhofer Verlag, Stuttgart
61. European Commission (2013) 2013/179/EU: Commission Recommendation of 9 April 2013 on the use of common methods to measure and communicate the life cycle environmental performance of products and organisations text with EEA relevance. https://eur-lex.europa.eu/eli/reco/2013/179/oj. Accessed 30 Jul 2019
62. Lehmann A, Bach V, Finkbeiner M (2015) Product environmental footprint in policy and market decisions: applicability and impact assessment. Integr Environ Assess Manag 11 (3):417–424
63. Manfredi S, Allacker K, Pelletier N, Schau E, Chomkhamsri K, Pant R, Pennington D (2015) Comparing the European Commission product environmental footprint method with other environmental accounting methods. Int J Life Cycle Assess 20(3):389–404
64. MA (2005) Ecosystems and human wellbeing: biodiversity synthesis. A report of the millennium ecosystem assessment. World Resources Institute, Washington
65. TEEB (2010) The economics of ecosystems and biodiversity: mainstreaming the economics of nature: a synthesis of the approach, conclusions and recommendations of TEEB. http://www.teebweb.org/our-publications/teeb-study-reports/synthesis-report/. Accessed 1 Oct 2017
66. Koellner T (2000) Species-pool effect potentials (SPEP) as a yardstick to evaluate land-use impacts on biodiversity. J Clean Prod 8(4):293–311
67. Koellner T, Scholz RW (2008) Assessment of land use impacts on the natural environment. Part 2: generic characterization factors for local species diversity in Central Europe. Int J Life Cycle Assess 13(1):32
68. Itsubo N, Inaba A (2012) LIME 2 life-cycle impact assessment method based on endpoint modeling chapter 1 – outline of LIME 2. In: JLCA News Letter English Edition No. 14
69. De Baan L, Alkemade R, Koellner T (2013) Land use impacts on biodiversity in LCA: a global approach. Int J Life Cycle Assess 18(6):1216–1230
70. Chaudhary A, Verones F, de Baan L, Hellweg S (2015) Quantifying land use impacts on biodiversity: combining species-area models and vulnerability indicators. Environ Sci Technol 49(16):9987–9995
71. Yamaguchi K, Ii R, Itsubo N (2016) Ecosystem damage assessment of land transformation using species loss. Int J Life Cycle Assess 23(12):1–12
72. Arrhenius O (1921) Species and area. J Ecol 9(1):95–99
73. Olson DM, Dinerstein E, Wikramanayake ED, Burgess ND, Powell GV, Underwood EC, D'amico JA, Itoua I, Strand HE, Morrison JC, Loucks CJ, Allnutt TF, Ricketts TH, Kura Y, Lamoreux JF, Wettengel WW, Hedao P, Kassem KR (2001) Terrestrial Ecoregions of the World: A New Map of Life on Earth: a new global map of terrestrial ecoregions provides an innovative tool for conserving biodiversity. Bioscience 51(11):933–938
74. IUCN Red list of threatened species. IUCN Global Species Programme Red List Unit, Cambridge. https://www.iucnredlist.org/
75. Michelsen O (2008) Assessment of land use impact on biodiversity. Int J Life Cycle Assess 13 (1):22
76. Coelho C, Michelsen O (2014) Land use impacts on biodiversity from kiwifruit production in New Zealand assessed with global and national datasets. Int J Life Cycle Assess 19(2):285–296

77. Lindner JP (2016) Quantitative Darstellung der Wirkungen landnutzender Prozesse auf die Biodiversität in Ökobilanzen. Fraunhofer Verlag, Stuttgart
78. Fehrenbach H, Grahl B, Giegrich J, Busch M (2015) Hemeroby as an impact category indicator for the integration of land use into life cycle (impact) assessment. Int J Life Cycle Assess 20 (11):1511–1527
79. Maia de Souza D, Flynn DF, DeClerck F, Rosenbaum RK, de Melo Lisboa H, Koellner T (2013) Land use impacts on biodiversity in LCA: proposal of characterization factors based on functional diversity. Int J Life Cycle Assess 18(6):1231–1242
80. Curran M, Maia de Souza D, Antón A, Teixeira RF, Michelsen O, Vidal-Legaz B, Sala S, Milà I, Canals L (2016) How well does LCA model land use impacts on biodiversity? A comparison with approaches from ecology and conservation. Environ Sci Technol 50(6):2782–2795
81. Teixeira R, Souza D, Curran M, Antón A, Michelsen O, Milà I, Canals L (2016) Towards consensus on land use impacts on biodiversity in LCA: UNEP/SETAC Life Cycle Initiative preliminary recommendations based on expert contributions. J Clean Prod 112:4283–4287
82. Finkbeiner M (2014) Indirect land use change – help beyond the hype? Biomass Bioenergy 62:218–221
83. Schmidt JH, Weidema BP, Brandão M (2015) A framework for modelling indirect land use changes in life cycle assessment. J Clean Prod 99:230–238
84. Pleissner D, Kümmerer K (2019) Green chemistry and its contribution to industrial biotechnology. In: Fröhling M, Hiete M (eds) Sustainability and life cycle assessment in industrial biotechnology. Advances in biochemical engineering/biotechnology. Springer, Berlin
85. UN (2009) Towards sustainable production and use of resources: assessing biofuels. United Nations Environment Programme, Paris
86. Rockström J, Steffen W, Noone K, Persson Å, Chapin FS, Lambin E, Lenton TM, Scheffer M, Folke C, Schellnhuber HJ, Nykvist B, de Wit CA, Hughes T, van der Leeuw S, Rodhe H, Sörlin S, Snyder PK, Costanza R, Svedin U, Falkenmark M, Karlberg L, Corell RW, Fabry VJ, Hansen J, Walker B, Liverman D, Richardson K, Crutzen P, Foley J (2009) Planetary boundaries exploring the safe operating space for humanity. Ecol Soc 14(2):32
87. Steffen W, Richardson K, Rockström J, Cornell SE, Fetzer I, Bennet EM, Biggs R, Carpenter SR, de Vries W, de Wit CA, Folke C, Gerten D, Heinke J, Mace GM, Persson LM, Ramanathan V, Reyers B, Sörlin S (2015) Planetary boundaries: guiding human development on a changing planet. Science 347(6223):736

Adv Biochem Eng Biotechnol (2020) 173: 255–280
DOI: 10.1007/10_2018_74

Published online: 23 September 2018

# Risk Assessment of Processes and Products in Industrial Biotechnology

**Chao Chen and Genserik Reniers**

## Contents



C. Chen
Faculty of Technology, Policy and Management, Safety and Security Science Group (S3G), TU Delft, Delft, The Netherlands

G. Reniers (✉)
Faculty of Technology, Policy and Management, Safety and Security Science Group (S3G), TU Delft, Delft, The Netherlands

National Institute for Public Health and the Environment (RIVM), Bilthoven, The Netherlands
e-mail: genserik.reniers@uantwerpen.be

**Abstract** Risk assessment has been used extensively as the main approach to prevent accidents in the chemical and process industry. Industrial biotechnology has many of the same hazards as chemical technology, but also encounters biological hazards related to biological agents. Employees in the biotechnology industry are susceptible to health risks because of different types of exposure to harmful agents. The external environment may also be affected by these agents in cases of accidental release. This chapter first presents several traditional risk assessment methods that may be used in industrial biotechnology after comparing differences between industrial biotechnology and chemical technology. Hazard identification in industrial biotechnology is then discussed, for biological as well as traditional hazards. Furthermore, risk assessment of occupational health and safety related to biological hazards is examined using exposure analysis and risk characterization. A two-stage risk assessment method is recommended to assess environmental and ecological risks in industrial biotechnology. Risk analysis of traditional accidents (fire, explosions, and toxic releases) in industrial biotechnology is also described.

**Graphical Abstract**

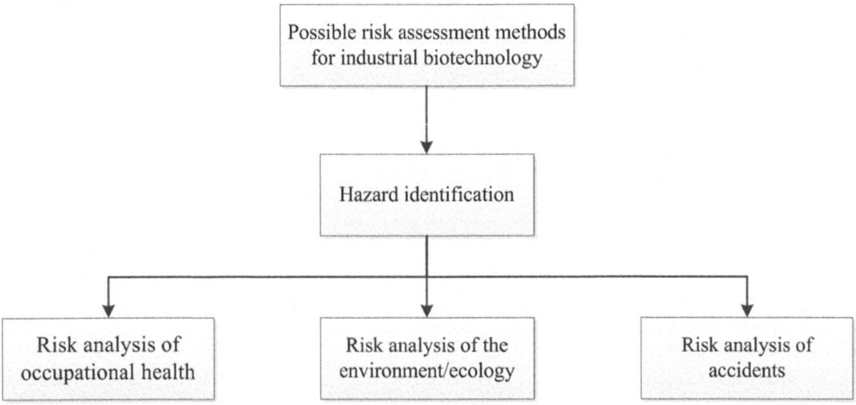


**Keywords** Environmental and ecological risks, Hazard identification, Industrial biotechnology, Occupational health and safety, Risk assessment

# 1 Introduction

Industrial biotechnology is one of the most promising technologies with many advantages, such as pollution prevention and resource conservation. Industrial biotechnology processes with both biological and chemical characteristics, make use of microorganisms, cells in culture, or enzymes to manufacture products or

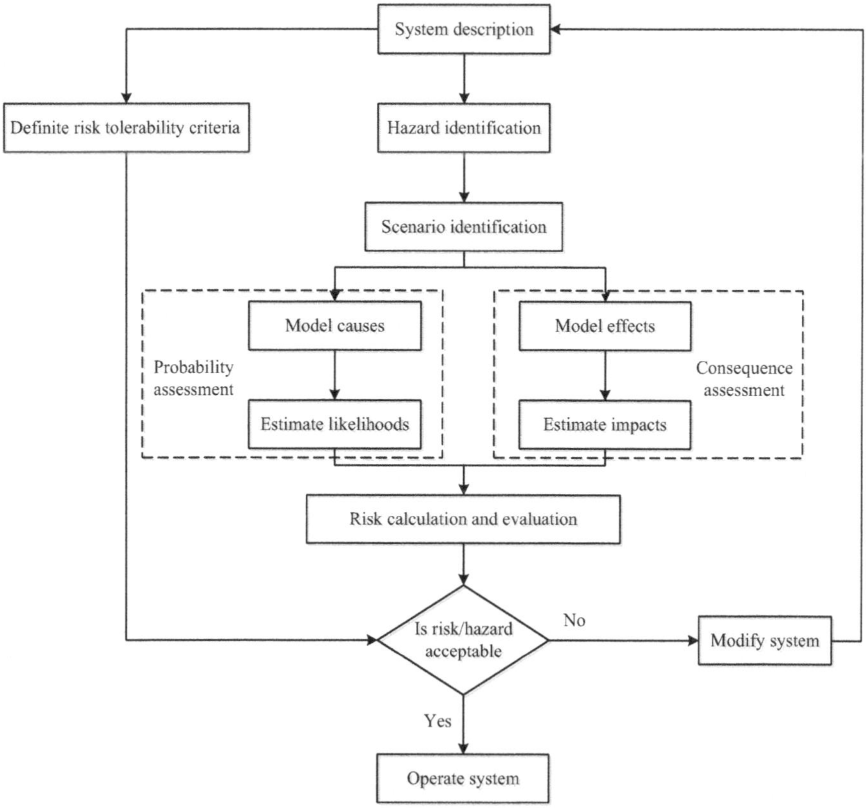


**Fig. 1** Risk assessment flowchart [2, 3]

complete chemical transformations. With the rapid development of industrial biotechnology over the past several decades, industrial biotechnology has gained widespread use in almost every aspect of our daily lives. Applications include the production of food, feed, drugs, plastics, bioenergy, mining, metallurgy, recycling, and waste treatment. Nevertheless, the potential risks from processes and products in industrial biotechnology cannot be overlooked because there are many hazards related to human health, safety, and the environment. An increase in serious accidents in bioenergy production has been witnessed with the rapid increase of bioenergy facilities, involving fires, explosions, and release of toxic gases [1]. Risk assessment of processes and products in industrial biotechnology should therefore gain more attention with the aim of reducing accidents. Figure 1 shows the process of risk assessment in the process industry.

The steps of risk assessment include hazard identification, scenario identification, probability assessment, consequence assessment, risk calculation, and risk evaluation, among others. After a brief description of the system, all hazards and accident scenarios in the system need to be identified. Further, the probability of accidents and

**Table 1** Comparison of the main characteristics between industrial biotechnology and chemical technology (incomplete enumeration)

| Types | Industrial biotechnology | Chemical technology |
|---|---|---|
| Raw materials | Microorganisms, plants, animals cells, or enzymes | Coal, oil, natural gas, and other chemicals |
| Products | Food, biomaterials, biofuels, biogas, and other chemicals | Fossil fuels, plastics, and other chemicals |
| Processes | Fluid dynamics, heat transfer, mass transfer, thermodynamics | Fluid dynamics, heat transfer, mass transfer, thermodynamics, high temperature and pressure |
| Waste materials | General liquid, solid, and gaseous waste; chemical waste; biological waste | General liquid, solid, and gaseous waste; chemical waste |
| Energy demand | Saves energy (enzymes are used to speed up biochemical reactions) | High energy consumption (chemical reactions usually take place under high-temperature, high-pressure conditions) |

the severity of the consequences are estimated. Finally, a qualitative ranking of risks (qualitative risk assessment) or quantified benefits or costs of risks (quantitative risk assessment) can be obtained via risk evaluation. The system can run normally if the risk is acceptable according to certain predetermined risk tolerability criteria. Otherwise, safety measures should be taken to reduce the risk. Risk assessment has been used extensively as the main approach to preventing accidents in the chemical process industry. It is necessary to make a comparison between industrial biotechnology and traditional chemical technology to reveal the risk features and explore possible risk assessment methods suitable for the former. The production processes in industrial biotechnology have many of the same hazards as in chemical technology, but there are also differences specifically related to biotechnology, as hazard identification mainly depends on raw materials, products, processes, and waste materials. An overview of the differences relating to risks between industrial biotechnology and chemical technology is given in Table 1.

According to Table 1, industrial biotechnology has many of the same hazards as chemical technology, along with other hazards specific to biotechnology. In addition to process safety risks related to toxic, flammable, and explosive substances, health risks and environmental risks may also be present in industrial biotechnology. With different types of exposure, contact is possible between employees and highly dangerous agents or products, and the presence of workers in processes of biotechnology is susceptible to health risks. The hazards of agents and products may be more significant in industrial biotechnology than traditional chemical risks, and the health risk to workers or the public is possibly higher in the case of accidental release. Furthermore, biological production or waste materials involve living organisms such as bacteria, yeasts, or plants may be harmful if the external environment is exposed to these organisms. Therefore, this chapter deals with risks in industrial biotechnology related to process safety, occupational health and safety, and environmental or ecological safety. First, we discuss some methods that have been widely used in the chemical industry and that may be applied to risk assessment in industrial biotechnology. Second, we expound on hazard identification in industrial biotechnology involving biological hazards and traditional hazards. Third, risk assessment of occupational health and safety related to biological hazards is discussed through exposure analysis and risk characterization. Finally, risk analyses of traditional accidents in industrial biotechnology are described.

## 2 Risk Assessment Methods

As there has been less experience with risk assessment for industrial biotechnology than for traditional chemical industrial processes, risk-related experiences can be taken from the chemical industry because of the many similarities. The most widespread risk identification and risk assessment methods in the chemical industry were investigated by means of a survey [4]. Many of the existing methods can also be used in industrial biotechnology. Methods such as the Checklist, the Hazard and Operability Study (HAZOP), the Fault Tree Analysis (FTA), and the Risk Matrix are therefore further briefly explained in this section.

### *2.1 Checklists*

The Checklist method is one of the most simplistic tools to identify hazards and are frequently used to indicate compliance with standard procedures. A Checklist consists of an enumeration of questions about safety concerns, for example, raw materials, agents, products, processes, and operation. Each item on the list can be physically verified. The effectiveness of a Checklist depends upon the expertise of its preparer and upon the qualifications of the personnel completing the list. A Checklist is easy to use and can provide results relatively quickly, leading to "yes or no" decisions about compliance with standard procedures. Checklists are even more effective if the questions cannot be answered by a simple "yes" or "no" and if they require some thought in formulating an answer [5]. The extent of the list can vary, but in general it is one of the quickest and least expensive risk analysis methods. It is highly cost-effective for common hazards.

### *2.2 Hazard and Operability Study (HAZOP)*

HAZOP is a formalized methodology to identify and to document hazards through imaginative thinking [6]. It involves a very systematic examination of design documents that describe the installation or the facility under investigation. The study is performed by a multidisciplinary team that analytically examines design intent deviations. The objective of the HAZOP study is to take a full description of the process and to question every part of it to discover what deviations from the intention of the design can occur and what the causes and consequences of these deviations may be. This method is often applied in the process industries and the main steps are [5, 7]:

- Definition of objectives and study scope
- Selection of multi-disciplinary team
- Preparation
- Conduct/facilitate
- Record/document study
- Preparation of HAZOP study report

### *2.3 Fault Tree Analysis (FTA)*

Fault Tree Analysis (FTA) is a deductive technique focusing on one particular accident event and providing a method for determining causes of that event. In a graphical way, the various combinations of equipment faults and failures that can result in the accident event are displayed. It is based on a specialized model that may be represented as a diagram of binary logic (yes-no). It is a structured topological method that can produce both quantitative and qualitative results. FTA is often used in situations where causes are made up of many parts [8].

### *2.4 Risk Matrix*

The "Risk Assessment Decision Matrix," often abbreviated as "The Risk Matrix," is a risk evaluation approach. The basis for the risk matrix is the standard conceptualization of risk as a combination of severity (consequences) and probability of a certain accident scenario [9]. This method provides a quick and simple priority sorting method, which is especially implemented in case of low probability–high severity hazards.

For further reading on the above methods, please refer to more detailed work [5, 7, 10].

## 3 Hazard Identification

In addition to various traditional physical and chemical hazards, biological hazards related to biological agents are present in workplaces devoted to industrial biotechnology. Hence, hazards in industrial biotechnology are divided into two categories:

biological hazards mainly involved in occupational health and environment/ecology, and "traditional" hazards (non-biological hazards) which may induce "traditional" safety accidents. Both biological hazards and traditional hazards are indispensable for risk assessment of processes and products in industrial biotechnology. These hazards depend on the characteristics of production plants using industrial technology, including:

- The type of products being manufactured
- The types of chemical-physical-bioprocesses used to manufacture the product and the types of raw materials and processing aids or solvents
- The specific type of biological organism or cellular products used, co-products or by-products produced, and the physical and chemical processes required for product recovery and purification
- The yields and waste streams generated per unit of product produced and the scale of operation and schedule for production [11]

## 3.1 Biological Hazards

Biological hazards are of primary consideration in industrial biotechnology hazard identification. This technology mainly depends on biological agents in raw materials, products, wastes, water streams, and processes. In the process of biological hazard identification, there are three types of factors that should be considered: agents, processes, and the environment.

According to the EEC directives 2000/54/EC and Safety Health and Welfare at Work (Biological Agents) Regulations 2013, biological agents are classified into four groups based on the severity of the harmful effects they may have on employees' health [12]. The classification was proposed on the basis of the following considerations [13]:

- The ability of the biological agents to cause disease in "healthy" people
- The mean severity of the disease
- The likelihood of the biological agents causing epidemics
- The existence of effective treatments and suitable prophylactic measures[1]

The four groups are elaborated in Table 2.

[1]Prophylactic measures include any measure taken to prevent disease before it occurs and procedures that help to prevent infection after exposure to a pathogen or to ease symptoms associated with an illness or health condition, such as vaccination and medical supervision.

**Table 2** Classification of biological hazards

| Group | Content |
|---|---|
| 1 | The biological agent that is unlikely to cause disease to employees |
| 2 | The biological agent that can cause human disease and might be a hazard to employees, although it is unlikely to spread to the community, and in respect of which there is usually effective prophylaxis or treatment available |
| 3 | The biological agent that can cause severe human disease and presents a serious hazard to employees and that may present a risk of spreading to the community, though there is usually effective prophylaxis or treatment available |
| 4 | The biological agent that causes severe human disease and is a serious hazard to employees and that may present a high risk of spreading to the community, and in respect of which there is usually no effective prophylaxis or treatment available |

**Table 3** Major pathological mechanisms for health effects associated with hazardous biological agents [14]

| Pathological type | Pathological mechanisms | Hazardous biological agents |
|---|---|---|
| Microbial infection | Infectious material | Hepatitis (A/B/C), Leptospira, Mycobacterium TB |
| | Opportunist pathogens | Legionella pneumophilia |
| | Zoonoses | *Bacillus anthracis*, *Chlamydia psittaci* |
| Allergic response | Microorganisms | Actinomycetes, Aspergillus |
| | Proteinaceous material | Pollen, dust, animal secretions |
| | Chemical compounds | Plicatic acid, gums, resins |
| Toxic/inflammatory response | Endotoxins (Gram-negative bacteria) | Stored grain, hay, cotton, swine and poultry confinement units |
| | Mycotoxins (fungi) and $\beta(1\rightarrow3)$-glucans | Stored fodder, grain, nuts |
| Carcinogenic | Wood dust | Hardwood (beech, oak), Softwood |
| | Mycotoxins (aflatoxin) | Stored nuts |

### 3.1.1 Biological Agents

Biological agents including cells, microorganisms, or cell cultures, either of natural origin or genetically modified may be associated with the risk of occupational illnesses such as infections or allergic or toxic reactions. For example, Table 3 shows the major pathological mechanisms for health effects associated with hazardous biological agents [14].

Pathogenic mechanisms vary with the hazardous biological agents. For a special agent, the factors listed here related to the agent should be considered [11]:

- Pathogenicity
- Infectious dose
- Virulence (primary or secondary communicability)
- Host factors (immunocompetence, pregnancy, underlying medical conditions, extreme age, or immunity)

- Sensitization reactions (allergens, toxins, or biologically active compounds)
- Incidence of laboratory acquired infections (LAI)
- Availability of vaccine and/or prophylactic treatment
- Environmental impact (agent stability – sensitivity to chemical and physical inactivation – survivability and dissemination in the environment)
- Respiratory: inhaling of contaminated particles
- Mucous membrane: splashing, spraying, or droplets in the eyes or mouth
- Parenteral: penetrations through the skin; for example, cuts, needle sticks, or abrasions
- Non-intact skin: contact with skin affected with dermatitis, chafing, hangnails, abrasions, acne, or other conditions that can alter the barrier properties of the skin
- Ingestion: swallowing contaminated material
- Adsorption: adhesion to a surface

A Checklist of questions for biological agents can be used to identify the hazards, as shown in Table 4.

**Table 4** Checklist for biological hazards related to biological agents [11]

| |
|---|
| What is the highest biosafety level[a] needed for containment of the agent(s) that will be used in the facility? |
| What is the mode of transmission? |
| What is the infectious dose? |
| How communicable is the agent? |
| Is the agent an opportunistic pathogen that could infect immunocompromised individuals? |
| Is the organism a select agent (an organism that is of particular concern to the U. S. federal government because of its potential use in biological weapons) or does it have characteristics that warrant increased security and oversight? |
| Has the disease that the organism causes been eradicated so that release could cause a serious public health threat by reintroducing it into the community? |
| Does the agent produce any toxic, biologically active, or allergenic compounds? |
| Is the agent susceptible to adventitious contaminants (that is, bacterial, fungal, mycoplasma, or viral) that may be harmful to humans? |
| Are vaccines, prophylaxis, or therapeutic measures available to prevent or mediate an infection? |
| Is the agent endemic in the area? |
| How well does the agent survive outside of the culture system? |
| Can the organism transfer genetic traits to other organisms in the environment? |
| How is the agent disseminated through vectors? |

[a]Biosafety levels (BSL) describe procedures, equipment and facilities, and levels of containment required to provide protection to employees, occupants of building, the environment, and the local community. The BSL ranges from the lowest biosafety level 1 (BSL-1) to the highest at level 4 (BSL-4) based on the agents or organisms that are being researched or worked on in any laboratory setting [15, 16]. CCPS defines four different levels for production setting: Good Large Scale Practices (GLSP), Biosafety Level 1 – Large Scale (BSL1-LS), Biosafety Level 2 – Large Scale (BL2-LS), and Biosafety Level 3 – Large Scale (BL3-LS) [11]

**Table 5** Checklist for hazards of processes and facilities [11]

| |
|---|
| Will the facility be dedicated to one agent or will a number of agents be used? |
| What volume(s) of active agents will be present in the facility? |
| Will the process be continuous or batch? |
| Will the equipment be stationary or movable? |
| What type of equipment will be used? |
| What types of manipulations need to be carried out |
| Does any of the equipment or manipulations generate aerosols? |
| Will the facility be required to comply with competent authority drug or device regulations, genetically modified organism (GMO) requirements, or other governmental regulations? |

### 3.1.2 Processes and Facilities

Possible biological hazards related to processes and facilities result from ventilation and laboratory design, laboratory procedures, containment equipment, personal protective equipment (PPE), training, facility sanitation, and medical surveillance. Details are as follows [11]:

- Ventilation and laboratory design: directional air, pressure gradients, airbreaks, separation of laboratories from offices, interlocking autoclave and airlock doors
- Laboratory procedures: use of inherently safer engineered sharps, containment of aerosols, and other means
- Containment equipment [17]: class II and III biological safety cabinets,[2] sealed centrifuges, cups, and rotors, gasket seals, and unbreakable tubes
- PPE: gloves, safety glasses, lab coats, face masks, respirators, or gowns
- Training: standard microbiological practices, aseptic practices, decontamination, spill cleanup, and handling of accidents
- Facility sanitation: decontamination, housekeeping, routine cleaning and disinfection, pest and rodent control programs
- Medical surveillance monitoring: as dictated by the risks present in the bioprocessing facility

A Checklist of questions about processes and facilities can also be employed, as shown in Table 5.

[2]A biological safety cabinet (BSC) is an enclosed, ventilated laboratory workspace for safely working. The U.S. Centers for Disease Control and Prevention (CDC) classifies BSCs into three classes based on the level of personnel and environmental protection provided and the level of product protection provided by the BSC. The Class I biological safety cabinets provide personnel and environmental protection, but no product protection. The Class II biological safety cabinets provide personnel, environmental and product protection. The Class III biological safety cabinets are designed for work with microbiological agents assigned to biosafety group 4, and provides maximum protection to the environment and the employee.

**Table 6** Checklist for hazards of the environment [11]

| |
|---|
| What are the climatic conditions in the area? |
| Has dispersion modeling data been compiled? |
| What is the geography of the site? |
| What are the native flora and fauna? |
| How close is the air supply intake or exhaust to other facilities |
| How near is the facility to private property? What is the usage of that property? |
| Is the site security adequate for the types of organisms handled? |
| How is waste treated prior to release to the local environment? |
| What type of waste is released (solid/liquid) and where it is released? |

### 3.1.3 The Environment

The last category of factors is related to the environment (which includes climate, geography, and proximity to people). A Checklist for hazards related to this category is provided in Table 6.

## *3.2 Traditional Hazards*

The presence of biological hazards in workplaces is the essential difference between industrial biotechnology and chemical technology, and this requires attention in risk assessment practices. As already mentioned, traditional hazards in industrial technology related to non-biological risks include the traditional chemical and mechanical hazards. These hazards at workplaces may induce serious accidents such as fires, explosions, leakages or releases of toxic or hazardous materials that can cause people illness, injury, disability, or death. Non-biological risk assessment is therefore not negligible in workplaces using industrial biotechnology, especially in the production of biogas. Occupational accidents in bioenergy production activities have occurred frequently in the past few decades because of poor safety culture and the lack of risk awareness. A more thorough data analysis indicates that major accidents have been increasing in recent years and the number is growing faster than bioenergy production [18]. For 13,171 European biogas stations, more than 800 accidents were found to have occurred in a 10-year period, and 3 of them were serious and life changing [19]. A total of 85 accidents were registered in the biodiesel industry from 2003 to 2013, and 14% of them were fatal [20].

### 3.2.1 Typical Scenarios Based on Historical Data

Typical scenarios in industrial biotechnology (mainly in bioenergy industry) are obtained by further analysis of historical data, which can be a guide to traditional hazard identification, as follows [1, 19, 21, 22]:

- Leakage caused by tank/reservoir rupture
- Leakage caused by pipe break
- Leakage caused by value/equipment failure
- Overpressure of digester caused by overflow/value freezing
- Release of $H_2S$ caused by improper handling of septic waste
- Overflowing sewage systems or storm-water control because of exceptional downpours or equipment failures in the event of massive influx of fire-water suppression
- Dangerous products in raw materials caused by improper detection
- Explosion caused by biogas and welding operations
- Explosion caused by improper repairing
- Explosion caused by inlet pipeline blocked
- Explosion caused by biogas production in an improper treatment process of wastewater
- Explosion caused by an improper operation procedure in raw material (vegetable and animal waste) unloading
- Water pollution caused by tank failure or effluent discharge
- Intoxication from gas fumes from organic waste
- Intoxication from improper operation of anaerobic digestion
- Ignition caused by electric sparks
- Spontaneous combustion or autoignition

#### 3.2.2 Checklist for Traditional Hazards

Integrating the above analysis with the Process Safety Management Guide [23], a Checklist of questions is obtained which can be used for identification of traditional safety hazards in industrial biotechnology, as shown in Table 7.

## 4 Risk Analysis of Occupational Health

Workplaces in industrial biotechnology can be associated with the risk of occupational illnesses such as infections, allergic reactions, or toxic reactions. Several legal considerations also require the safe handling of biological agents. A risk assessment has to consider the adverse health effects derived from both biological agents' presence and possible exposure in the workplace [24].

### *4.1 Risk Assessment Method*

A four-step method has been widely used for assessing the risks of cancer and other health risks that result from exposure to chemicals [25]. This risk assessment method

**Table 7** Checklist for traditional hazards in industrial technology procedures

| |
|---|
| What is/are the equipment/components within the scope of the analysis? |
| What are the failure forms of the equipment and components? |
| What are the substances present in what quantity? |
| What are the combustibles present in the main quantity? |
| What are the characteristics related to flammability and explosiveness of the combustibles? |
| What are the toxics present in the main quantity? |
| What are the toxicity characteristics of the toxics? |
| What is the phase of the combustible/ toxics? |
| What is the frequency of use? |
| Are any products hazardous from a toxic or fire standpoint? |
| Could their quantity introduce an additional hazard to the process? |
| Are hazardous reactions possible because of mistakes or contaminations/impurities? |
| What could be the other products of the reactions? |
| In what cases could unwanted reactions develop? |
| What deviation of operating parameters can introduce a hazard? |
| Is hazard possible from loss of utility? |
| Does the process work in subatmospheric pressure? |
| Are the following present: relief systems, flare systems, vents, drains, or other process equipment? |
| Are liquid seals protected against freezing? |
| Does the process work in or near the flammable range? |
| Can the process reach a temperature lower than the ductile/brittle transition temperature? |
| Are gas detectors used? |
| Are all employees required to use personal protective equipment when handling raw materials or products? |
| Are eyewash fountains and safety showers present in the working area? |
| Are operators included in a medical surveillance program which is appropriate for the types of chemicals to which they are exposed? |
| Do the operators have any medical conditions or take any drugs that might interact with raw materials or products? |
| Is any medical test recommended? |
| Are operators trained in the use of first aid procedures? |
| Are proper storage methods used to minimize the risk of fire and spontaneous combustion? |
| Have practices and procedures been established to control potential fire hazards/ignition sources? |
| Is there any ATEX (explosive atmospheres) zone? |
| Are there preventive or protective measures? |
| Does the company have a written fire prevention plan? |
| Is the local fire department well-acquainted with company facilities, location, and special hazards? |
| Are operators trained in the use of extinguishers and fire protection procedures? |

consists of four steps: hazard identification, dose-response assessment, exposure assessment, and risk characterization. These steps are adapted for the purpose of assessing hazards in industrial technology, as follows:

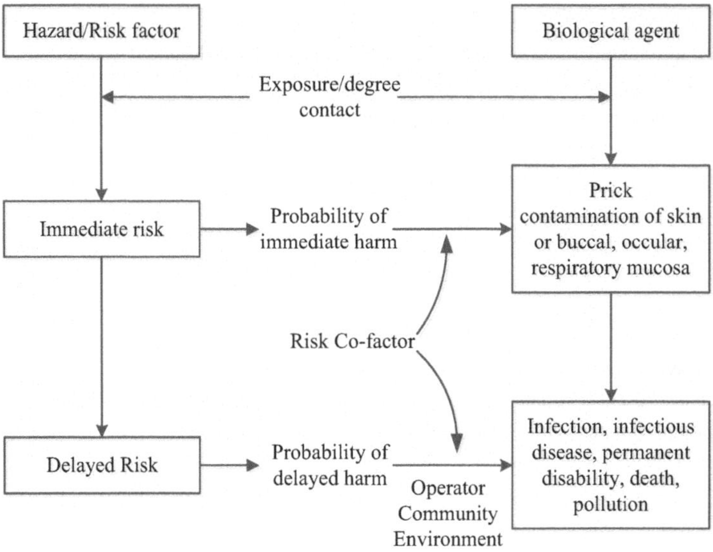


**Fig. 2** Risk assessment method proposed by Caucheteux and Mathot [13]

- Hazard identification: determination of all biological hazards related to health effects
- Dose-response assessment: determination of the relation between the magnitude of exposure and the probability of occurrence of the health effects in question
- Exposure assessment: determination of the extent of human exposure before or after application of regulatory controls
- Risk characterization: description of the nature and often the magnitude of human risk, including exposure uncertainty

Another risk assessment method for biological risk divides the risk into two categories: immediate risk and delayed risk, as shown in Fig. 2 [13].

The occurrence of immediate effects depends upon the probability of immediate harm and some cofactors.[3] The occurrence of delayed effects is determined by the probability of immediate effects and other cofactors. Both methods include hazard identification and exposure risk assessment. Hazard identification related to industrial biotechnology is discussed in Sect. 3.1. Section 4.2 expounds on exposure assessment.

[3]Risk cofactors are characteristics that may affect the risk or the probability of the occurrence of a harmful effect, such as work condition parameters and characteristics of operators.

## *4.2 Exposure Assessment*

Exposure within the occupational safety and health area means exposure of an employee to a biological agent at a certain location in the workplace. It can be evaluated in terms of the duration of exposure or the frequency with which the operator is exposed to the hazard [13]. This approach shows that employees may be exposed continuously or intermittently to a biological agent. The agent may reach the right specific point of entry for it to adhere, multiply, and produce harmful effects. The different points of entry are divided as follows [13]:

- Unprotected, injured or weakened skin
- The natural orifices: eyes, nose, mouth, urogenital area

The object of exposure assessment is to determine in which cases there can be the transmission of the biological agent to exposed people and what immediate and possibly delayed damage (contamination, infection, disease, and death) can follow. In other words, the probability that a given biological agent gets transmitted and arrives at a specific port of entry depends on the activity or operation. Special risks derive from procedures with an increased likelihood of direct contact or with procedures involving a corresponding risk of accident [24]:

- The use of glass tubes, pipettes, and syringes with the risk of injury
- Direct contact with biological agents in great volumes (waste treatment)
- Handling of infected animals/treatment of infected patients
- Risk of inhalation of bioaerosols

Considering these characteristics, exposures are classified into five different groups, which can be used for qualitative analysis. The ratings are as follows [24]:

Group 1, no exposure: no contact (remote operations)
Group 2, low exposure: infrequent contact, low concentration of biological agents (closed systems)
Group 3, moderate exposure: frequent contact at low concentration/infrequent contact at high concentration, engineering control in place (biological safety cabinet)
Group 4, high exposure: frequent contact at high concentration, limited engineering control in place (bioaerosols)
Group 5, very high exposure: frequent contact at a very high concentration (wastewater, waste product handling) or increased risk of accidental exposure

With biological agents of risk group 2 or higher in use, one needs to consider whether remedial measures (e.g., vaccinations) are available and whether they should be offered to employees.

The quantification of exposure for biological agents is difficult and has not been studied extensively by experts. Risk assessment is seriously hampered by the lack of valid quantitative exposure assessment methods and the limited dose-response relationships. International authorities therefore do not require the quantification of the exposure but do require that its nature, the degree, and the duration of the workers' exposure be determined to assess any health and safety risk and to take measures [26]. An employer must retain an occupational exposure list of employee (s) who may be exposed to group 3 or group 4 biological agent(s). This requirement also applies to the following group 2 agents: human herpesvirus type 8 (HHV8), BK polyomavirus, JC polyomavirus, and human papillomaviruses. An exposure record is also required where there is a likelihood of exposure, not just when there has been a known incident or accident related to a biological agent according to the regulation "Safety Health and Welfare at Work (Biological Agents) Regulations 2013" [27].

## *4.3 Risk Characterization*

The hazard identification and exposure assessment related to biological agents have been discussed separately in Sects. 3.1 and 4.2. The risk assessment for employees' activities with biological agents should combine the two aspects to obtain the final results related to occupational safety and health.

The principles of the assessment are summarized as follows [13]:

- Characterizing the biological agent and classifying it in one of the four hazard groups
- Identifying the potential immediate and delayed harmful effects on humans and the environment
- Defining the work situation and looking for procedures that favor the agent's transmission or contact with it
- Evaluating the probability of the occurrence of immediate harm in view of the work situation
- For contained uses (containment measures to minimize or prevent the release of biological agents): defining their usage risk class (maximum risk class of usage)

Equation (1) concerns the calculation approach for the risk which may be employed to quantify risk as recommended by the Health and Safety Authority [28]. However, the quantification of the risk related to biological agents is difficult because the severity of a hazard and the probability of exposure are variable and are possibly affected by many factors. For example, for the employees exhibiting obvious signs of infection, what is important is the probability of transmission to other employees and the employees' contact with the specific point of entry. This will vary according to the employees' pathology, the transmission's period of the agent, and the effectiveness of the treatment. The probability is obviously variable [13].

$$\text{Risk} = \text{Severity of hazard} \times \text{Probability of exposure} \tag{1}$$

A risk matrix combining the biological hazard and the exposure of employees is therefore proposed as shown in Fig. 3. According to the risk matrix, the biological risk to human safety and health can be obtained. It should also be noted that all of the

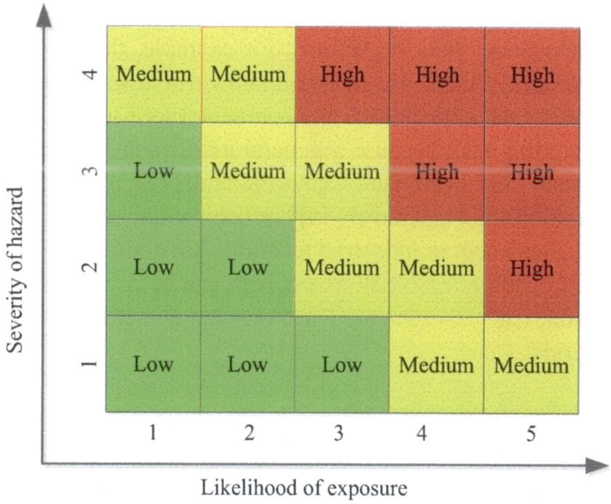


**Fig. 3** Risk matrix for occupational health risk assessment

classifications of the biological hazard, the exposure, and the risk should be adjusted according to the practical situation at hand.

## 5 Risk Analysis of the Environment/Ecology

Life cycle assessment (LCA) shows that the conventional risk to human health in industrial biological for bio-based products is lower than for petrochemical products [29]. Five products are considered in the assessment: plastics polytrimethylene terephthalate (PTT), polyhydroxyalkanoates (PHA), polyethylene terephthalate (PET), polyethylene (PE), and ethanol. The assessment is developed by a comparative analysis between bio-based products and petrochemical products using statistics on technological disasters, accidents, and work-related illnesses, ignoring potential hazards related to biological agents in the external environment. Risk related to human health with potential exposure to hazardous biological agents has attracted much attention, but the potential environmental risk from accidental or deliberate releases in industrial biotechnology is rather neglected and only little research can be found. Although biological agents are applied under contained conditions, that is, physically, chemically and sometimes biologically separated from the external environment, they may be released after accidents, such as failure of equipment or leakages. Biological agents can also enter the external environment through the process of waste discharge and product transmission. These releases may result in contamination of nature or consumable products for humans or animals, especially with the release of GMOs. The contamination of nature may directly interfere with natural populations by production of toxins or indirectly by competition for nutrients, water, light, and other important factors essential for survival of various species. Animal or human risks in the external environment include food contamination and infection by hazardous biological agents. The consequences may be serious, for example, the extinction of entire species. Most importantly, there are many unknown or unexpected side effects associated with these agents in the natural environment. Environmental risk assessment is therefore a demanding task because a general assessment covering all biological agents is not possible [30]. Fortunately, environmental risk assessment for biological agents in the area of biological control [31] has been widely investigated, which can be helpful for risk assessment in industrial technology by combining it with the risk of release. The result of the release risk assessment can be a precursor probability for the environmental risk assessment. The two-stage method is elaborated in the following sections.

## *5.1 Release Risk Assessment of Biological Agents*

The release risk assessment of both biological agents and chemical agents in plants share obvious similarities, involving quantifying the extent to which a risk source releases or otherwise introduces risk agents into the external environment. Methods for release assessment of biological agents should focus on describing the characteristics of the agents and processes, or systems that have the potential for creating risk. Available methods for release risk assessment are mainly divided into five categories: monitoring, performance testing, accident investigation, statistical methods, and modeling [8]. The method being appropriate for release assessment depends on what information about releases is required for the risk assessment, that is, what aspects of the risk source, risk agents, and release processes need to be quantified. The five release risk assessment methods are simply summarized as follows:

- Monitoring can be applied for release assessment with the purpose of collecting data characteristics of a risk source, and the information related to releases or potential releases. It focuses on current and past status, mainly used for current risk assessment and calibrating release models. However, monitoring has a significant limitation when applied to rare events or events that are difficult to detect.
- Performance testing collects data about a system under controlled, usually stressful, conditions. It is especially useful in characterizing risk sources that contain common electrical and mechanical components, but it is difficult to determine and simulate the conditions that are of greatest concern in risk assessment.
- Accident investigation is to reconstruct the accident based on postaccident information, which can provide a lot of useful data which are not readily obtainable from standard monitoring methods for risk assessment. However, it is inappropriate to use this method in accidents with obscured or obliterated evidence because, in that case, inference is difficult.

- Statistical methods provide a means for converting repeated measurements of a risk source. It can be extremely effective to quantify uncertainty about a risk source in situations where a great amount of data has been collected (for instance, automobile accidents and fires).
- Modeling is to establish models for obtaining important characteristics of a risk source from more fundamental factors and their relationships. The model is an abstraction of physical reality, which can be used to estimate the release rate of hazardous agents and the probability of an accident. However, inaccuracy or errors may occur because of the lack of understanding of cause-effect processes and elements contained in the model.

For a more detailed discussion of the method of release risk assessment we refer to the book by Covello and Merkhoher [8].

## 5.2 *Environmental/Ecological Risk Analysis*

After release assessment, the environmental risk assessment in industrial biotechnology can be regarded as adverse effect (consequence) analysis. In that case, the release assessment methods used in biological control can be employed in the area of industrial biotechnology. Hickson proposed a method for environmental risk management in New Zealand [32], which is the starting point for the development of risk evaluation for biological control agents [31]. The basis of the method is that the release risk of agents used in biological control is the product of the likelihood (probability) and the magnitude (consequence). Five groups of risks related to the release of biological agents are considered in this method, that is, the establishment, the dispersal, the host specificity, the direct effects, and the indirect (non-target) effects. For the purpose of using this method in industrial technology, the five groups of risks are redefined as follows [31]:

- Establishment: the potential of a biological agent to establish (no reproduction or reproduction) after release to the external environment
- Dispersal: the potential for dispersal of a biological agent after release to the external environment
- Host range: testing for host specificity of a biological agent in the plant using industrial technology or, in other words, determining the host range
- Direct effects: direct adverse effects of a biological agent on other organisms in the external environment, including attack, enrichment, and vectoring;
- Indirect effects: indirect adverse effects of a biological agent on other organisms in the external environment, including competition, intraguild predation, and enrichment

**Table 8** Qualitative scales for likelihood [32]

| Rank | Likelihood | Description |
|---|---|---|
| 1 | Very unlikely | Not impossible but only occurring in exceptional circumstances |
| 2 | Unlikely | Could occur but is not expected to occur under normal conditions |
| 3 | Possible | Equally likely or unlikely |
| 4 | Likely | Will probably occur at some time |
| 5 | Very likely | Is expected to occur |

**Table 9** Qualitative scales for magnitude [32]

| Rank | Magnitude | Description |
|---|---|---|
| 1 | Minimal | Insignificant (repairable or reversible) environmental impact |
| 2 | Minor | Reversible environmental impact |
| 3 | Moderate | Slight effect on native species |
| 4 | Major | Irreversible environmental effects but no species loss and remedial action available |
| 5 | Massive | Extensive irreversible environmental effects |

The likelihood and the magnitude of adverse effects are subsequently estimated separately according to Tables 8 and 9.

Finally, the environmental risk is obtained by putting the likelihood and magnitude results into a risk matrix as shown in Fig. 4. By combining with the result of release assessment, the result of environmental risk assessment related to biological agents release can be obtained. Certainly, this two-stage method as a preliminary model can be further improved. For example, the qualitative scales can be detailed more in depth based on the characteristics of the agent used in industrial biotechnology.

## 6 Risk Analysis of Accidents

There are many traditional hazards found both in industrial biotechnology and in chemical technology, as shown in Sect. 3. Raw materials and products used in industrial biotechnology processes may be toxic, flammable, or explosive. For example, acetone, with hazardous characteristics of flammability and toxicity, often used as an organic solvent in the production of biopharmaceuticals, may induce serious accidents. In industrial biotechnology, the main products (e.g., biogas, bio-alcohol, and biodiesel) with flammability features can cause fires or explosions during production, storage, and transportation. In addition, some inadvertent products (e.g., $H_2S$) emerging in the production chain may be harmful to

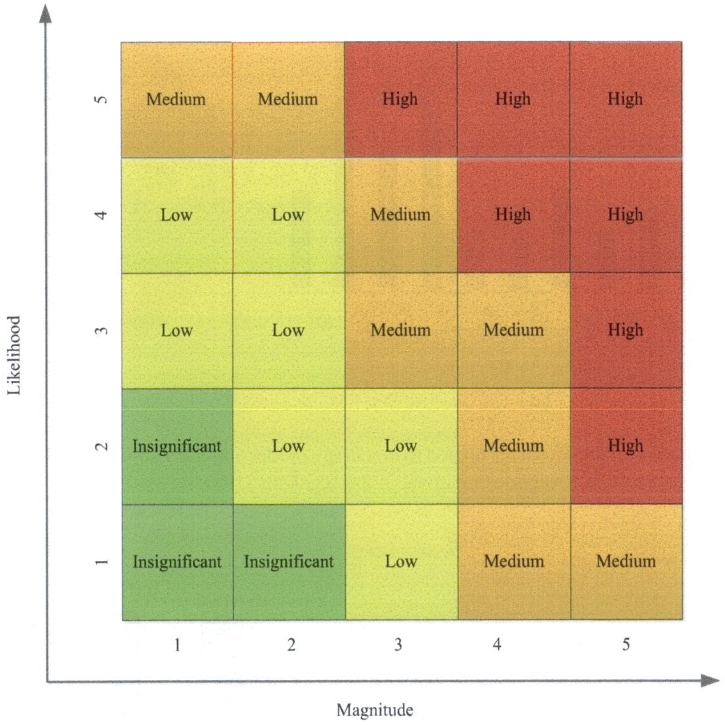


**Fig. 4** Risk matrix for environmental/biological risk assessment

humans. Consequently, traditional accidents found in the chemical industry also occur in the process of industrial biotechnology.

Industrial biotechnology, with some advantages compared to traditional chemical technology, has developed rapidly in recent decades. The number of recorded occupational accidents in the former industry is, however, increasing and mainly related to bioenergy production [33]. From 2004 to 2013, a total of 57 major accidents were recorded involving the production of biomass, biofuels, and biogas, demonstrating an increasing trend in the 10-year period, as shown in Fig. 5a. The accident categories include toxic release, fire, and explosion as illustrated in Fig. 5b. Fire has the highest proportion (53%), followed by explosion (40%) [18]. Figure 6 shows that the number of major accidents seems to be growing faster than the bioenergy production output in recent years.

In addition, Moreno conducted research on cause-consequence analysis (CCA) in bioenergy production to find the characteristics of accidents in bioenergy production [18, 22, 34]. For example, a fishbone diagram was proposed based on available data, providing a qualitative representation of direct causes, as shown in Fig. 7. The effect

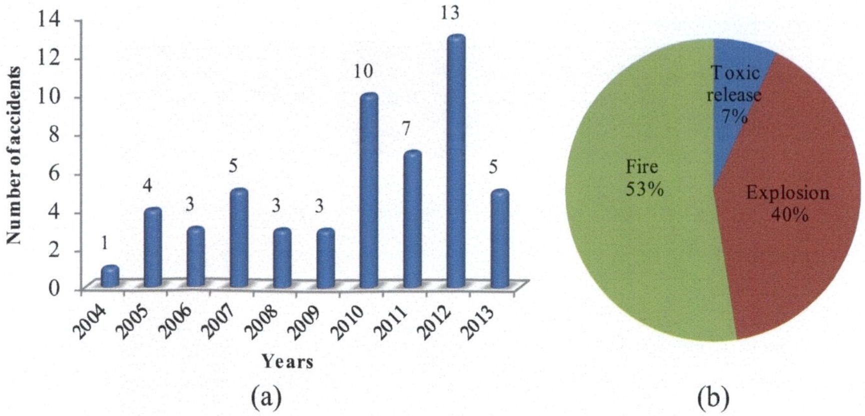


**Fig. 5** Statistics of major accidents related to bioenergy production from 2004 to 2013: (**a**) number of accidents per year; (**b**) categories involved in the accidents [18]

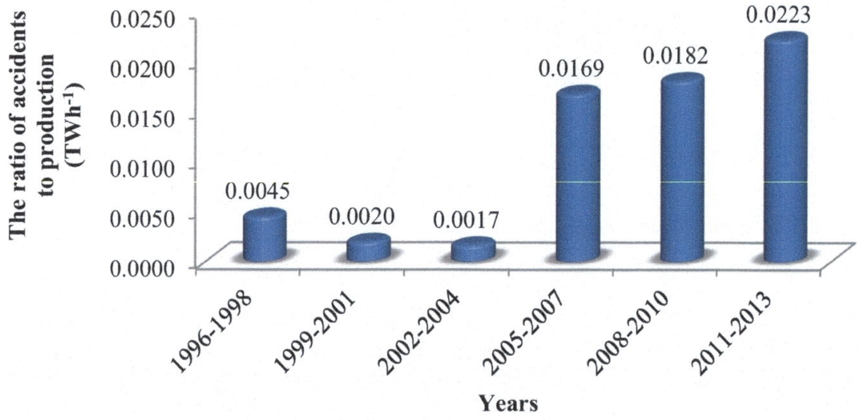


**Fig. 6** The ratio of major accidents to bioenergy production from 1996 to 2013 [18]

(accident) is shown as the fish's head, facing to the right, with the causes extending to the left as a fishbone; five ribs branch off the backbone for major causes, with sub-branches for root-causes. The following major categories of causes were identified via the fishbone diagram method [18]:

- Maintenance errors, defined as operations carried on during maintenance that caused an incident, or the lack of maintenance itself
- Operational errors
- Equipment failures
- Component failures
- Design errors

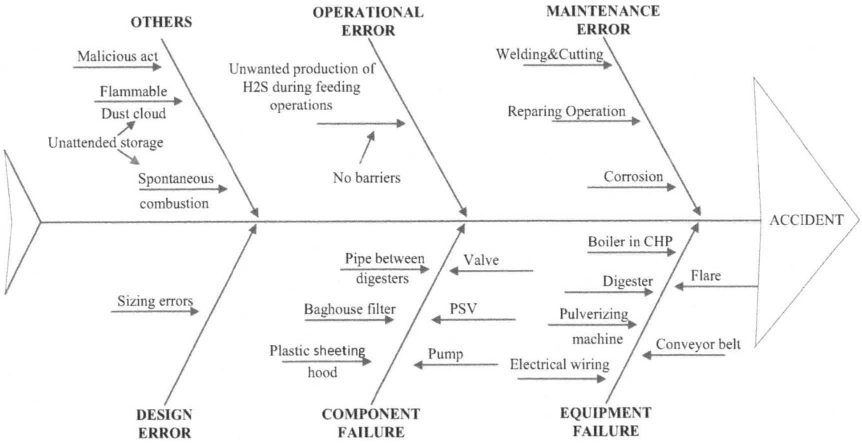


**Fig. 7** Fishbone diagram of accident causes in bioenergy plants [18]

# 7 Summary and Conclusions

This chapter discusses the risk assessment of processes and products in industrial biotechnology. Industrial biotechnology has both biological and chemical process characteristics, this being the essential difference with the traditional process industry. Hazards in industrial biotechnology can therefore be divided into two categories: biological hazards and "traditional" hazards. Risks related to human health, safety, and the environment cannot be ignored in this industrial area because of the increasing number of industrial biotechnology plants.

Although there is not much research and experience with respect to risk assessment in industrial biotechnology, the methods of risk assessment used in the chemical industry can be used as the basis. Therefore, at the beginning of this chapter, traditional tools for hazard identification and risk analysis are discussed, which may also be used in the area of industrial biotechnology. Hazard identification related to biological and traditional risk has been discussed, and Checklists which can be used to identify hazards are provided. Then occupational health and safety risks are discussed mainly by exposure analysis and risk characterization to account for the potential exposure to hazardous biological agents. A two-stage risk assessment method combining release assessment and environmental/ecological risk analysis is proposed to study environmental and ecological risks in industrial biotechnology. Finally, the characteristics of traditional safety accidents in industrial biotechnology are described based on available literature.

This chapter presents an overview analysis of how to identify and assess risks in industrial biotechnology, providing guidance on risk assessment practices in that industry. Conventional risks of industrial biotechnology may be lower than those of fossil-fuel-derived chemical technology. Nevertheless, it should be noted that risk awareness and safety culture need to be strengthened in industrial biotechnology to respond to the growing number of accidents. Besides, many uncertainties related to possible safety, health, environmental, and ecological effects in industrial biotechnology should be further studied, such as the effects of genetically modified agents on other organisms in the external ecosystem. Furthermore, quantitative risk assessment may be a potential research issue in future in this industrial area.

## References


1. Rivera SS, Olivares RDC, Baziuk PA, McLeod JEN (2015) Assessment of biofuel accident risk: a preliminary study. In: Proceedings of the world congress on engineering
2. Arendt J (1990) Management of quantitative risk assessment in the chemical process industry. Process Saf Prog 9(4):262–268
3. Flynn AM, Theodore L (2001) Health, safety, and accident management in the chemical process industries. Marcel Dekker Inc, New York
4. Reniers GLL, Dullaert W, Ale BJM, Soudan K (2005) Developing an external domino accident prevention framework: Hazwim. J Loss Prev Process Ind 18(3):127–138. https://doi.org/10.1016/j.jlp.2005.03.002
5. Lees F (2012) Lees' loss prevention in the process industries: hazard identification, assessment and control. Butterworth-Heinemann, Boston
6. CCPS (1992) Guidelines for hazard evaluation procedures2nd edn. American Institute of Chemical Engineers, New York
7. Meyer T, Reniers G (2016) Engineering risk management. Walter de Gruyter GmbH & Co KG, Berlin
8. Covello VT, Merkhoher MW (2013) Risk assessment methods: approaches for assessing health and environmental risks. Springer Science & Business Media, New York
9. Markowski AS, Mannan MS (2008) Fuzzy risk matrix. J Hazard Mater 159(1):152–157
10. Kletz TA (1999) HAZOP and HAZAN: identifying and assessing process industry hazards. IChemE, Warwickshire
11. CCPS (2011) Guidelines for process safety in bioprocess manufacturing facilities. AIChE, New York
12. Council E (2000) EU Council Directive 2000/54/EC on the protection of workers from risks related to exposure to biological agents at work. Off J Eur Communities L 262:221–245
13. Caucheteux D, Mathot P (2005) Biological risk assessment: an explanation meant for safety advisors in Belgium. Appl Biosafety 10(1):10–29
14. Jeebhay MF (2002) An approach to hazardous biological agents in the workplace – legal provisions and practical considerations. Occup Health South Afr 8(2):8–13
15. Chosewood LC, Wilson DE (1999) Biosafety in microbiological and biomedical laboratories. U.S.GPO, Washington DC
16. Directive Council (1990) 90/679/EEC of 26 November 1990 on the protection of workers from risks related to exposure to biological agents at work. Off J Eur Communities L 374:1–12
17. Richmond J, McKinney R (2000) Primary containment for biohazards: selection, installation and use of biological safety cabinets. United States Department of Health and Human Services/Centers for Disease Control and Prevention/National Institutes of Health, Washington DC
18. Moreno VC, Cozzani V (2015) Major accident hazard in bioenergy production. J Loss Prev Process Ind 35:135–144. https://doi.org/10.1016/j.jlp.2015.04.004
19. Kotek L, Trávníčekb P, Blechaa P (2015) Accident analysis of European biogas stations. Chem Eng Trans 43:1933–1938
20. Calvo Olivares RD, Rivera SS, Núñez Mc Leod JE (2014) Database for accidents and incidents in the biodiesel industry. J Loss Prev Process Ind 29:245–261. https://doi.org/10.1016/j.jlp.2014.03.010

21. Moreno VC, Giacomini E, Cozzani V (2016) Identification of major accident hazards in industrial biological processes. Chem Eng Trans 48:679–684
22. Moreno VC, Papasidero S, Scarponi GE, Guglielmi D, Cozzani V (2016) Analysis of accidents in biogas production and upgrading. Renew Energy 96:1127–1134. https://doi.org/10.1016/j.renene.2015.10.017
23. CSChE (2012) Process safety management guide4th edn. Canadian Society for Chemical Engineering, Ottawa
24. Eckebrecht T (2000) Occupational standards for the protection of employees in biotechnology. Int Arch Occup Environ Health 73:4–7
25. National Research Council (1983) Risk assessment in the Federal Government: managing the process. National Academies Press, Washington DC
26. Douwes J, Thorne P, Pearce N, Heederik D (2003) Bioaerosol health effects and exposure assessment: progress and prospects. Ann Occup Hyg 47(3):187–200
27. Authority HaS (2013) Safety health and welfare at work (biological agents) regulations 2013. The Stationery Office, Dublin
28. HSA (2014) Guidelines to the safety, health and welfare at work (biological agents) regulations 2013. Health and Safety Authority, Dublin
29. Roes AL, Patel MK (2007) Life cycle risks for human health: a comparison of petroleum versus bio-based production of five bulk organic chemicals. Risk Anal 27(5):1311–1321. https://doi.org/10.1111/j.1539-6924.2007.00959.x
30. Patel M, Crank M, Dornburg V, Hermann B, Roes A, Huesing B, Overbeek L, Terragni F, Recchia E (2006) Medium and long-term opportunities and risks of the biotechnological production of bulk chemicals from renewable resources - the potential of white biotechnology. UU CHEM NW&S (Copernicus), Utrecht
31. Van Lenteren JC, Babendreier D, Bigler F, Burgio G, Hokkanen HMT, Kuske S, Loomans AJM, Menzler-Hokkanen I, Van Rijn PCJ, Thomas MB (2003) Environmental risk assessment of exotic natural enemies used in inundative biological control. BioControl 48(1):3–38
32. Hickson R, Moeed A, Hannah D (2000) HSNO, ERMA and risk management. New Zealand Sci Rev 57(3–4):72–77
33. Saracino A, Moreno VC, Antonioni G, Spadoni G, Cozzani V (2016) Application of a self-assessment methodology for occupational safety to biogas industry. Chem Eng Trans 53:247–252. https://doi.org/10.3303/CET1653042
34. Moreno VC, Guglielmi D, Cozzani V (2018) Identification of critical safety barriers in biogas facilities. Reliab Eng Syst Saf 169:81–94. https://doi.org/10.1016/j.ress.2017.07.013

Adv Biochem Eng Biotechnol (2020) 173: 281–298
DOI: 10.1007/10_2018_73

Published online: 1 October 2018

# Green Chemistry and Its Contribution to Industrial Biotechnology

**Daniel Pleissner and Klaus Kümmerer**

## Contents



**Abstract** Sustainable chemistry is a broad framework that starts with the function that a chemical product is offering. Not only chemical but also economic and ethical aspects come into focus throughout the complete lifecycle of chemical products. Green chemistry is an important building block for sustainable chemistry and addresses the issue of greener synthesis and, to a certain degree, the more benign properties of chemicals. The principles of green chemistry clearly aim at making chemical reactions and processes more environmentally friendly. Aspects such as atom efficiency, energy efficiency, harmless reactants, renewable resources, and pollution prevention are considered. Despite the progress made toward a "greener" chemistry, biotechnological processes, as processes for the conversion of biomass into value-added products, have not been properly adapted to new developments. Processes used in industrial biotechnology are predominantly linear. This review elaborates on the potential contributions of green chemistry to industrial

D. Pleissner (✉) and K. Kümmerer
Institute of Sustainable and Environmental Chemistry, Leuphana University of Lüneburg, Lüneburg, Germany
e-mail: daniel.pleissner@leuphana.de

biotechnology and vice versa. Examples are presented of how green chemistry and biotechnology can be connected to make substrate supply, upstream and downstream processing, and product formation more sustainable. The chapter ends with a case study of adipic acid production from lignin to illustrate the importance of a strong connection between green chemistry and biotechnology.

**Graphical Abstract**

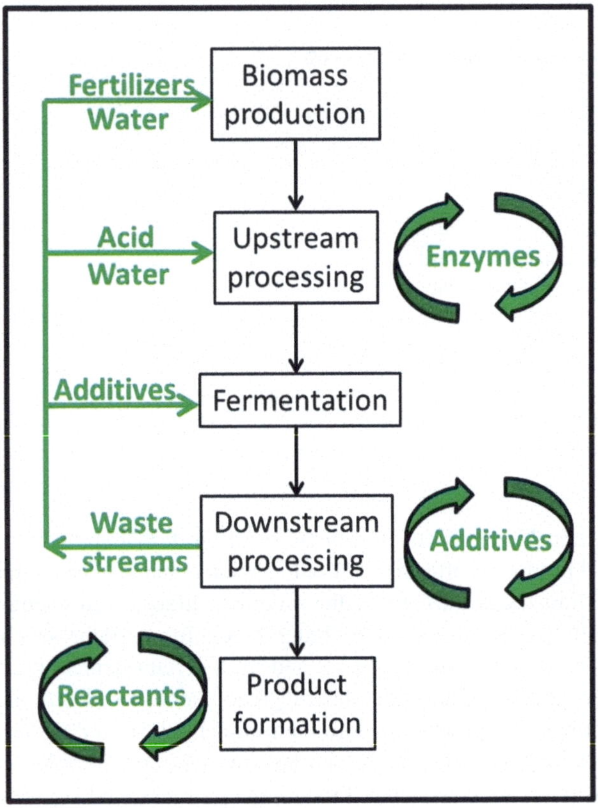


**Keywords** Adipic acid, Downstream processing, Fermentation, Renewable resources, Sustainable chemistry, Upstream processing

# 1 Introduction

Chemistry as a science – and an industry – is an indispensable prerequisite for sustainable development. Many of the Sustainable Development Goals (SDGs) of the United Nations are unavoidably linked to chemistry [1]. One important issue is resources. Fossil resources, such as crude oil, are finite and will sooner or later be depleted, as illustrated by the oil peak shown in Fig. 1 [2]. Consequently, the responsible use of bioresources grows in importance.

A characteristic of fossil-based chemistry is usage of alkanes as a starting point for almost all organic chemicals. Alkanes are chemicals of low complexity (structure, stereochemistry) and reactivity. For most products of the chemical and pharmaceutical industries, however, more complex chemicals are needed in order to offer the required functions, and this applies, to a certain degree, even to fuels. Most of these molecules therefore have to be chemically modified by changing their basic structure to form, for instance, isomers and molecules of different molecular sizes bearing functional groups. Most often, molecules of medium or high polarity are needed for chemical applications and in products. Alkanes, however, are of low polarity and reactivity, and need to be activated for proper chemical reactions to occur. This activation – and most often oxidation – requires a catalyst, some of them expensive metals. Bioresources offer a chance to escape this challenging situation using molecules of low complexity (e.g., fatty acids, plant oils) and high complexity (e.g., sugars, secondary plant metabolites) as feedstocks. Microorganisms used in industrial biotechnology can modify these molecules through their versatile metabolic pathways and biocatalysts at ambient temperature and pressure.

Even though organic materials contain molecules with high functionalization, complexity, and stereochemistry, a predominant amount is "wasted" in bioenergy production. The reduction of oxidized chemical compounds, for instance, to biomethane is associated with a loss of carbon and with the functionalization and stereochemistry of molecules. In contrast, material utilization aims to preserve functional groups and carbon. For example, the fermentative production of lactic acid from sugars is considered efficient as no carbon dioxide is formed during conversion. However, it needs to be clearly stated that material use is, to a certain

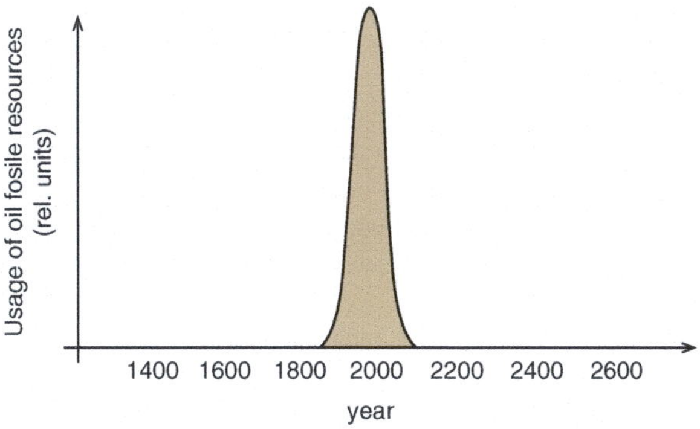


**Fig. 1** The fossil age [2]

degree, associated with loss of functionalization and stereochemistry. Such aspects of resource preservation need to be considered when industrial biotechnological processes are developed, evaluated, and implemented.

## 2 Green and Sustainable Chemistry

At the Earth Summit held in Rio de Janeiro in 1992 it was stated that it is important for research to intensify the development of safe substitutes for chemicals (Agenda 21, #19.21). Principles that address a more integrative view were subsequently established in the European Union in 1996 by the Integrated Pollution Prevention and Control (IPPC) Council Directive (96/61/EC; an amendment came into force in 2010 as Industrial Emissions Directive (IED) 2010/75/EU of the European Parliament and of the Council). In general, use of the best available techniques, efficient use of energy, and prevention of accidents and limitations of their consequences were addressed. In Annex IV of the directive, specific measures were specified (Table 1). Furthermore, in 1998 the 12 principles of green chemistry were published [3] (Table 2), and at the Johannesburg World Summit in 2002, as part of the millennium goals set-up, it was agreed (among other things) to increase resource efficiency. This resulted in the establishment of a Strategic Approach to International Chemicals Management (SAICM). Since then it has become apparent that an even broader approach is urgently needed – sustainable chemistry. In general, only rarely are aspects addressed by green chemistry that go beyond the chemicals themselves and their technical issues, whereas sustainable chemistry generally includes all

**Table 1** Measures stated in Annex IV of Industrial Emissions Directive 2010/75/EU of the European Parliament and of the Council for best available techniques, efficient energy use and prevention of accidents and limitations of their consequences

| |
|---|
| Use of low-waste technology |
| Use of less hazardous substances |
| Fostering of recovery and recycling of substances generated and used in the process, and of waste, where appropriate |
| Comparable processes, facilities or methods of operation, which have been tried with success on an industrial scale |
| Technological advances and changes in scientific knowledge and understanding |
| Nature, effects, and volume of the emissions concerned |
| Commissioning dates for new or existing installations |
| Length of time needed to introduce the best available technique |
| Consumption and nature of raw materials (including water) used in the process, and their energy efficiency |
| Need to prevent or reduce to a minimum the overall impact of the emissions on the environment and the risks to it |
| Need to prevent accidents and to minimize the consequences for the environment |

**Table 2** The 12 principles of green chemistry [3]

| Principles | Explanation |
|---|---|
| 1. Prevention | It is better to prevent waste than to treat or clean up waste after it has been created |
| 2. Atom economy | Synthetic methods should be designed to maximize the incorporation of all materials used in the process into the final product |
| 3. Less hazardous chemical syntheses | Wherever practicable, synthetic methods should be designed to use and generate substances that possess little or no toxicity to human health and the environment |
| 4. Designing safer chemicals | Chemical products should be designed to affect their desired function while minimizing their toxicity |
| 5. Safer solvents and auxiliaries | The use of auxiliary substances (e.g., solvents, separation agents, etc.) should be made unnecessary wherever possible and innocuous when used |
| 6. Design for energy efficiency | Energy requirements of chemical processes should be recognized for their environmental and economic impacts and should be minimized. If possible, synthetic methods should be conducted at ambient temperature and pressure |
| 7. Use of renewable feedstocks | A raw material or feedstock should be renewable rather than depleting whenever technically and economically practicable |
| 8. Reduce derivatives | Unnecessary derivatization (use of blocking groups, protection/deprotection, temporary modification of physical/chemical processes) should be minimized or avoided if possible, because such steps require additional reagents and can generate waste |
| 9. Catalysis | Catalytic reagents (as selective as possible) are superior to stoichiometric reagents |
| 10. Design for degradation | Chemical products should be designed so that at the end of their function they break down into innocuous degradation products and do not persist in the environment |
| 11. Real-time analysis for pollution prevention | Analytical methodologies need to be further developed to allow for real-time, in-process monitoring and control prior to the formation of hazardous substances |
| 12. Inherently safer chemistry for accident prevention | Substances and the form of a substance used in a chemical process should be chosen to minimize the potential for chemical accidents, including releases, explosions, and fires |

aspects of a product related to sustainability, including social and economic aspects related to the use of resources, the stakeholders, and the consumers [4].

Both green and sustainable chemistry are associated with the full life cycle of chemicals and not just one stage of it: (1) raw materials, (2) synthesis, (3) production, (4) use, and (5) fate after use ("end of life"). Whereas green chemistry focuses on issues related to the products themselves, such as workplace and process safety, toxicity, and other properties of chemicals, sustainable chemistry also includes economic, social, and other aspects related to manufacturing and application of chemicals and products (Fig. 2). Sustainable chemistry also requires chemistry and

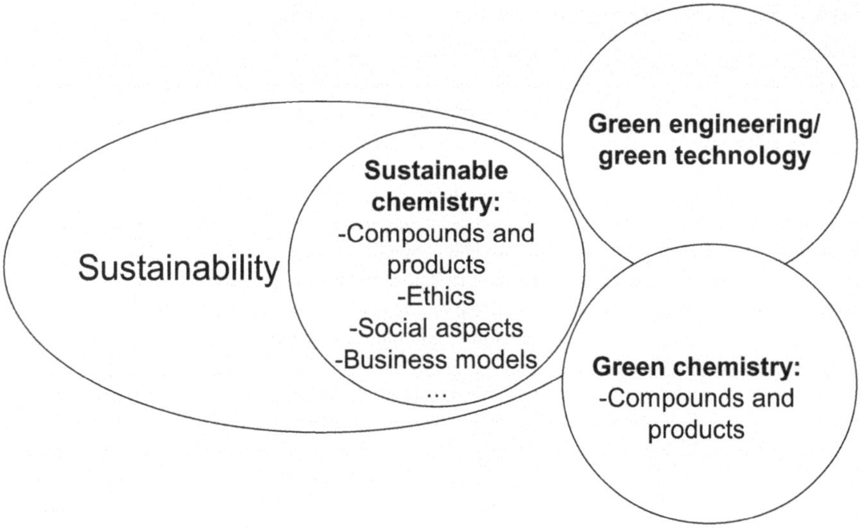


**Fig. 2** Green chemistry, green engineering, green technology, sustainable chemistry, and their relation to sustainability ([2], modified)

the chemical industry and its products to contribute to sustainability. Sustainable chemistry was acknowledged as a major framework to promote sound chemicals management by the United Nations Environment Program (UNEP) second general assembly held in Nairobi in May 2016. However, that is only a starting point. Sustainable chemistry is a guiding principle going much further than that. It also seeks to identify and to avoid rebound effects at the very early stages of process and product development [4].

Green chemistry principle #7 (Table 2) addresses the use of renewable resources. As for conventional chemistry, it is also of importance for biotechnology not only to link itself to green chemistry and green processing principles and technology but also to take into consideration the broader picture of sustainability and to avoid, for instance, material flows that are too big. The negative environmental and social impacts as well as impacts on biodiversity from growing of thousands of hectares of palm trees for generating plant oils as a "renewable bioresource" or sugarcane for bioethanol are well-documented, and are not in line with SDGs (see also [5, 6]). However, responsible use of bioresources can help to reduce or even avoid such effects. The share of organic matter originating from growing food and other plants not usable as food offers a sounder resource for biotechnology and chemistry. Such material is not a waste but a precious resource for old and new chemicals.

## 3 Biotechnology and Green Chemistry

The strategy of industrial biotechnology using new approaches for the production of food and feed, healthcare and fine chemicals, drugs, flavors, bio-based monomers and polymers as well as biofuels means that fossil resources are conserved [7]. Industrial biotechnology makes use of the biochemical versatility of microorganisms to convert carbon- and nitrogen-rich organic materials into microbial biomass and value-added products under controlled conditions in bioreactors [8]. In first-generation biotechnological processes, pure nutrients mostly in form of organic carbon compounds, such as glucose, and organic nitrogen compounds, such as amino acids, are used. Carbon compounds originate from food and feed streams, such as corn, sugarcane, and sugar beet [9], and nitrogen compounds from nitrogen-rich biomass, such as soy. The critical debate about the needs in terms of land (see also [5, 10]), fertilizer and possible groundwater contamination by pesticides and surplus fertilizer, and the use of potential food and feed for the production of chemicals, materials and even fuels initiated the development of second generation biotechnological processes. These are based on residual lignocellulosic materials not useable as food and feed, such as straw from agriculture [9]. The ability of microorganisms not only to produce various value-added compounds but also to utilize a broad range of organic residues contributed to the development of various valorization processes [7, 9].

Resource efficiency of industrial biotechnological processes depends on volumetric productivity, meaning that as much product as possible is formed in a given time. Furthermore, titer and yield of product are of relevance. The use of all fractions of organic materials further contributes to efficiency. Use of all fractions is achieved by cascade use [11], which considers first the production of food and feed, then material use and, finally, energetic use of remaining fractions.

Industrial biotechnology leads to the production of alcohols, short- and long-chain fatty acids, and polymers [7, 8]. Those compounds should be considered not only as final products but also as intermediates to be used as feedstocks for chemical industries searching for greener reactants in chemical reactions. It should also be mentioned that, despite the advantages that come with it, industrial biotechnology implementation is challenging. Industrial biotechnological processes are complex and rely on the several factors: (1) supply of biomass with predictable quantity and quality, (2) logistics, (3) upstream processing, (4) fermentation, (5) downstream processing, and (6) final product formation, isolation, and purification [12, 13]. Every process step requires energy and sometimes adjuvants. The production of biomass also requires water, fertilizer, and arable land. Furthermore, upstream and downstream processing needs additional chemicals, such as acids and bases. The complexity of each process step necessitates separate investigation regarding efficiency, sustainability, and economic feasibility.

Green chemistry aims to provide safer and more environmentally benign chemical reactions and processes. It considers aspects such as waste prevention, energy efficiency, safety and harmless reactants, renewable resources, and pollution prevention (see the 12 principles of green chemistry in Table 2 [3]). During the last few decades, new chemical processes, technologies, solvents, and chemicals have been developed that make chemical reactions and products greener. The principles of green chemistry, however, should apply not only to chemical reactions and chemical products but also to processes used in industrial biotechnology and the resulting products. Technologies developed and investigated to make chemical reactions greener can also contribute to the "greening" of biotechnological processes.

## 4 Biomass Production

The demand for biomass is a result of the limitation in fossil resources and ongoing climate change, and the aim to establish a green, sustainable, and bio-based society. Expenditures on biomass production, such as water, fertilizers, land, and energy demands, however are, critically, often not discussed enough when it comes to an assessment of sustainability. The same applies to the competition for food and feed, impact on water use and quality, land-use, soil carbon stock balance, fertility, greenhouse gas emissions, biodiversity, toxicological risks, and energy efficiency [14] (see also [5, 6, 10]).

There are several studies available elaborating on the demand for biomass for energy production. For example, a predominant fraction of biomass produced in Poland is used for energy production. The conversion of biomass into biogas or bioethanol is expected to contribute 32.4%, corresponding to 157 PJ, to the energy from renewable energy sources to reach the 15% EU target (contribution of 15% of renewable energy to total energy) by 2020 [15]. Around 28%, corresponding to 135 PJ, comes from energy crops (e.g., willow and miscanthus). The total demand for cereals and rape for energy purposes is 1,915,050 tons and 4,494,550 tons, respectively. The demand for molasses for the production of bioethanol is 165,693 tons. Baum et al. [15] concluded that such development would compete with food production for soil and water. Interestingly, the study of Baum et al. did not consider the material utilization of biomass before energy production, which is necessary to cover the future demand for bio-based compounds and materials. Consequently, biomass available for energy production is much less than estimated [15] and competition for soil and water is even more severe.

Establishing a bio-based society relies on fertilizers, crucial for both current and future production of biomass. As illustrated by the study of Baum et al. [15], enormous amounts of biomass are needed to produce a significant amount of energy and even more is necessary when biomass is further used for bio-based materials and chemicals production. The challenge that needs to be solved is the sustainable production of fertilizers, but also the effect of run-off and transfer into groundwater, as well as volatilization. For instance, for maize production in Nigeria, around 15–16 kg per ha phosphorus and 40–47 kg per ha nitrogen were applied annually (data from 2010/2011 and 2012/2013) [16]. The input data came from 1,200 maize plots. Phosphorus is a limited resource. A possible contribution of green chemistry to biomass production and consequently industrial biotechnology could be the implementation of effective methods to recover phosphate from wastewater and organic waste streams.

There is more than one approach to recover phosphorus. One approach is the TetraPhos process currently studied by Remondis in Hamburg (Germany) at pilot-scale [17]. First, phosphoric acid is added to the combusted sewage sludge acting as a leaching agent. The remaining ash residues are disposed of. The second step involves the addition of sulfuric acid to the phosphoric acid-rich stream. After filtration, the filtrate is purified to remove the metal salts added during sewage sludge precipitation and the phosphoric acid is stream concentrated. By-products of this process are gypsum and aluminum and iron salts. Gypsum is subsequently used for constructions as drywall, wallboard, or plasterboard.

Zhang et al. [18] investigated the production of monoammoniumphosphate and diammoniumphosphate from phosphate rock in a life cycle assessment study [18]. The investigated scenarios included sulfuric and phosphoric acid syntheses, sludge concentration, and wastewater treatment. Both scenarios created gas emissions such as $SO_2$, $NH_3$, dust, $NO_2$, and HF, solid waste such as sulfur, catalysts, cinder, phosphogypsum, sludge, and garbage, as well as wastewater. The study revealed that, compared to diammoniumphosphate, fewer emissions were produced and fewer raw materials were consumed when monoammoniumphosphate was produced. Phosphogypsum production, however, was higher. Generally, side-products should be utilized and phosphogypsum, which up to now has only rarely been utilized, should be considered for the cement, fertilizer, and sulfuric acid industry.

## 5 Upstream Processing

Upstream processing in industrial biotechnology processes is essential to (1) make biomass available as substrate, (2) separate high value biomass constituents from low value constituents and from nutrients, and (3) separate compounds with inhibiting effects on microbial performance. Upstream processing can be skipped when easily digestible biomass streams, such as starch- and protein-rich materials, are used. Here, the ability of microorganisms to secrete amylases and proteases can be utilized to degrade biomass. Food waste, for instance, is easily digestible and various approaches have been developed, making upstream processing unnecessary. The strain *Streptococcus* sp., for instance, is able to recover sugars from food waste and simultaneously convert the sugars into lactic acid. The performance is thereby highly comparable to fermentation processes where hydrolysis has been carried out separately [19, 20].

Upstream processing is essential when complex and recalcitrant biomass streams, such as wood, spent sulfite liquor (a process stream of the sulfite pulping process), and lignocellulosic-rich agricultural residues, are used. Wood consists of cellulose, hemicellulose, and lignin. The pretreatment is mostly carried out using acids and high temperatures separate cellulose and hemicellulose from lignin and to release fermentable sugars. A drawback is that lignin, depending on the pretreatment, is also partly degraded and phenolic compounds are released.

Alexandri et al. [21] investigated the extraction of phenolic compounds from spent sulfite liquor containing 12.4 g/L of phenolic compounds originating from delignification of wood and a considerable amount of sugars. In their study a liquid-liquid extraction method to (1) separate phenolic compounds, such as gallic acid, isorhamnetin, syringic acid, syringaldehyde, vanillic acid, vanillin, and catechin with antioxidant activity and (2) convert spent sulfite liquor into a fermentable nutrient source was used. Using ethyl acetate as solvent at a solvent-to-spent sulfite liquor ratio of 3.67:1 and a pH of 2.22, 7.5 g gallic acid equivalents (a sum parameter for stating phenol concentration) per L spent sulfite liquor was extracted. The extract showed a strong antioxidant activity of 3.64 based on an antioxidant activity index [21]. Subsequently, the dephenolized liquor was successfully used as source of carbon for succinic acid production with *Actinobacillus succinogenes* and *Basfia succiniciproducens*. Compared to the untreated spent sulfite liquor, the productivity of both strains was two to three times higher.

Chemical hydrolysis of agricultural residues to convert cellulose and hemicellulose into fermentable sugars can be carried out using concentrated or dilute sulfuric acid and hydrochloric acid. Acid hydrolysis results in high yields of sugars (up to 80%), but acids also cause serious corrosion of equipment, such as piping. More serious is the necessity to neutralize acids [22] after use, which both requires considerable amounts of base and also creates salt solutions. A further drawback is the need for temperatures of more than 200°C to hydrolyze lignocellulosic materials effectively [23].

Recent approaches consider the application of ionic liquids. Zhou et al. [24] investigated six novel ionic liquids based on 2-phenyl-2-imidazoline. They found "superior catalytic activity" regarding the hydrolysis of cellulose when 1-propyl sulfonic acid-2-phenyl imidazoline hydrogensulfate was used together with water. The total yield of reducing sugars was 85.1% within 1 h at 100°C. In 1-butyl-3-methyl imidazolium chloride the yield of reducing sugars was 72.1% at 110°C after 3 h. Even though ionic liquids show promising properties for the hydrolysis of biomass constituents, economic aspects and particularly their fate in the environment still need to be evaluated. Haiß et al. [25] investigated phenylalanine-based ionic liquids and found no biodegradation in the sense of OECD guidelines. Even though ionic liquids can be recycled, release into the environment cannot be excluded, and thus the use of fully degradable ionic liquids, in accordance with the principles of green chemistry (Table 2), should have priority. In this sense, pyridinium-substituted phenylalanine-derived ionic liquids represent, according to Haiß et al. [25], a promising structure for further degradation optimization.

In addition to acids and ionic liquids, the degradation of biomass can be carried out using enzymes. The application of hydrolytic enzymes, such as cellulases and hemicellulases, can significantly decrease process temperature and reduce the use of chemicals. The drawbacks of enzymatic hydrolysis are product inhibition, lignin adsorption, low solid loading, and high costs [26]. Furthermore, it might be necessary to perform a pretreatment of the biomass. The highest yields, for instance, were found when biomass is pretreated using acids or ionic liquids and later the residues are enzymatically hydrolyzed after neutralization of acids or removal of ionic liquids. Alternatively, biomass pretreated with ultrasonication increased the hydrolytic efficiency of enzymes [27].

The contribution of green chemistry results from the challenges associated with upstream processing of biomass streams. The first target should be a minimization of side-streams, such as salts and other streams that require a further treatment. Every side-stream that requires further treatment negatively affects the overall sustainability, requires energy, and contributes to process costs. The second target should be a recovery and recycling of all reactants. Because of natural denaturation, enzymes can be recovered and recycled only a couple of times. The same applies to acid streams because of the accumulation of inhibitory compounds. The third target should be the application of mechanical pretreatments that do not require high temperature and pressure. That would decrease the energy demand of the whole process and minimize the appearance of compounds that may affect fermentation carried out subsequently.

## 6 Downstream Processing and Product Formation

The most crucial process in industrial biotechnology is the downstream processing needed to separate the product of interest from impurities and remaining nutrients, and to recover additives added during upstream processing and fermentation. According to substance properties, Sasiradee et al. listed and clustered a couple of separation technologies either depending on (1) mechanical unit operations, such as sedimentation, filtration, chromatography, electrophoresis, and centrifugation, (2) mass transfer unit operations, such as sorption and desorption, distillation, extraction, crystallization, and lyophilization, and (3) reactive separation techniques, such as reactive distillation, chromatography reactors, and reactive membrane separation [28].

Illustrations of a conventional, linear industrial biotechnology process and a process improved by green chemistry are shown in Fig. 3. The advantage of the improved, circular process is that waste streams are minimized and additives, as well as water, reactants, and enzymes, are recovered during downstream processing and recycled later on. In contrast, conventional, linear biotechnological processes result in the formation of various liquid and solid side-streams, which may require neutralization or other treatment before discharge into municipal wastewater treatment facilities.

Enzymes needed in biomass hydrolysis and solvents, as well as reactants for extraction and product formation, can be designed in such a way as to achieve almost complete recovery and recycling. The amounts of additives, such as acids and bases, for pH regulation during upstream processing, fermentation, and downstream

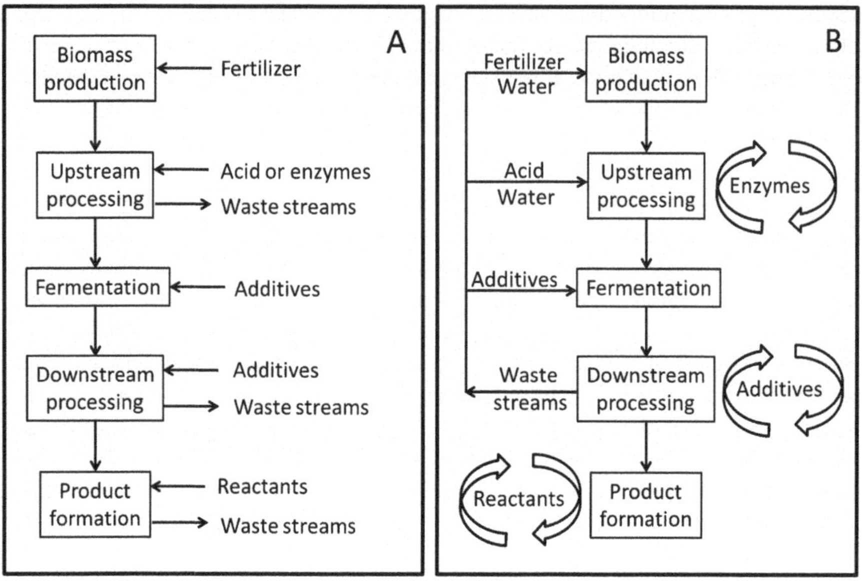


**Fig. 3** Illustration of a conventional industrial biotechnology process (**a**) resulting in various liquid and solid side-streams, and by a green chemistry improved process considering the recirculation of side-streams and recovery of water, fertilizer, additives and enzymes (**b**)

processing, however, are highly diluted, which challenges their recovery. The concentration of phosphate in fermentation broth can be below 50 mg/L [29], which seems too low to allow a feasible phosphate recovery step on an industrial scale. Nevertheless, considering wastewater streams of 50 $m^3$ or more from fermentation processes, within only 1 or 2 days a significant amount of phosphorus can be recovered and reused for the production of biomass. There is currently no effective and efficient process available for the recovery of phosphate from highly diluted fermentation broths integrated in downstream processing. In wastewater treatment plants, the formation of struvite ($MgNH_4PO_4{\cdot}6H_2O$) is used to recover phosphate from liquid streams by adding magnesium salts. Another approach is the incineration of wastewater sludge and recovery of phosphorus from ashes. In sustainable biotechnological processes the recovery of additives and water should have priority, and thus dosing of additional reactants should be avoided to ease downstream processing and product recovery.

For additives recovery, Pleissner et al. [30] investigated the downstream processing of a lactic acid-rich fermentative broth for pure lactic acid production and recovery of acids and bases. They studied first the Amberlite resins FPA 53 and CR 5550 for separating lactic acid from salt ions using water or sulfuric acid as eluent. Second, monopolar and bipolar electro-dialysis were used to concentrate and convert salt ions into the corresponding acid or base, respectively. They further investigated the recycling of recovered base and acid in fermentation and downstream processing, respectively. Using FPA 53 and 12.5 mM sulfuric acid, 90% of lactic acid was separated from salt ions, whereas the Amberlite CR 5550 and water as eluent did not result in a satisfactory separation. Subsequently, electrodialysis of salt-rich streams resulted in 1.1 M sodium hydroxide and 0.8 M hydrochloric acid. The sodium hydroxide solution was concentrated to 5.3 M and used for regulation of pH in lactic acid fermentation. An advantage of this approach is that, except for eluent and resin material, no extra chemicals are needed. However, the use of sulfuric acid introduces further ions into the salt stream, which complicates the production of pure hydrochloric acid by electrodialysis.

To separate organic acids from fermentation broths, the use of liquid-liquid extraction with solvents has been investigated. The hydrophilic character of short-chain carboxylic acids prevents the direct extraction in organic solvents and a reagent is needed to trap the compound in the organic phase [31]. Aliphatic amines interacting with acidic groups have been tested for the recovery of short-chain carboxylic acids, such as malic, citric, glycolic, propionic, pyruvic, and lactic acids [31]. Solvent screening experiments carried out on 3-hydroxypropionic acid revealed that H-bond donor characteristic and polarity have a significant impact on recovery yield. Using solvents such as *n*-decanol and oleyl alcohol, low extraction yields at high tri-*n*-octylamine concentrations were observed. An increase in initial concentration of tri-*n*-octylamine from 0.003 up to 0.13 mol/L resulted in an increase in extraction yield although a further increase from 0.1 to 0.56 mol/L led to a reduction of the extraction yield caused by amine transfer in the aqueous phase [31]. Furthermore, the authors observed the release of *n*-octylamine and di-noctylamine from commercial tri-*n*-octylamine into the aqueous phase, which negatively influenced extraction yields at low initial acid concentrations. It is important for industrial biotechnological processes that released amines might be toxic to microorganisms when water streams are reused in fermentation processes [31].

# 7 Case Study: Biotechnological Production of Adipic Acid from Lignin

The production of adipic acid from lignin and the eventual conversion of adipic acid into nylon-6,6 is a good case study to illustrate the potential of green chemistry to contribute to industrial biotechnology. Lignin is the second most abundant bio-polymer, after cellulose, in the terrestrial ecosystem [32]. Lignin is part of the so-called lignocellulosic biomass. Although cellulose and hemicellulose from lignocellulosic biomass have been extensively investigated as carbon sources in various fermentations, the efficient use of lignin is still in its infancy. The complexity of the polymer consisting of *p*-coumaryl, coniferyl, and sinapyl alcohol, and *p*-coumaraldehyde, coniferaldehyde, and sinapaldehyde monomers [33] makes a controlled degradation for the formation of specific products challenging. Unspecific oxidative enzymes, such as peroxidases, formed by bacteria and fungi can cleave the chemical bonds between monomers and can release various phenolic compounds. However, because of the recalcitrant character of lignin, enzymatically carried out degradation is rather slow compared to chemical and thermal degradations. In chemical oxidation processes, oxidants, such as oxygen, hydrogen peroxide, and metal oxides [34], are needed to break bonds between monomers. Chemical oxidation results in a conversion of more than 80% of the lignin polymer and, depending on the catalysts, the reaction can last up to 48 h at a temperature of 80–135°C. The reaction mixture is most often highly heterogeneous. For instance, when 2-phenoxy-1-phenylethanol is applied as substrate and the treatment carried out at 80°C for 8 h in the presence of $VO(acac)_2$/acetic acid, a conversion efficiency of 94.2% can be obtained. The mixture, however, consists of 1-(3,5-dimethoxyphenyl)-3-hydroxy-2-(2-methoxy phenoxy)propan-1-one (65%), phenol (53%), and 2-phenoxy-1-phenylethanone (2.3%) [35]. An improvement of the lignin degradation process may come with the use of ionic liquids. Ionic liquids are able to dissolve lignin selectively and contribute to its degradation. For instance, kraft lignin was degraded at 75°C for 1 h in the presence of pyridinium formate, pyridinium acetate, and pyridinium propionate. The yield was 95% and the ionic liquids were recyclable.

Despite the problems associated with the degradation of lignin and purification of products, studies have been conducted using lignin-derived aromatic compounds in biotechnological processes. Various bacteria can oxidize aromatic compounds, such as benzoates, toluene, phenol, aniline, anthranilates, mandelates, and salicylates, and form catechol as the central aromatic intermediate [36]. One example is the use of aromatic compounds for the fermentative production of muconic acid ((2*E*,4*E*)-hexa-2,4-dienedioic acid) [37]. Muconic acid can be chemically converted into adipic acid (hexanedioic acid), a precursor of nylon-6,6. Van Duuren et al. [38] investigated a pH-stat fed-batch process for the production of *cis,cis*-muconic acid from benzoic acid by an engineered *Pseudomonas putida* KT2440-JD1 strain. The strain oxidizes benzoic acid to benzoate diol and subsequently to catechol, and finally forms *cis,cis*-muconic acid. They reported that the *P. putida* strain used was unable to grow in the presence of more than 50 mM benzoic acid or 600 mM *cis,cis*-muconic acid. By feeding glucose as growth substrate, a final titer of *cis,cis*-muconic acid of 18.5 g/L was reached. The maximum volumetric productivity was 0.8 g/L/h. Johnson et al. used other *P. putida* strains derived from strain KT2440 to produce muconic acid from either aromatic compounds or sugars [39]. The production of muconic acid from either aromatic compounds or sugars requires a decarboxylase to convert protocatechuate to catechol. For the production of muconic acid the protocatechuate decarboxylase has been identified as metabolic bottleneck. The optimized engineered strains produced in bioreactor experiments more than 15 g/L muconic acid from *p*-coumarate and 4.9 g/L from glucose at a productivity of 0.21 and 0.09 g/L/h, respectively [39].

The formation of final products is only partly achieved using biotechnological processes alone. To produce nylon-6,6, the conversion of muconic acid into adipic acid is necessary. Capelli et al. [40] worked on a catalyzed hydrogenation of muconic acid under mild operation conditions. They used commercial Pt/C as catalyst reduced for 3 h at 200°C and 6 bar of static hydrogen. The reaction was tested using an artificial fermentation broth containing 28 g/L *cis,cis*-muconic acid, 50 g/L $Na_2HPO_4$, 15 g/L $KH_2PO_4$, 2.5 g/L NaCl, and 40 g/L NaOH at 70°C. In contrast to experiments carried out using defined media, they did not find a complete conversion to adipic acid, probably because of adsorption of salts present in the artificial fermentation broth on the catalyst surface [40]. A downstream processing and purification of muconic acid prior to catalytic conversion is therefore recommended.

The case study of biotechnological production of adipic acid from lignin illustrates the potential but also the challenges that come along with the utilization of biomass. Lignin is a source of phenolic compounds that can find application in various chemical and biotechnological processes. However, the heterogeneity of phenolic compounds obtained after cleavage of lignin makes a utilization of all side-streams challenging. Furthermore, the case study illustrates the difficulties associated with the use of a complex feedstock such as lignin. According to the state-of-the-art, rough chemical pretreatments are necessary to degrade lignin. Nevertheless, the microbial conversion of released phenolic compounds seems straightforward and, with further improvements in strain development regarding titer and productivity of muconic acid, better efficiency can be achieved.

## 8 Conclusions and Future Perspectives

In the past, various approaches have been developed to convert biomass biotechnologically into new value-added products. Even though some processes exist which consider the efficient utilization of the whole potential of biomass, a large number of processes do not. The focus is predominantly on the production of pure products, not taking into account the appearance of side-streams, neither a complete balance of pretreatment and upstream steps nor post-treatment and downstream steps. The principles of green chemistry should be carefully considered by industrial biotechnology. In biotechnological processes, side-streams can be acidified wastewater and waste organic material coming from fermentation as well as up- and downstream processing. In the future, processes are required that (1) consider all biomass fractions as resource and (2) minimize the production of unwanted side-streams. This would not only contribute to a holistic use of biogenic resources, but also strengthen the economic feasibility of industrial biotechnology. The goal of "green biotechnology" can be achieved, for instance, by recirculation of side-streams, such as water, to the fermenter and reuse of phosphate as fertilizer in biomass production. The former would contribute to a conversion of all nutrients and the latter would contribute to the formation of new biomass, which subsequently can be used as feedstock in biotechnological processes. It is expected that future biotechnological processes and biorefineries will operate in a circular way. The development of those processes – not losing any water, nutrients, and reactants, and going far beyond the formation of pure products – makes a rethinking of conventional applied downstream processes necessary. Green chemistry can give guidance on how biotechnology can unfold its potential much more effectively and can move toward sustainability.

## References


1. Matlin SA, Mehta G, Hopf H, Krief A (2015) The role of chemistry in inventing a sustainable future. Nat Chem 7:941. https://doi.org/10.1038/nchem.2389
2. Kümmerer K, Clark J (2016) Green and sustainable chemistry. In: Heinrichs H, Martens P, Michelsen G, Wiek A (eds) Sustainability science: an introduction. Springer Netherlands, Dordrecht, pp 43–59. https://doi.org/10.1007/978-94-017-7242-6_4
3. Anastas PT, Warner JC (1998) Green chemistry: theory and practice. Oxford University Press, Oxford
4. Kümmerer K (2017) Sustainable chemistry: a future guiding principle. Angew Chem Int Ed 56:16420–16421. https://doi.org/10.1002/ange.201709949
5. Festel G (2018) Economic aspects of industrial biotechnology. Adv Biochem Eng Biotechnol. https://doi.org/10.1007/10_2018_70
6. MacLean H, Saville B (2018) Environmental aspects of industrial biotechnology. Adv Biochem Eng Biotechnol
7. Koutinas AA et al (2014) Valorization of industrial waste and by-product streams via fermentation for the production of chemicals and biopolymers. Chem Soc Rev 43:2587–2627. https://doi.org/10.1039/C3CS60293A
8. Pleissner D, Qi Q, Gao C, Rivero CP, Webb C, Lin CSK, Venus J (2016) Valorization of organic residues for the production of added value chemicals: a contribution to the bio-based economy. Biochem Eng J 116:3–16. https://doi.org/10.1016/j.bej.2015.12.016
9. Moncada JB, Aristizábal VM, Cardona CA (2016) Design strategies for sustainable biorefineries. Biochem Eng J 116:122–134. https://doi.org/10.1016/j.bej.2016.06.009
10. Benoît Norris C (2018) Social life cycle assessment for industrial biotechnology. Adv Biochem Eng Biotechnol
11. Geldermann J et al (2016) Improved resource efficiency and cascading utilisation of renewable materials. J Clean Prod 110:1–8. https://doi.org/10.1016/j.jclepro.2015.09.092
12. Ahmadi N, Khosravi-Darani K, Mortazavian AM (2017) An overview of biotechnological production of propionic acid: from upstream to downstream processes. Electron J Biotechnol 28:67–75. https://doi.org/10.1016/j.ejbt.2017.04.004
13. Özdenkçi K et al (2017) A novel biorefinery integration concept for lignocellulosic biomass. Energy Convers Manag 149:974–987. https://doi.org/10.1016/j.enconman.2017.04.034
14. de Jong E, Jungmeier G (2015) Biorefinery concepts in comparison to petrochemical refineries. https://doi.org/10.1016/B978-0-444-63453-5.00001-X
15. Baum R, Wajszczuk K, Pepliński B, Wawrzynowicz J (2013) Potential for agricultural biomass production for energy purposes in Poland: a review. Contemp Econ 7:63–74
16. Liverpool-Tasie LSO, Omonona BT, Sanou A, Ogunleye WO (2017) Is increasing inorganic fertilizer use for maize production in SSA a profitable proposition? Evidence from Nigeria. Food Pol 67:41–51. https://doi.org/10.1016/j.foodpol.2016.09.011
17. Gorazda K et al (2017) Fertilisers production from ashes after sewage sludge combustion – a strategy towards sustainable development. Environ Res 154:171–180. https://doi.org/10.1016/j.envres.2017.01.002
18. Zhang F, Wang Q, Hong J, Chen W, Qi C, Ye L (2017) Life cycle assessment of diammonium- and monoammonium-phosphate fertilizer production in China. J Clean Prod 141:1087–1094. https://doi.org/10.1016/j.jclepro.2016.09.107
19. Demichelis F, Pleissner D, Fiore S, Mariano S, Navarro Gutiérrez IM, Schneider R, Venus J (2017) Investigation of food waste valorization through sequential lactic acid fermentative

production and anaerobic digestion of fermentation residues. Bioresour Technol 241:508–516. https://doi.org/10.1016/j.biortech.2017.05.174
20. Pleissner D, Demichelis F, Mariano S, Fiore S, Navarro Gutiérrez IM, Schneider R, Venus J (2017) Direct production of lactic acid based on simultaneous saccharification and fermentation of mixed restaurant food waste. J Clean Prod 143:615–623. https://doi.org/10.1016/j.jclepro.2016.12.065
21. Alexandri M, Papapostolou H, Vlysidis A, Gardeli C, Komaitis M, Papanikolaou S, Koutinas AA (2016) Extraction of phenolic compounds and succinic acid production from spent sulphite liquor. J Chem Technol Biotechnol 91:2751–2760. https://doi.org/10.1002/jctb.4880
22. Galbe M, Zacchi G (2002) A review of the production of ethanol from softwood. Appl Microbiol Biotechnol 59:618–628. https://doi.org/10.1007/s00253-002-1058-9
23. Ji W, Shen Z, Wen Y (2015) Hydrolysis of wheat straw by dilute sulfuric acid in a continuous mode. Chem Eng J 260:20–27. https://doi.org/10.1016/j.cej.2014.08.089
24. Zhuo K, Du Q, Bai G, Wang C, Chen Y, Wang J (2015) Hydrolysis of cellulose catalyzed by novel acidic ionic liquids. Carbohydr Polym 115:49–53. https://doi.org/10.1016/j.carbpol.2014.08.078
25. Haiß A, Jordan A, Westphal J, Logunova E, Gathergood N, Kümmerer K (2016) On the way to greener ionic liquids: identification of a fully mineralizable phenylalanine-based ionic liquid. Green Chem 18:4361–4373. https://doi.org/10.1039/C6GC00417B
26. Fenila F, Shastri Y (2016) Optimal control of enzymatic hydrolysis of lignocellulosic biomass. Resour Effic Technol 2:S96–S104. https://doi.org/10.1016/j.reffit.2016.11.006
27. Revin V, Atykyan N, Zakharkin D (2016) Enzymatic hydrolysis and fermentation of ultradispersed wood particles after ultrasonic pretreatment. Electron J Biotechnol 20:14–19. https://doi.org/10.1016/j.ejbt.2015.11.007
28. Jantasee S, Kienberger M, Mungma N, Siebenhofer M (2017) Potential and assessment of lactic acid production and isolation – a review. J Chem Technol Biotechnol 92(12):2885–2893. https://doi.org/10.1002/jctb.5237
29. Pleissner D, Lau KY, Schneider R, Venus J, Lin CSK (2015) Fatty acid feedstock preparation and lactic acid production as integrated processes in mixed restaurant food and bakery wastes treatment. Food Res Int 73:52–61. https://doi.org/10.1016/j.foodres.2014.11.048
30. Pleissner D, Schneider R, Venus J, Koch T (2017) Separation of lactic acid and recovery of salt-ions from fermentation broth. J Chem Technol Biotechnol 92:504–511. https://doi.org/10.1002/jctb.5023
31. Chemarin F et al (2017) New insights in reactive extraction mechanisms of organic acids: an experimental approach for 3-hydroxypropionic acid extraction with tri-n-octylamine. Sep Purif Technol 179:523–532. https://doi.org/10.1016/j.seppur.2017.02.018
32. Roopan SM (2017) An overview of natural renewable bio-polymer lignin towards nano and biotechnological applications. Int J Biol Macromol 103:508–514. https://doi.org/10.1016/j.ijbiomac.2017.05.103
33. Sonoda T, Ona T, Yokoi H, Ishida Y, Ohtani H, Tsuge S (2001) Quantitative analysis of detailed lignin monomer composition by pyrolysis-gas chromatography combined with preliminary acetylation of the samples. Anal Chem 73:5429–5435. https://doi.org/10.1021/ac010557c
34. Dai J, Patti AF, Saito K (2016) Recent developments in chemical degradation of lignin: catalytic oxidation and ionic liquids. Tetrahedron Lett 57:4945–4951. https://doi.org/10.1016/j.tetlet.2016.09.084
35. Ma Y, Du Z, Liu J, Xia F, Xu J (2015) Selective oxidative C-C bond cleavage of a lignin model compound in the presence of acetic acid with a vanadium catalyst. Green Chem 17:4968–4973. https://doi.org/10.1039/C5GC00645G
36. Harwood CS, Parales RE (1996) The β-ketoadipate pathway and the biology of self-identity. Annu Rev Microbiol 50:553–590. https://doi.org/10.1146/annurev.micro.50.1.553
37. Kohlstedt M et al (2018) From lignin to nylon: cascaded chemical and biochemical conversion using metabolically engineered Pseudomonas putida. Metab Eng 47:279–293. https://doi.org/10.1016/j.ymben.2018.03.003

38. van Duuren JBJH, Wijte D, Karge B, Martins dos Santos VAP, Yang Y, Mars AE, Eggink G (2012) pH-stat fed-batch process to enhance the production of cis, cis-muconate from benzoate by *Pseudomonas putida* KT2440-JD1. Biotechnol Prog 28:85–92. https://doi.org/10.1002/btpr.709
39. Johnson CW, Salvachúa D, Khanna P, Smith H, Peterson DJ, Beckham GT (2016) Enhancing muconic acid production from glucose and lignin-derived aromatic compounds via increased protocatechuate decarboxylase activity. Metab Eng Commun 3:111–119. https://doi.org/10.1016/j.meteno.2016.04.002
40. Capelli S et al (2017) Bio-adipic acid production by catalysed hydrogenation of muconic acid in mild operating conditions. Appl Catal B Environ 218:220–229. https://doi.org/10.1016/j.apcatb.2017.06.060

Adv Biochem Eng Biotechnol (2020) 173: 299–324
DOI: 10.1007/10_2018_82

Published online: 19 February 2019

# Application Potentials of Geobiotechnology in Mining, Mineral Processing, and Metal Recycling

**Franz Glombitza, Rene Kermer, and Susan Reichel**

## Contents



**Abstract** This chapter highlights the huge and manifold possibilities of reactions which result from the interactions between microorganisms and the geosphere and which are used for mining, mineral processing, and metal recycling. Besides the introduction (Sect. 1) the contribution is divided into five different sections describing the mobilization (Sect. 2) and immobilization (Sect. 3) of valuable substances, the processes of biosorption and bioaccumulation (Sect. 4), as well as transformation of metals into metal organic compounds (Sect. 5). A special topic (Sect. 6) addresses the application of $CO_2$ as an important component for the formation of energy-rich

F. Glombitza (✉), R. Kermer, and S. Reichel
G.E.O.S. Ingenieurgesellschaft mbH, Halsbrücke, Germany
e-mail: f.glombitza@geosfreiberg.de; r.kermer@geosfreiberg.de; s.reichel@geosfreiberg.de

compounds and chemicals. Each section starts with an overview of the relevant reactions and an explanation of the reaction condition. Afterward information about applications and different technological processes as well as sustainability aspects are provided.

**Graphical Abstract**

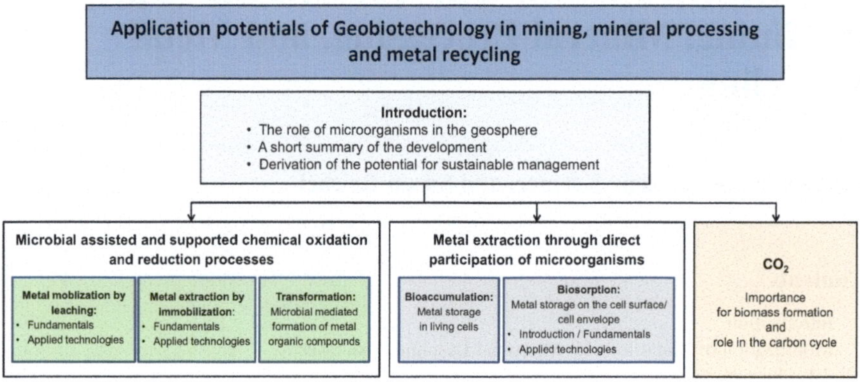


**Keywords** Bacterial leaching, Bioaccumulation, Biosorption, Geobiotechnology, Metal extraction, Mine water treatment

## Abbreviations

| | |
|---|---|
| AMT | Advanced mineral technology |
| ATP | Adenosine triphosphate |
| BCF | Biological concentration factors |
| Da | Dalton (unit for the mass of molecules) |
| E | Energy |
| $H^+$ | Hydrogen ion |
| hv | Symbol for sunlight energy |
| L | Organic ligands |
| Me | Metal |
| µM | Micromole (unit for molarity) |
| MO | Microorganisms |
| NS | Nutrient salts (mineral nutrients) |
| ppm | Parts per million |
| REE | Rare earth element |

# 1 Introduction

During the earth's development, microorganisms (MO) adapted to different ecosystems by means of evolution, influenced their states, and determined their further development. As a result, they participate in numerous natural reactions today leading to mobilization and immobilization of elements and compounds in the geosphere. These reactions represent essential components of the closed cycles of matter in nature which, however, are far away from being fully understood in their complexity until now, as recently demonstrated [1].

The identification and elucidation of these reactions and the underlying processes as well as the analysis of the involved microorganisms are essential components of the scientific field of geomicrobiology. The technologies which were developed and applied based on this knowledge are subject of the field of geobiotechnology offering manifold possibilities for the extraction of metals from ores and its processing residues but also for metal refinement, purification, and further utilization options by special reactions or syntheses [2–6]. The geobiotechnology may be regarded as a part of the industrial biotechnology.

While past studies and considerations in this field were mainly focused on processes for the extraction of copper, zinc, cobalt, nickel, uranium, and gold from the ores, present research in particular concentrates on recycling processes which could notably contribute to sustainability and on the extraction of rare earth and trace elements like gallium, indium, or lithium [7].

In the past, huge amounts of (different) metal-bearing industrial residues were accumulated. These include (1) residues from smelters, normally as slags; (2) residues from ore and mineral processing plants, stored/dumped as flotation residues and tailings; (3) iron-containing red muds from aluminum production; (4) filtration residues from titan dioxide production; (5) sludges from galvanic/electrolysis plants; (6) ashes and residues from the different combustion processes in gas, oil, hard coal, and lignite-fired power plants and from municipal waste incineration plants; and (7) dusts from de-dusting and filtration plants. Furthermore the diverse residues from wastewater treatment and biogas-producing plants as well as sludges from rivers and harbors have to be added, which are mainly oxides and hydroxides, phosphates, carbonates, and silicates containing a number of essential metals and trace elements. The residues mentioned do not only have to be considered as relics of the past, because they are still generated and accumulated today by the diverse combustion, cleaning, and valuable substance-producing processes.

The latter residues still produced and accumulated today especially include sludges from drinking water and water treatment plants, ashes from the different combustion processes, and slags of running smelters.

In addition, there are an increasing number of goods which have to be recycled and reintroduced into the economic cycle comprising spent catalysts, electronic scrap, composite materials, and various residues from new and modern technologies such as products from the photovoltaic and the chip industry. In contrast, the residues from other processes such as television- and cathode-ray tube-producing factories are decreasing or even do no longer accumulate.

The concentrations of valuable substances in the residues often are much higher than the corresponding concentrations in the existing ore bodies or can be increased to such magnitudes by means of biotechnological enrichment or concentration processes. For example, it is possible to achieve concentrations of 140 mg U/g biomass, 170 mg Cu/g biomass, or 100 mg REE/g biomass by such processes [8–10]. Political decisions, in particular those targeting the closure of material cycles, play a crucial role in accelerating and/or influencing the development of utilization processes [11, 12].

This contribution aims to demonstrate the manifold possibilities and starting points for the application of geobiotechnological processes. Besides the variety of metal extraction processes, the formation of $CO_2$ and its reduction to methane, both related to processes of the carbon cycle, are mentioned within this chapter.

The contribution also comprises considerations on sustainability of the applied technologies. According to its classical sense termed by Carlowitz, sustainability means that the production, application, and consumption of goods from a specific amount of material should happen in the same amount of time required for the growth or renewal of this material amount [13]. This definition, however, cannot be used for ores, metals, or similar raw materials. Here, the processes of production, application, and consumption vs. the renewal and growth have to be considered in a different way. For ore and metal-like raw materials, the term sustainability is usually connected with the idea of environmental protection targeting specific goals. For instance, a technology relating to ore and metal-like raw material processing should be operable at lower costs; consume less energy and/or water; accumulate less residues, etc.; and form closed cycles during its later operation. The latter is of particular importance for sustainability, as the production and the recycling/renewal of a specific metal will happen by different processes. For example, the production of a metal happens by a specific ore leaching process, while after application and consumption of the same, other technologies and processes are necessary for recycling/renewal of the metal or ore from the residues, wastes, or by-products. Only with such a complexity of different reactions closed material cycles and sustainability can be fulfilled.

## 2 Mobilization of Metal Ions by Microbial Leaching Processes

### *2.1 Introduction*

The fundamentals of all reactions carried out by microorganisms rely on anabolic and catabolic processes, which provide, firstly, the energy needed for growth and secondly, the building blocks required for reproduction. The provided energy is used in a subsequent step to perform various cell substance syntheses. Therefore, these reactions are always connected to an energy-consuming and an energy-delivering

$$\text{microorganisms} + e^-(\text{donor}) - e^-(\text{acceptor}) + C + O_2 + H_2 + \text{nutrients} \longrightarrow \text{cell substances (biomass)} + \text{primary metabolites} + \text{secondary metabolites}$$

**Fig. 1** Representation of a schematic cell growing process

process. The energy delivery takes place by an oxidation process during which a donor provides energy in the form of electrons. A second adequate process must exist in which these electrons are taken up from an acceptor. The total process is described by the following schematic equation (Fig. 1).

Besides energy, microorganisms also require a carbon source (C), oxygen ($O_2$), and hydrogen ($H_2$). Nutrients like nitrogen, potassium, magnesium, phosphorous, and sulfur are additionally required and used for the formation of the cell substances (biomass). The occurring metabolites are side products of these processes. For instance, $CO_2$ and fatty acids are formed as primary metabolites, while amino acids and/or other organic substances are produced as secondary metabolites. All components of these reactions are potential starting points for geobiotechnological processes.

## *2.2 Fundamentals of Microbial Reactions Important for Leaching Processes*

Microbial leaching processes are based on the ability of microorganisms to dissolve a solid substance. First, this can be accomplished by a growth connected oxidation or reduction process resulting in a specific modification of the solid's components and finally to an increase of its solubility. A second microbial strategy involves the formation of side products which can dissolve the solid substance or change its solubility in a second reaction [14].

Reactions according to the first strategy are known for many years and are denoted as autotrophic leaching processes because the microorganisms are able to use $CO_2$ present in the air as carbon source [15].

Leaching processes relying on a reaction product are usually heterotrophic processes, because the microorganisms utilize an organic carbon source for formation of the side products and for growth [16].

Another approach for classification of the mineral dissolution is based on analysis of the leaching processes and the occurring reactions. Accordingly, these are denoted redoxolyses, acidolyses, or complexolyses [17].

Important oxidation reactions are:

The oxidation of sulfides with the final products sulfur or sulfate:

$$S^{2-} \rightarrow S^{\circ} + 2e^- \tag{1}$$

$$S^{2-} + 4H_2O \rightarrow SO_4{}^{2-} + 8H^+ + 8e^- \quad (2)$$

The oxidation of iron 2+ ions:

$$Fe^{2+} \rightarrow Fe^{3+} + e^- \quad (3)$$

The oxidation of chromium 3+ ions:

$$Cr^{3+} \rightarrow Cr^{6+} + 3e^- \quad (4)$$

The oxidation of uranium ions:

$$U^{4+} \rightarrow U^{6+} + 2e^- \quad (5)$$

The oxidation of manganese ions:

$$Mn^{2+} \rightarrow Mn^{4+} + 2e^- \quad (6)$$

But also the nitrification:

$$N^{3-}\,H_4{}^+ + 2.0\,O_2 \rightarrow N^{5+}\,O_3{}^- + 2H^+ + H_2O + 8e^- \quad (7)$$

Numerous oxidation reactions exist besides these processes, but they have minor importance for leaching processes so far.

Part of the reactions taking place during these processes as well as the participating enzyme systems, especially those for the dissolution of sulfides, have already been subject of continuous and comprehensive research [18–20]. They allowed elucidating the interaction between mineral and microorganisms and the involved metabolic pathways, which can lead to the formation of sulfuric acid in some cases [2, 21, 22]. Accordingly, the oxidation of metal sulfides is known to take place after two different mechanisms, the thiosulfate mechanism and the polysulfate mechanism [20].

Besides oxidation processes carried out by microorganisms, also chemical processes between minerals and formed reaction products exist. An important example is the oxidation of sulfide by microbially formed iron III ions. So, the reactions taking place during leaching processes are a mixture of chemical and microbiological processes.

The dissolution of pyrite is an appropriate example to illustrate the described correlations between microbial and chemical processes.

During the first step, sulfidic sulfur is oxidized to $SO_4{}^{2-}$ by the existing microorganisms which are usually attached to the mineral surface.

$$FeS_2 + O_2 + H_2O \rightarrow Fe^{2+} + 2H^+ + 2SO_4{}^{2-} \tag{8}$$

In the second step, oxidation of the previously formed $Fe^{2+}$ to $Fe^{3+}$ is performed by iron-oxidizing microorganisms.

$$2Fe^{2+} + 0.5\ O_2 + 2H^+ \rightarrow 2Fe^{3+} + H_2O \tag{9}$$

The produced $Fe^{3+}$ ions act as oxidizing agents and accelerate the further dissolution of pyrite.

$$14Fe^{3+} + FeS_2 + H_2O \rightarrow 15Fe^{2+} + 16H^+ + 2SO_4{}^{2-} \tag{10}$$

Subsequent oxidation of $Fe^{2+}$ to $Fe^{3+}$ ions takes place by microorganisms again.

According to the latter reactions, it is obvious that the presence of iron plays an important role in leaching processes. Iron-free minerals such as zinc blende (sphalerite) or galenite are not able to form sulfuric acid. In such a case, the presence of microbially oxidizable sulfide is needed, e.g., pyrite, $FeS_2$. When no sulfide is present, as in case of pure oxides, the leaching process can be initialized by adding sulfur, iron, or pyrite.

Reduction processes usable for leaching are:

The reduction of iron 3+ ions:

$$Fe^{3+} + e^- \rightarrow Fe^{2+} \tag{11}$$

The reduction of sulfate anions, when the release of barium, radium, or lead is targeted and they are bound as sulfates:

$$SO_4{}^{2-} + 8e^- + 4H_2O \rightarrow S^{2-} + 8OH^- \tag{12}$$

The reduction of manganese 4 ions:

$$Mn^{4+} + 2e^- \rightarrow Mn^{2+} \tag{13}$$

The reduction of $As^{5+}$ to $As^{3+}$, which in many cases is an undesirable process due to the higher solubility of $As^{3+}$ compounds, but which at the same time is necessary for the leaching of arsenic-containing minerals:

$$As^{5+} + 2e^- \rightarrow As^{3+} \tag{14}$$

Important leaching processes carried out by heterotrophic microorganisms rely on the formation of organic acids from an organic carbon source through blocking of the successive metabolic pathways and prevention of a further degradation after primary oxidation.

$$C_nH_pO_x + e^- \underset{MO}{\rightarrow} C_nH_pO_x^- + H^+ \tag{15}$$

Such acid formation processes are:

- The formation of citric acid and/or isocitric acid [23]
- The oxidation of glucose and formation of gluconic acid [24]

The process of gluconic acid formation has been achieved by cultivation of methanol-utilizing bacteria, which were able to produce gluconic acid from glucose in a nitrogen-free medium. The formed gluconic acid is able to dissolve oxidic and siliceous ores and slags, if its production takes place in the presence of the minerals. An example is the release of rare earth elements during leaching of zirconium or slags of phosphorous production [25–27].

During this process gluconic acid forms a hardly soluble gluconate with the calcium ions which enables the subsequent separation of the rare earth elements from the leach solution [28].

- The formation of fatty acids [29] and amino acids [30] based on oxidation of $n^-$ paraffins and iso-alkenes

In particular, studies with fatty acids were conducted to leach non-sulfidic ores with high carbonate content containing nickel or copper. However, it was not possible to carry out this process in a bigger scale outside the laboratory so far [31, 32].

Because these processes are often connected with the formation of metal organic compounds, they could be assigned to the reaction class of complexolyses taking place after the following general equation:

$$Me^+(\text{mineral}) + H^+L^- \rightarrow H^+(\text{mineral}) + LMe \tag{16}$$

with

$H^+$ Hydrogen ion

L Abbreviation for organic ligands capable of forming a complex

$L^-$ Anion of organic compounds, Me Abbreviation for metal

Such a complex can be electrically charged or have an electro neutral character as in case of chelating agents.

Another technology which is generally important but hardly utilized by industry so far involves leaching processes based on further complex formations. Among these are, first, the formation and application of siderophores and, second, the microbial formation of cyanide as a ligand for complex formation [17, 33–36].

Siderophores are substances which are produced by a cell and delivered into the environment with the aims to bind iron and to transport it through the cell membrane into the cell. They are produced only by bacteria, fungi, and different plant roots. About 200 natural siderophores are known, which can be found in soil or in seawater partially at high concentrations. The reuptake of the loaded siderophores into the cell is carried out more or less specifically through special channels in the cell membrane and by energy consumption in the form of ATP. Many kinds of bacterial strains are able to take up a broad range of siderophores, also such not formed by themselves.

Because iron exists only in hardly soluble compounds, siderophores have to possess an extremely high complex formation constant, which is achieved by their characteristic to be a strong chelating ligand.

The known siderophores produced by microorganisms are divided into three groups – the catecholates, the hydroxamates, and the α-hydroxy/keto carboxylates [37, 38]. The most important ones are enterobactin or enterochelin, a derivate of catechol; aerobactin, a derivate of hydroxam; and rhodotorulic acid, also a derivate of hydroxam [2]. An amount of 1 mol with the molar weight of 300–2,000 Da is able to bind 1–2 mol of iron, corresponding to 56–112 g of iron.

The produced amounts of siderophore per 1 g of microorganism are quite differing. For example, *Staphylococcus* sp. produce siderophore amounts of 3–383 μM, relating to a maximum of 2 mg/g biomass. This amount is able to bind 0.1 mg Fe and 1 g of siderophore can bind 50 mg Fe, respectively.

Hydrocyanic acid or cyanides represent exceptionally reactive ligands for complex formation. Cyanide forms stable water-soluble complexes with different metals, which even can be chromatographically separated [33, 34].

A number of bacteria (in particular strains of *Pseudomonas*) are able to form hydrocyanic acid for which the amino acid glycine is used as immediate starting product. In contrast to acidophilic microorganisms, cyanide-forming bacteria are particularly active in the alkaline region (pH 8–9), therefore allowing the leaching and separation of metals at neutral or alkaline pH [17].

## 2.3 Overview of Microbial Leaching Applications

Today a variety of materials is subjected to leaching processes including ores, technically unusable low-grade ores, residues of the mineral processing, sludges, sediments, and different industrial wastes like dusts, slags, catalysts, and electronic scrap.

The technological design and implementation of leaching processes were and will be influenced and conducted by economical aspects and are mainly determined and controlled by the market price of the metal(s) of interest. Leaching processes are carried out as tank leaching, dump leaching, and heap leaching but also as in situ leaching and underground leaching [5, 39].

Besides pure economical aspects, the requirement of sustainability moved into focus during the last years due to political pressure by the government to establish a closed circular economy. This led to the integration of residues and reusable wastes into the consideration and to the development of suitable technologies. So, besides various metal-containing residues, sludges, ashes, and dusts were integrated into the processes for metal recovery/production.

Most importantly, these include the residues from old mining activities, ashes from waste incineration, sludges of former copper smelters as Theisen sludge, and tailings of flotation processes from tin ore processing.

The first microbial leaching processes were mentioned after World War II. In Bulgaria the leaching of the copper mining dumps of the mine Elshitza from the 1920s started. Furthermore, the extraction of uranium was initiated by means of dump leaching, underground leaching, and different in situ leaching processes. Roughly since the 1970s, various leaching processes were established in the USA, in South America especially in Chile, in Canada, in Australia, in the former Soviet Union, and in the former German Democratic Republic to name only the most important countries [40–42].

The applied technologies were adapted to the minerals to leach and are continuously advanced to date. Parts of these technologies are largely to be considered as standard processes, while newer technologies still wait for their industrial application.

The most important processes include:

- Leaching of copper-containing sulfides with aerobic processes and with sulfide- and iron-oxidizing microorganisms [43]
- Leaching of iron-rich and low sulfide copper ores by means of a redox potential controlled process for minimizing undesirable iron precipitations [44]
- Leaching of iron-rich and low sulfide-containing nickel ores (laterite) by iron-reducing microorganisms for the dissolution of limonite or goethite with simultaneous adjustment of an acidic pH value through oxidation of added sulfur under anaerobic conditions [45]
- Leaching of oxide ores, in particular rare earth element-containing materials, with microbially formed organic acids like gluconic acid, e.g., leaching of residues from the phosphor production [46]
- Leaching of uranium 4 containing minerals like pitchblende through leaching with $Fe^{3+}$-containing sulfuric acid solutions and regeneration of the leaching solution by oxidation of the formed $Fe^{2+}$ ions in a separate process [47]
- Leaching of sulfide-containing spent catalysts for nickel and molybdenum recovery [48]

# 3 The Separation of Metal Ions from Liquids by Immobilization

## *3.1 Introduction*

Metal extraction by immobilization is based on the formation of hardly soluble metal compounds, which can be separated from the liquid phase afterward.

Different (microbial) leaching processes can potentially be utilized for metal extraction. However, at the same time, suitable technologies for separation of the dissolved metals from the produced leach liquids are also required.

Besides treatment of process water, a treatment of natural water, such as drainage water, groundwater, or surface water, is needed if they contain either valuable metals or toxic substances.

Applicable technologies are manifold and among other things include:

- Physical technologies, e.g., different membrane, ion exchange, and evaporate processes
- Chemical technologies, e.g., precipitations accomplished through solubility changes, oxidation and reduction processes, as well as solvent extraction techniques and electrochemical methods such as electrolyses
- Processes which are induced and carried out by microorganisms during their life and growth processes and which are parts of the natural elemental cycles

## 3.2 *Microbial Reactions Relevant for Metal Precipitation*

Several microbial reactions can be applied for metal extraction by immobilization. First, reduction processes can be used, such as the reduction of sulfate to sulfide leading to the formation of hardly soluble metal sulfides.

$$SO_4{}^{2-} + Me^{(+n)} + 4H_2O \rightarrow S^{2-} \rightarrow Me\,S_n + 8OH^- \quad (17)$$

All these processes usually take place at neutral or slightly alkaline pH ranges, but there is also a sulfate reduction occurring at acidic pH conditions which is intensively studied in recent years [49, 50]. For the formation of metal sulfides, however, this type of sulfate reduction currently has no importance due to the considerably lower solubility of the sulfide ions.

Another possibility for immobilization is based on the simultaneous microbial oxidation of iron 2+ to iron 3+ and of manganese 2+ to manganese 4+ ions. The generated iron 3+ is able to precipitate and can be separated as hydroxysulfates, usually as schwertmannite or jarosite, if an acidic pH exists and sulfate is present. Manganese precipitates as manganese dioxide.

$$\begin{aligned} Fe^{2+} - e^- + SO_4^{2-} + nH^+ \rightarrow {} & (Fe^{3+})_3O_8(OH)_x(SO_4)_y \\ & x = 3.5 - 6.0 \\ & y = 1 - 1.5 \\ & \text{(schwertmannite)} \end{aligned} \quad (18)$$

$$\begin{aligned} & KFe_3\left[(OH)_6(SO_4)_2\right] \\ & \text{(jarosite)} \end{aligned} \quad (19)$$

$$Mn^{2+} - 2e + O_2 \rightarrow MnO_2 \quad (20)$$

## *3.3 Application Examples Using Microbially Mediated Metal Precipitation*

Water treatment processes in which a sulfate reduction takes place are suitable examples for the application of immobilization [51, 52]. The company Paques developed different variants of such processes and implemented them for different treatment purposes. In a separate process, sulfide is generated from sulfur or sulfate and subsequently used for the separation of metals [53, 54]. The metals are precipitated as sulfides with the aid of the microbially produced sulfide. An overview of the process is shown in Fig. 2.

The advantages of this technology comprise a smaller sludge volume, a decrease of the sulfate concentration, and an increasing of the pH value.

This so-called Thiopaq®–Thiomet® technology was used, among others, in the Caribou Mine located in New Brunswick in Canada for the selective recovery of Cu and Zn and the separation of Cd and Pb from acidic drainage water with a flow rate of 700 $m^3$/day [55, 56]. Some published data of this process is shown in Fig. 3.

The separation of iron as hydroxysulfate usually takes place in an iron-containing water at a pH value of 2.5–3.5 after microbial oxidation of $Fe^{2+}$ to $Fe^{3+}$ and in the presence of sulfate. The so-formed crystals are shown in Fig. 4. If a carrier material is available in the water, deposits as shown in Fig. 5 are formed.

This process is used for the preparation of particles from schwertmannite which are applied in the paint industry as well as for water cleaning technologies based on the particles' special adsorption capacity [57, 58]. It is also applied for the separation of excess iron from leaching solutions after cementation processes, used for copper recovery with iron scrap, and a further microbial oxidation of the formed $Fe^{2+}$, which is known under the name BACFOX process [59].

Due to the presence of polar groups, the formed schwertmannite particles have a high adsorption capacity. So, they are well suited for the separation of cations from water [60].

This application has been successfully established and tested for the removal of arsenic from different mine waters [61].

# 4 Biosorption and Bioaccumulation

## *4.1 Introduction*

Microorganisms are able to store large amounts of different cations which are fixed on the cell surface or stored inside the cell. This behavior in some cases is simply caused by the structure of the cell wall consisting of different polar chemical groups. In other cases it can be explained by the microbe's attempt to prevent toxic concentrations inside the cell. In general, the metal storage processes are divided into bioaccumulation and biosorption. In both processes the stored metal

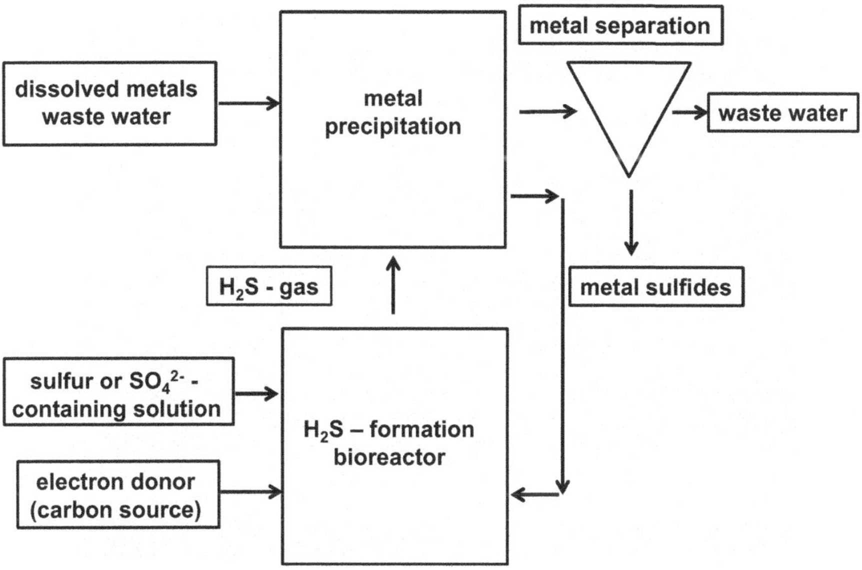


**Fig. 2** Scheme of the Paques Thiopaq®-Thiomet® process

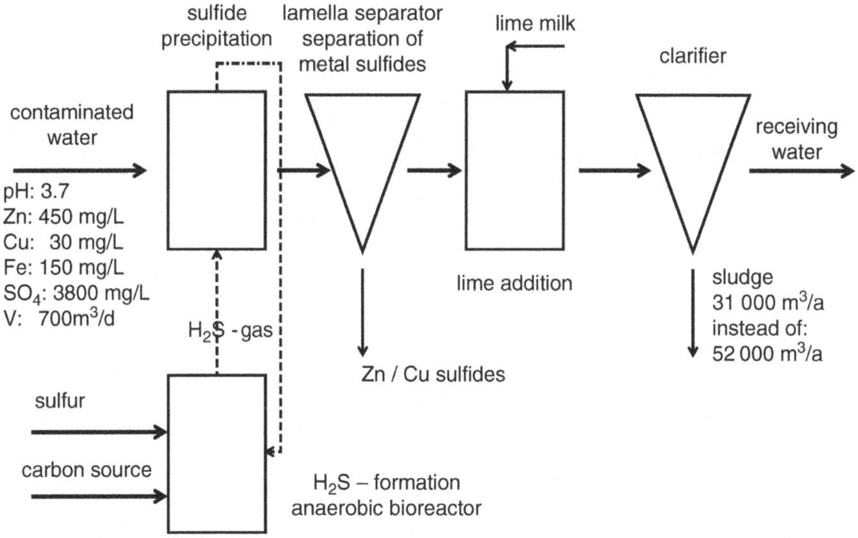


**Fig. 3** Flow diagram of the Paques Thiomet® process in the Caribou Mine

concentration is considerably higher than in the surrounding environment or medium when the concentration in the cell volume is compared with the concentration in the same volume of the surrounding liquid. Due to these different mechanisms, microorganisms are used as single or immobilized living or dead cells.

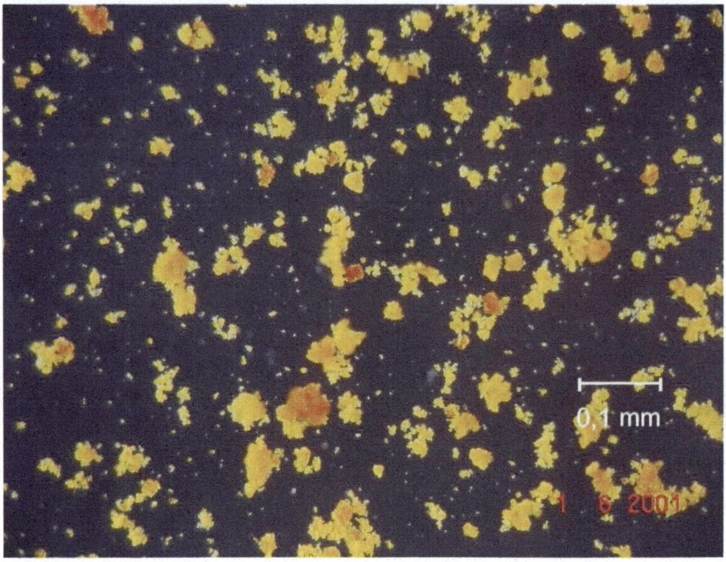


**Fig. 4** Flakes of schwertmannite in a mine water (photograph by G.E.O.S.)

## *4.2 Bioaccumulation*

Bioaccumulation is a process by which a substance, especially a metal ion, is taken up by a living cell and stored in its inside due to different reasons. The probably most famous example is the storage of iron by distinct microorganisms, which are known as magnetotactic bacteria and keep iron as magnetite in the inside of their cells [62]. Besides iron the storage of manganese is also known.

Such metal storage processes usually take place in municipal wastewater treatment plants in which activated sludge (comprising a huge number of microbes and different microbial communities) collects and stores metal ions [63]. A relatively new and thus unknown process of mineral formation is the synthesis or formation of nanoparticles, respectively [64, 65].

## *4.3 Biosorption*

### 4.3.1 Fundamentals

The term biosorption is used, if microorganisms, as whole cells or as cell components, are used for the binding and storage of elements or metals, respectively. The occurring processes are similar to the processes taking place at ion exchange resins.

The cell wall of microorganisms contains a set of functional groups, such as the $OH^-$ group, the COOH group, the $PO_4^{3-}$ group, or the $SH^-$ group. These groups are

**Fig. 5** Deposits of schwertmannite on a carrier material (photograph by G.E.O.S.)

able to dissociate at a distinct pH value and form electrically charged particles. In addition, they can already generate negatively or positively charged zones on the cell wall due to their existing charge. Metal cations or anions are able to attach to such polar groups and so can be removed from a distinct water or aqueous solution.

The achievable concentrations depend on various parameters involving the process conditions, especially the pH value, the respective metal concentration in the medium, the biomass concentration, and the kind of process management. In general, the concentrations achievable by biosorptive processes are much higher than concentrations in the ambient water or medium. Table 1 shows the so-called BCF values indicating the concentration factors that are attainable through the (biosorptive) storage.

More recent research is based on the knowledge that the storage predominantly takes place at the functional groups of the cell wall proteins and that the preparation of these proteins and their application in biosorption processes lead to remarkably higher enrichment levels [66, 67].

**Table 1** Compilation of specifically stored concentrations (BCF values)

| Strain/DSM – number | Zn | Cd | Hg | Cu |
|---|---|---|---|---|
| *Bacillus cereus* 2301 | 2,220 | 3,180 | 111,000 | 51,150 |
| *Bacillus subtilis* 387 | 1,160 | 1,920 | 102,000 | 8,760 |
| *Bacillus subtilis* 402 | 4,830 | 7,840 | 77,400 | 2,300 |
| *Bacillus subtilis* 2109 | 19,500 | 11,500 | 119,000 | 880 |
| *Citrobacter freundii* 30047 | 1,500 | 1,630 | 25,700 | 18,100 |
| *Escherichia coli* 498 | 1,120 | 3,120 | 48,500 | 5,670 |
| *Micrococcus luteus* 1790 | 2,710 | 3,550 | 46,300 | 6,200 |
| *Pseudomonas putida* 291 | 3,240 | 3,760 | 138,000 | 15,000 |
| *Saccharomyces cerevisiae* | 700 | 190 | 8,700 | 2,350 |
| *Rhodotorula putidis* | 210 | 120 | 3,100 | 3,460 |
| *Hansenula anomala* | 500 | 700 | 16,400 | 500 |

### 4.3.2 Processes and Technologies Using Biosorption

The development and application of biosorption processes is strongly connected with the prevailing economic situation. In the 1980s the development concentrated on exploitation of uranium, platinum group elements, and precious metals. The AMT process developed by Brierley for the recovery of metals by using *Bacillus subtilis* biomass probably was the first process applying biosorption [68, 69]. Following research focused on the testing of various technologies for the pretreatment of the used biomass and on analysis of the storage capability of the different microbial strains and biomasses [70].

The pretreatment consisted of spray drying, immobilization with alginate, sulfone, or another suitable substance for immobilization [71, 72]. The aims were to achieve a high loading capacity at long lifetimes and to allow a reutilization after regeneration by means of desorption. Besides the aspect of an economical metal production, the aspect of environmental protection was more and more taken into account due to the separation of harmful substances [73, 74]. So a process was developed by Tsezos for the separation of radionuclides based on *Rhizopus arrhizus* biomass and successfully applied [75] (Table 2).

Taking into consideration that microbial biomass is able to store and to enrich metals, processes in wetlands and biological water treatment processes are of particular importance, because they already provide an enrichment of metals taking place as a side process. Accordingly, a biomass from a wastewater treatment plant can already contain considerable metal amounts which even can be significantly increased by a methane formation process and the connected mass reduction [88]. In Table 3 the metal contents in the ash of a biomass from a municipal wastewater treatment plant are shown [89].

The processes of storage and mineralization of organic substances in a wetland can lead to the formation of new ore bodies to a certain extent, and, even more importantly, clean water is obtained by this kind of treatment; both are the reasons for constructing artificial wetlands by mining companies [90]. Depending on the

**Table 2** Compilation of selected, tested, and applied processes

| Name | Process | Principle | Quote |
|---|---|---|---|
| AMT–BIOCLAIM™ | Granulated biomass with a diameter of about 0.1 mm for heavy metal separation from wastewaters | Application of *Bacillus subtilis* after alkaline treatment | Brierley et al. [68, 69] |
| AlgaSORB™ | The material was well suitable for heavy metal separation in diluted waters with low concentrations in a range of 1–100 mg/L | Biomass of *Chlorella vulgaris* immobilized on a silica gel matrix | Developed by the Bio-Recovery Systems Inc. [73, 76, 77] |
| BIO-FIX™ | Treatment of industrial wastewaters and mine drainage waters for heavy metal separation | A mixture of cyanobacteria, yeasts, and plants immobilized in a porous organic matrix, immobilization with polysulfon, polyethylene, or polypropylene for the formation of porous balls | Jeffers et al. [78] |
| METEX® | The METEX® – Process was developed for heavy metal elimination and extraction from industrial wastewaters | The technology was applied with a microbial population from an activated sludge under anaerobic conditions in a countercurrent reactor with a conical cylinder | Comys [79], Fürst and Morper [80], and Morper [81] |
| WISMUT | Radionuclide separation, especially uranium and radium | Application of a granulated and spray-dried biomass of a methylotrophic bacterium MB 127 | Glombitza et al. [82], Glombitza and Eckard [83], and Glombitza and Eckardt [84] |
| BV-SORBEX™ | Metal separation from diluted and highly concentrated waters with high efficiency (>99%) | A granulated powder is used as biosorbent consisting of different algae (*S. natans*, *A. nodosum*, *H. opuntia*, *P. pamata*, *C. crispus*, *C. vulgaris*) | Volesky [85] |
| MetaGeneR and RAHCO Bio-Beads | Metal separation from electroplating wastewaters and from mine waters | Two commercial and highly effective biosorption processes | Atkinson et al. [86] and Chojnacka [87] |

kind of metals and their concentrations, wetlands can serve as a kind of ore-forming plants in which microorganisms are used for the separation of metals under oxidizing and reducing conditions [91]. The accompanying organic substances are always degraded and transformed to $CO_2$ after metal uptake took place. The metals are stored in the residues as oxides, sulfides, hydroxides, or carbonates.

**Table 3** Metal contents in the ash of an activated sludge from a municipal wastewater treatment plant

| Compound/ element | Concentration (%) | Compound/ element | Concentration (ppm) | Compound/ element | Concentration (ppm) |
|---|---|---|---|---|---|
| $P_2O_5$ | 20 | Fe | 48,000 | Ni | 1,900 |
| $NO_3$ | 5 | Al | 31,600 | Pb | 1,450 |
| $SiO_2$ | 7 | Zn | 11,400 | Mn | 472 |
| Ca | 3.4 | Co | 10,900 | Ga | 300 |
| Mg | 0.5 | Cr | 2,180 | Cd | 190 |
| | | Ag | 600 | Au | 30 |
| | | Pd | 4 | Pt | 1 |

**Table 4** Metal(loid) organic compounds from different sources

| Waste disposal | Sewage sludge | Contaminated soil |
|---|---|---|
| $(CH_3)_2AsH$ | $(CH_3)AsH_2$ | $(CH_3)AsH_2$ |
| $(CH_3)_3As$ | $(CH_3)_2AsH$ | $(CH_3)_2AsH$ |
| $(CH_3)_2AsC_2H_5$ | $(CH_3)_3As$ | $(CH_3)_3As$ |
| $(CH_3)_3Sb$ | $(CH_3)_3Sb$ | $(CH_3)_2BiH$ |
| $(CH_3)_3Bi$ | $(CH_3)_3Bi$ | $(CH_3)_3Bi$ |
| $(CH_3)_2Te$ | $(CH_3)_2Te$ | $(CH_3)_2Se$ |
| $(CH_3)_2Hg$ | $(CH_3)_4Sn$ | $(CH_3)_2Se_2$ |
| $(CH_3)_4Sn$ | | $(CH_3)_2Te$ |
| $(CH_3)_4Pb$ | | $(CH_3)_2Te$ |
| | | $(CH_3)_2Hg$ |
| | | $(CH_3)_4Sn$ |
| | | $(CH_3)_4Pb$ |

## 5 Transformations

Transformations are reactions, by which metals or metalloids are combined with organic molecules. The resulting compounds have totally different properties in comparison to the starting substances. The best known reactions are methylation and ethylation of metals and metalloids like arsenic and selenium [92, 93], during which methyl and ethyl groups are transferred under anaerobic conditions.

Table 4 contains a compilation of different methylated and ethylated compounds which were identified in different habitats and can be attributed to microbial activity [94].

In case of mercury, these are undesirable processes because the formation of methyl mercury or ethyl mercury leads to volatile and toxic products [93]. In contrast, methylation of a series of metals seems to open new perspectives for their extraction and recovery. This especially applies to tin because tin extraction from oxidic ores is only possible by expensive thermal reducing processes.

# 6 $CO_2$ as an Important Component in Carbon Cycle

$CO_2$ exhibits a particular importance due to its influence on the climate. Closed process cycles in which $CO_2$ is transformed and utilized and thereby removed again are of remarkable importance and interest [95, 96]. Two processes are available for microbial $CO_2$ transformation and utilization.

The first one is related to lithoautotrophic processes for which $CO_2$ serves as the source of carbon and so enables microbial growth.

$$MO_1 + Fe^{2+} + CO_2 \rightarrow MO_{(1+n)} + Fe^{3+} \quad (21)$$

Microorganisms use the $CO_2$ and form new biomass. This is indicated in Eq. 21 by the term $MO_{(1+n)}$. One gram of dry biomass contains an average amount of about 450 mg carbon. Accordingly, an amount of ca. 1,650 mg $CO_2$ has to be consumed for the synthesis and production of 1 g of biomass. This is a minimum value because the degree of efficiency is always lower than 100%. Besides $Fe^{2+}$ another energy-supplying substrate can be used, e.g., $H_2$. This category of $CO_2$-utilizing processes also includes the formation of biomass through phototrophic processes as to be found during the cultivation of algae.

$$CO_2 + \text{energy} \underset{hv}{\rightarrow} \text{biomass} + O_2 \quad (22)$$

The second $CO_2$-utilizing process is the enzymatic reduction to $CH_4$.

$$CO_2 + 4H_2 + E \rightarrow CH_4 + 2H_2O \quad (23)$$

In this way the biological conversion of $CO_2$ to $CH_4$ in natural gas fields and coal mines takes place [97–99] and likewise the microbial synthesis of $CH_4$ from $CO_2$ and $H_2$ [100].

# 7 Summary

The contribution demonstrates the manifold possibilities of applying microbiological processes and their interactions with the abiotic geosphere for metal extraction technologies in mining, mineral processing, and recycling processes. Oxidation and reduction reactions are compiled in an overview first, and possible leaching processes are subsequently derived.

The use of microorganisms for the separation of metals from different kinds of water by means of precipitation reactions is shown in a further section. Next, metal separation processes with the aid of biosorption and bioaccumulation are considered, and the best-known applications are reported.

A short explanation of microbial metal transformations for the formation of metal organic compounds as well as the representation of $CO_2$ application and the resulting technological options complete the contribution.

By applying these processes, it is possible to decrease the costs for the production of metals, to increase the degree of their reutilization, and to enlarge the amount of potentially usable resources.

## References


1. Müller H, Meckenstock RU (2017) Kabelbakterien ermöglichen neue geochemische Prozesse. BIOspektrum 23(4):388–390
2. Ehrlich HL (1996) Geomicrobiology, 3rd edn. Marcel Dekker, New York
3. Karavaiko GI, Groudev SN (1985) Biogeotechnology of metals. In: Proceedings of the international seminar and training course "modern aspects of microbiological hydrometallurgy and training course on microbiological leaching of metals from ores", 24 May–25 June 1982, Moscow – Sophia. Centre of International Projects GKNT, Moscow
4. Karavaiko GI, Rossi G, Avakyan ZA (1990) Biohydrometallurgy. In: International seminar on dump and underground bacterial leaching metals from ores. Centre of International Projects, USSR State Committee for Environment Protection, Moscow
5. Rossi G (1990) Biohydrometallurgy. McGraw-Hill, Hamburg
6. Torma AE, Bosecker K (1982) Bacterial leaching. Prog Ind Microbiol 16:77–118
7. Ali SH, Giurco D, Arndt N, Nickless E, Brown G, Demetriades A, Durrheim R, Enriquez MA, Kinnaird J, Littleboy A, Meinert LD, Oberhänsli R, Salem J, Schodde R, Schneider G, Vidal O, Yakovleva N (2017) Mineral supply for sustainable development requires resource governance. Nature 543:367–372
8. Brierley CL, Brierley JA, Davidson MS (1989) Applied microbial processes for metal recovery and removal from waste water. In: Beveridge TJ, Doyle RJ (eds) Metal ions and bacteria. Wiley, New York, pp 359–382
9. Bullmann M, Iske U, Glombitza F, Dittrich B (1987) Wechselwirkungen zwischen seltenen Erdmetallen und Mikroorganismen. Acta Biotechnol 7:409–412
10. Norberg A (1983) Production of extracellular polysaccharide by Zoogloea ramigera and its use as an adsorbing agent for heavy metals. Doctoral dissertation, LUTKDH (CTKMB-1003)/131 Lund University, Lund
11. Bundesministerium für Justiz und Verbraucherschutz (2012) Gesetz zur Förderung der Kreislaufwirtschaft und Sicherung der umweltverträglichen Bewirtschaftung von Abfällen (Kreislaufwirtschaftsgesetz – KrWG) 24 Feb 2012
12. Europäische Kommission Brüssel (2011) Communication from the Commission to the European Parliament, the Council, the European Economic and Social Committee and the Committee of the Regions, closing the loop – an EU Action Plan for the Circular Economy, COM (2015) 614/2
13. Sächsische Hans-Carl-von-Carlowitz-Gesellschaft (ed) (2013) Die Erfindung der Nachhaltigkeit. Oekom Verlag, Munich
14. Corrans IJ, Harris B, Ralph BJ (1972) Bacterial leaching: an introduction to its application and theory and a study on its mechanism of operation. J South Afr Inst Min Metall 72:221
15. Boujard D, Anselme B, Cullin C, Raguénès-Nicol C (2014) Zell- und Molekularbiologie im Überblick: Kapitel Energetische Vorgänge in der Zelle. Springer Spektrum, Berlin
16. Fuchs G, Eitinger T, Schlegel H-G (2007) Allgemeine Mikrobiologie. Georg Thieme, p 158ff

17. Brandl H, Lehmann S, Faramarzi MA, Martinelli D (2008) Biomobilization of silver, gold and platinum from solid waste materials by HCN-forming microorganisms. Hydrometallurgy 94:14–17
18. Bosecker K (1997) Bioleaching: metal solubilization by microorganisms. FEMS Microbiol 20:591–604
19. Sand W, Gehrke T, Hallmann R, Schippers A (1995) Sulfur chemistry, biofilm, and the (in) direct attack mechanism – a critical evaluation of bacterial leaching. Appl Microbiol Biotechnol 43:961–966
20. Vera M, Schippers A, Sand W (2013) Progress in bioleaching: fundamentals and mechanisms of bacterial metal sulfide oxidation – part A. Appl Microbiol Biotechnol 97:7529–7541
21. Rohwerder T, Sand W (2007) Mechanisms and biochemical fundamentals of bacterial metal sulfide oxidation. In: Donati ER, Sand W (eds) Microbial processing of metal sulfides. Springer, Amsterdam, pp 35–58
22. Schippers A, Glombitza F, Sand W (eds) (2014) Geobiotechnology I – metal related issues. Advances in biochemical engineering/biotechnology, vol 141. Springer, Heidelberg
23. Stottmeister U, Behrens U, Weissbrodt E, Barth G, Franke-Rinker D, Schulze E (1982) Nutzung von Paraffinen und anderen Nichtkohlenhydrat-Kohlenstoffquellen zur mikrobiellen Citronensäuresynthese. Z Allg Mikrobiol 22:399–424
24. Uhlig H, Karbaum K, Steudel A (1986) Acetobacter methanolicus sp. nov. an acidophilic facultatively methylotrophic bacterium. Int J Syst Bacteriol 36(2):317–322
25. Iske U, Bullmann M, Glombitza F, Babel W, Miethe D, Dietze H-J, Becker S (1985) Verfahren zur mikrobiellen Laugung von Seltene Erdmetalle enthaltenden Phosphatmineralien und Abprodukten, Patent DD 249 156 A3, 04.04.1985, 02.09.1987
26. Iske U, Bullmann M, Dietze H-J, Becker S (1985) Verfahren zum mikrobiellen Aufschluss metallhaltiger silikatischer und/oder oxidischer Rohstoffe hoher chemischer Resistenz, Patent DD 250 235 A3, 04.04.1985, 08.10.1987
27. Iske U, Bullmann M, Glombitza F (1987) Organoheterotrophe laugung resistenter materialien. Acta Biotechnol 7:401–407
28. Dittrich B, Bullmann M, Iske U, Glombitza F (1987) Verfahren zur Abtrennung von Seltenen Erdelementen aus calciumhaltigen Fermentationslösungen, Patent DD 259 212 A1, 30.03.1987; 17.08.1988
29. Stichel E (1978) Untersuchungen zum Einfluss individueller *n* – Alkane auf den Prozess der Erdölkonvertierung durch Hefen. Dissertation, Akademie der Wissenschaften, Institut für technische Chemie, Leipzig
30. Hermann T (2003) Industrial production of amino acids by coryneform bacteria. J Biotechnol 104:155–172
31. Coto O, Galizia F, Hernandez I, Marrero J, Donati E (2008) Cobalt and nickel recoveries from lateritic tailings by organic and inorganic bio-acids. Hydrometallurgy 94:18–22
32. Karsson S, Sjöberg V, Grandin A (2011) Heterotrophic leaching of LD slag – formation of organic ligands. In: Rüde T, Freund A, Wolkersdorfer C (eds) IMWA 2011 – Aachen, mine water – managing the challenges, pp 371–374
33. Brandl H, Stagars M, Faramarzi MA (2003) A novel type of microbial metal mobilization: cyanogenic bacteria and fungi solubilize metals as cyanide complexes. In: Tsezos M, Hatzikioseyian A, Remoundaki E (eds) Proceedings of the 15th international symposium on biohydrometallurgy IBS, 14–19 Sept 2003, Athens. Biohydrometallurgy: a sustainable technology in evolution. National Technical University of Athens, Electronic means of production – N. J. Zographakis, pp 457–463
34. Faramarzi MA, Stagars M, Pensini E, Krebs W, Brandl H (2004) Metal solubilization from metal-containing solid materials by cyanogenic Chromobacterium violaceum. J Biotechnol 113:321–326
35. Miethke M, Marahiel MA (2007) Siderophore-based iron acquisition and pathogen control. Microbiol Mol Biol Rev 71:413–451

36. Snow GA (1970) Mycobactins: iron-chelating growth factors from mycobacteria. Bacteriol Rev 34(2):99–125
37. Müller-Graf K (2005) Siderophore, $Fe^{3+}$ Aufnahme durch Bakterien, Pilze und Pflanzenwurzeln. In: Vortrag im Rahmen des anorganisch chemischen Fortgeschrittenenpraktikums, 6 May 2005
38. Neilands JB (1995) Siderophores: structure and function of microbial iron transport compounds. J Biol Chem 270:26723–26726
39. Karavaiko GI, Torma AE (1985) Microbiological processes for the leaching of metals from ores – state of the art review. Centre of International Projects GKNT UNEP, Moscow
40. Donati ER, Sand W (2007) Microbial processing of metal sulfides. Springer, Dordrecht
41. Hedrich S, Schippers A (2017) Metallgewinnung mittels Geobiotechnologie. Chem Ing Tech 89:29–39
42. Johnson DB, Roberto FF (1997) In: Rawlings DE (ed) Biomining: theory, microbes and industrial processes. Springer, Berlin
43. Rawlings DE (2002) Heavy metal mining using microbes. Annu Rev Microbiol 56:65–91
44. Ranjbar M, Esmailbagi M, Schaffie M (2017) Bioelectrochemical leaching of copper sulfide minerals. In: Hedrich S, Rübberdt K, Glombitza F, Sand W, Schippers A, Vera Veliz M, Willscher S (eds) Proceedings of the 22nd international biohydrometallurgy symposium, vol 262. Trans Tech Publications, Solid State Phenomena, pp 16–19
45. Hallberg KB, Grail BM, du Plessis CA, Johnson DB (2011) Reductive dissolution of ferric iron minerals: a new approach for bio-processing nickel laterites. Miner Eng 24:620–624
46. Glombitza F, Iske U, Bullmann M (1988) Mikrobielle Laugung von Seltenen Erdmetallen und Spurenelementen. Bio-Engineering/Forschung und Praxis, München 4:37–43
47. Autorenkollektiv (1999) Chronik der Wismut copyright by Wismut GmbH, Teil 2 Beschreibung der Lagerstätten und der Betriebe Teil 2.2 Beschreibung der Gewinnungsbetriebe 2.2.11 Uranerzbergbau in der Lagerstätte Königstein, 2.2.14 Uranerzbergbau in der Lagerstätte Ronneburg
48. Iske U, Glombitza F, Huber H, Stetter K-O (1991) Leaching of carbonate and/or alkali containing ore materials by chemolithoautotrophic moderate acidophilic bacteria. In: Verachtert H, Verstraete W (eds) Environmental biotechnology. Proceedings of the international symposium, Ostende, Part II, VZEW, Koninklijke Vlaamse Ingenieursvereniging, pp 501–504
49. Ñancucheo I, Johnson DB (2012) Selective removal of transition metals from acidic mine waters by novel consortia of acidophilic sulfidogenic bacteria. Microb Biotechnol 5:34–44
50. Santos AL, Johnson DB (2017) The effects of temperature and pH on the kinetics of an acidophilic sulfidogenic bioreactor and indigenous microbial communities. Hydrometallurgy 168:116–120
51. Glombitza F (2001) Treatment of acid lignite mine flooding water by means of microbial sulphate reduction. Waste Manag 21:197–203
52. Oyekola OO, van Hille RP, Harrison STL (2010) Kinetic analysis of biological sulphate reduction using lactate as carbon source and electron donor: effect of sulphate concentration. Chem Eng Sci 65:4771–4781
53. Barnes LJ, Jannsen FJ, Sherren J, Versteegh JH, Koch RO, Scheeren PJH (1991) A new process for the microbial removal of sulphate and heavy metals from contaminated waters extracted by a geohydrological control system. Trans Inst Chem Eng 69:184–186
54. Gadd GM, White C (1993) Microbial treatment of metal pollution – a working biotechnology? Bioremediation trends. In: BioTECH, vol 11. Elsevier Science, London, pp 353–359
55. Huiman E (2005) Biological produced sulphide for purification of process streams, effluent treatment and recovery of metals in the metal and mining industry. In: Harrison STL, Rawlings DE, Petersen J (eds) Proceedings of the 16th international biohydrometallurgy symposium, pp 445–453
56. Reidl H (2000) Elimination of sulfur compounds and metals from waste water and off gas using biotechnology. In: ACHEMA 2000 international meeting on chemical engineering, environmental protection and biotechnology abstracts of the lecture groups environmental technology, pp 83–86

57. Burgos WD, Borch T, Troyer LD, Luan F, Larson LN, Brown JF, Lambson J, Shimizu M (2012) Schwertmannite and Fe oxides formed by biological low – pH Fe(II) oxidation versus abiotic neutralization: impact on trace metal sequestration. Geochim Cosmochim Acta 76:29–44
58. Janneck E, Arnold I, Koch T, Meyer J, Burghardt D, Ehinger S (2010) Microbial synthesis of schwertmannite from lignite mine water and its utilization for removal of arsenic from mine waters and for production of iron pigments. In: Wolkersdorfer C, Freund A (eds) Mine water and innovative thinking. Nova Scotia (CBU Press), Sydney, pp 131–135
59. Livesey-Goldblatt E, Tunley TH et al (1977) Pilot plant bacterial film oxidation (BACFOX process) of recycled acidified uranium plant ferrous sulfate leach solution. In: Schwartz W (ed) Conference bacterial leaching, GBF monograph no. 4. Verlag Chemie, Weinheim
60. Reichel S, Janneck E, Burghardt D, Peifer S, Kießig G, Koch T, Arnold I, Laubrich J (2017) Microbial production of schwertmannite: development from microbial fundamentals to marketable products. In: Hedrich S, Rübberdt K, Glombitza F, Sand W, Schippers A, Vera Veliz M, Willscher S (eds) Proceedings of the 22nd international biohydrometallurgy symposium, vol 262. Trans Tech Publications, Solid State Phenomena, pp 568–572
61. Burghardt D, Richter J, Simon E, Reichel S, Janneck E, Laubrich J (2016) Treatment of seepage water from a tailings pond of uranium mining: column tests with a novel schwertmannite adsorbent. In: Drebenstedt C, Paul M (eds) Proceedings IMWA 2016, mining meets water – conflicts and solutions, Freiberg, Leipzig, pp 889–893
62. Dondero NC (1975) The Sphaerotilus-Leptothrix group. Annu Rev Microbiol 29:407–428
63. Matlakowska R, Sklodowska A (2003) Isolation and evaluation of indigenous iron- and sulphur oxidizing bacteria for heavy metal removal from sewage sludge. In: Tsezos M, Hiatzikioseyian A, Remoundaki E (eds) Biohydrometallurgy: a sustainable technology in evolution (first edition 2004), Proceedings of the 15th international biohydrometallurgy symposium (IBS), National Technical University of Athens, Athens, 14–19 Sept 2003, pp 265–276
64. Castro L, Blazquez ML, Munoz JA, Gonzalez FG, Ballester A (2014) Mechanism and applications of metal nanoparticles prepared by bio-mediated process. Rev Adv Sci Eng 3:199–216
65. Dahl J, Maddux BI, Hutchison JE (2007) Toward greener nanosynthesis. Chem Rev 107:2228–2269
66. Merroun ML, Raff J, Rossberg A, Hennig C, Reich T, Selenska-Pobell S (2005) Complexation of uranium by cells and S-layer sheets of Bacillus sphaericus JG-A12. Appl Environ Microbiol 71(9):5532–5543
67. Vogel M, Matys S, Lehmann F, Drobot B, Günther T, Pollmann K, Raff J (2017) Use of specific metal binding of self-assembling S-layer proteins for metal bioremediation and recycling. Solid State Phenom 262:389–393
68. Brierley JA, Brierley CL, Goyak GM (1985) AMT-BIOCLAIM®, a new wastewater treatment and metal recovery technology. In: Lawrence RW, Branion RMN, Ebner HG (eds) Fundamental and applied biohydrometallurgy. Elsevier, Amsterdam, pp 291–303
69. Brierley JA, Brierley CL, Decker RF, Goyak GM (1988) Metal recovery U.S. Patent, 4,789,481
70. Kushwah A, Srivastav JK, Palsania J (2015) Biosorption of heavy metal: a review. Eur J Biotechnol Biosci 3(11):51–55
71. Brady D, Duncan JR (1993) Bioaccumulation of metal cations by Saccharomyces cerevisiae. In: Torma AE, Apel ML, Brierley CL (eds) Biohydrometallurgical technologies, vol 2. Proceedings of the international biohydrometallurgical symposium, Jackson Hole, A Publication of TMS, 22–25 Aug 1993, pp 711–724
72. Klein J, Vorlop K-D (1985) Immobilization techniques – cells. In: Moo-Young M (ed) Comprehensive biotechnology, vol 2. Pergamon Press, Oxford, pp 203–224
73. Glombitza F, Eckardt L, Hummel A (1995) In: Slate S, Baker R, Benda G (eds) Biosorption for the separation of radionuclides from drainage and process waters of the uranium mining industry. ASME, New York, pp 1451–1454

74. Tsezos M (1990) Engineering aspects of metal binding by biomass. In: Ehrlich HL, Brierley CL (eds) Microbial mineral recovery. McGraw-Hill, New York, pp 325–340
75. Tsezos M, McReady RGL, Salley J, Cuif JP (1989) The use of immobilized biomass for the recovery of uranium from Elliot Lake ore. Pilot plant results. In: Salley J, Ron GL, Mc Cready RGL, Wichlaz P (eds) Biohydrometallurgy, Proceedings of the international symposium, Jackson Hole, CANMET SP 89-10, pp 279–288
76. Barkley NP (1991) Extraction of mercury from groundwater using immobilized algae. J Air Waste Manage Assoc 41:1387–1393
77. Darnall DW, McPherson RA, Gardea-Torresdey J (1989) Metal recovery from geothermal waters and groundwater's using immobilized algeae. In: Salley J, Ron GL, McCready RGL, Wichlaz P (eds) Biohydrometallurgy, Proceedings of the international symposium, Jackson Hole, CANMET SP 89-10, pp 341–348
78. Jeffers TH, Ferguson CR, Seidel DC (1989) Biosorption of metal contaminants using immobilized biomass. In: Salley J, Ron GL, McCready RGL, Wichlaz P (eds) Biohydrometallurgy, Proceedings of the international symposium, Jackson Hole, CANMET SP 89-10, pp 317–328
79. Comys AE (2014) Enzyklopädisches Wörterbuch genannter Prozesse in der chemischen Technologie, 4. Auflage, CRC Press/Taylor and Francis Group
80. Fürst P, Morper MR (1990) METEX® – a biological process for removal of heavy metals. In: Proceedings of the DECHEMA biotechnology conferences. Lectures held at the seventh DECHEMA annual meeting of biotechnologists, Frankfurt, 30–31 May 1989, pp 1019–1023
81. Morper MR (2004) Biologische Abwasserreinigung. In: P & A Kompendium, Engineering und Anlagenbau, Realisierung und Optimierung, pp 110–113
82. Glombitza F, Eckardt L, Hummel A (1994) Cleaning of an uranium mining drainage water by means of biosorption. In: Bioengineering 2/94, VAAM annual meeting, Hannover, Mar 1994, p 90, poster 453
83. Glombitza F, Eckard L (1993) BMBF Projekt, Abschlussbericht zum BMBF Projekt: Anwendung des Prinzips der Biosorption zur Abtrennung und Rückgewinnung von metallischen Inhaltsstoffen aus kontaminierten Abwässern und Erprobung im Pilotmaßstab, DFA Fertigungs- und Anlagenbaugesellschaft GmbH, FE-Vertrag-Z/A-78, FKZ 02 WA 9173, Auftraggeber Wismut GmbH
84. Glombitza F, Eckardt L (1997) Untersuchung zum Einsatz und Erprobung eines dynamischen Sandfilters mit immobilisierten Mikroorganismen zur Reinigung radionuklidhaltiger Haldensickerwässer. Auftraggeber: Wismut GmbH 15.05.1997 Vertrag 1/116838/0/97/3.1/7858
85. Volesky B (1990) Biosorption and biosorbents. In: Volesky B (ed) Biosorption of heavy metals. CRC Press, Boca Raton, pp 3–5
86. Atkinson BW, Bux F, Kasan HC (1989) Considerations for application of biosorption technology to remediate metal-contaminated industrial effluents. Water SA 24:129–135
87. Chojnacka K (2010) Biosorption and bioaccumulation – the prospects for practical applications. Environ Int 36:299–307
88. WABAG Technologiereport (2012) Nachhaltige Klärschlammbehandlung. http://www.wabag.com/wp-content/uploads/2012/04/SludgeDigestion_2012_v32_WEB.pdf
89. McLean VA (1978) Verfahren zur Gewinnung von ausgewählten Metallen aus Abwasser. Patentschrift:DE 2722 561 Offenlegungstag 24.05.1978 Anmeldetag 18.05.1977
90. Kunze C, Glombitza F, Gerth A, Kiessig G, Küchler A (2002) Long-term stability and resilience of passive mine water treatment facilities: a joint experimental and simulation approach. In: Merkel BJ, Planer-Friedrich B, Wolkersdorfer C (eds) Uranium in the aquatic environment. Proceedings of the UMH III, 15–21 Sept 2002. Springer, Freiberg, pp 597–604

91. Glombitza F, Karnatz F, Fischer H, Pinka J, Janneck E (2003) Fundamentals of the uranium separation in constructed wetlands. In: Tsezos M, Hatzikioseyian A, Remoundaki E (eds) Proceedings of the 15th international symposium on biohydrometallurgy IBS, 14–19 Sept 2003, Athens. National Technical University of Athens, Electronic means of production – N. J. Zographakis, pp 575–583
92. Feldmann J, Hirner AV (1995) Occurrence of volatile metal and metalloid species in landfill and sewage gases. Int J Environ Anal Chem 60:339–359
93. Kaschak E (2013) Untersuchungen zur biotischen Bildung von Methylquecksilber aus anorganischem Quecksilber durch intestinale Sulfat-reduzierende Bakterien und die Bestimmung alkylierter Quecksilberverbindungen mittels GC-ICP-MS und Isotopenverdünnungsanalyse, Dissertation, Fachbereich Biologie der Johannes Gutenberg – Universität Mainz
94. Thomas F, Diaz-Bone RA, Meyer A, Michalke K, Hensel R (2012) Methylierte Metall(oid)e. BIOspektrum 18(5):487–489
95. Hoth N, Gniese C, Rakoczy J, Weber A, Kümmel S, Reichel S, Freese C, Hache M, Kassahun A, Schulz A, Fischer H, Mühling M, Starke R, Kahnt R, Vogt C (2015) $CO_2$ Bio Perm – influence of biogeochemical $CO_2$ transformation processes on the long-term permeability. In: Liebscher A, Münch U (eds) Geological storage of CO2 – long term security aspects, Science report no 22. Springer, Cham, pp 73–96
96. Kotelnikova S (2002) Microbial production and oxidation of methane in deep subsurface. Earth-Sci Rev 58:367–395
97. Behr A, Kuhlmann R (2018) Chemische Umsetzung von Kohlendioxid. Chem Ing Tech 90:593–601
98. Gniese C, Bombach P, Rakoczy J, Hoth N, Schlömann M, Richnow H-H, Krüger M (2014) Relevance of deep – subsurface microbiology for underground gas storage and geothermal energy production. In: Schippers A, Glombitza F, Sand W (eds) Geobiotechnology II. Springer, Berlin, pp 95–121. ISBN: 978-3-662-44474-3
99. Mayer F, Enzmann F, Lopez AM, Holtmann D (2017) Die mikrobielle Elektrosynthese von Methan. BIOspektrum 23:471–473
100. Zurbel A, Kraft M, Kavurucu-Schubert S, Bertau M (2018) Methanol synthesis by $CO_2$ hydrogenation over $Cu/ZnO/Al_2O_3$ catalysts under fluctuating conditions. Chem Ing Tech 90:1721–1724

Adv Biochem Eng Biotechnol (2020) 173: 325–358
DOI: 10.1007/10_2019_102

Published online: 2 August 2019

# Assessing Industrial Biotechnology Products with LCA and Eco-Efficiency

**Peter Saling**

## Contents



**Abstract** Biotechnology is applied in many industrial areas and uses microorganisms, enzymes, or precursors replacing chemicals to produce goods, including chemicals, plastics, food, agricultural and pharmaceutical products, and energy carriers from renewable raw materials and increasingly also from waste from agriculture and forestry (BIOPRO Baden-Württemberg GmbH, Facts and Figures. Biotechnologie.de. https://www.biooekonomie-bw.de/en/articles/dossiers/industrial-biotechnology-biological-resources-for-industrial-processes/, 2013). In

P. Saling (✉)
BASF SE, Sustainability Strategy, Ludwigshafen, Germany
e-mail: peter.saling@basf.com

comparison with conventional processes, industrial biotechnology processes often run under relatively mild reaction conditions, moderate temperatures, and the use of aqueous media. They might reduce in general the energy requirements and the number of by-products. Since product concentration and formation rate are often very low, the resulting products need to be purified and recovered in marketable quantities in a process that is referred to as downstream processing. Product quantity can also be increased by optimizing the manufacturing processes or biocatalysts used (OECD, the application of biotechnology to industrial sustainability. www.oecd.org/sti/biotechnology, 2001).

In this context, developing a sustainable bio-based economy that uses eco-efficient processes is one of the key strategic challenges for the twenty-first century. Decisions in the technology development are often supported by sustainability assessment results using different types of sustainability assessment methods. In the last decades, we developed different types of sustainability assessment methods evaluating aspects of economy, ecology, and society to support decision-making processes. We show in this chapter how different types of questions can be answered, how more sustainable solutions can be identified, and how this information can be used for marketing and research activities.

## Graphical Abstract

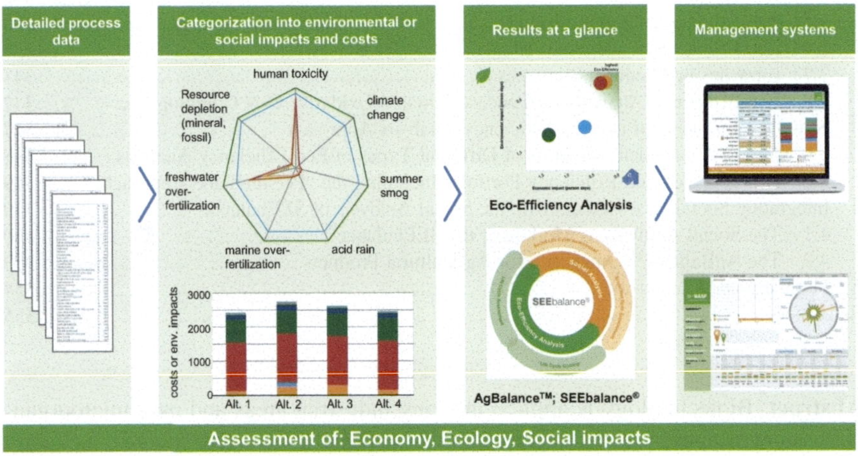


**Keywords** AgBalance™, Eco-efficiency, LCA, SEEbalance®, Social analysis, Sustainability evaluation, Sustainability management tools

# 1 Sustainability Assessment Methods

Potential risks and impacts to the environment associated with chemical production and chemical products should be analyzed critically. The Eco-Efficiency Analysis (EEA) [1] considers the economic and life cycle environmental effects of a product or process, giving them equal weighting.

The instrument provides early recognition and systematic detection of economic and environmental opportunities and risks in existing and future business activities. In BASF, this tool has become integral part of the decision-making process for new investments as well as in the interaction, collaboration, and communication with customers and suppliers [2, 3]. In this approach, technological details can be assessed and linked with the analysis of possible improvement potentials. Inventions and potential innovations can be assessed before realizing them by the preparation of holistic life cycle analysis with different levels of details. The new process under investigation can be ranked in comparison to the existing process and to process alternatives due to their higher sustainability. This information can be used in the decision-making process to reduce the risk of bigger investment activities.

Proving early recognition and systematic detection of economic and environmental opportunities for production processes in the chemical industry has been used in BASF since 1997.

In addition to the Eco-Efficiency Analysis (EEA), some years ago, the SEEbalance® was developed [4, 5] and has been used by BASF and its customers to assist strategic decision-making, facilitate the identification of product and process improvements, enhance product differentiation as well as to support the dialogue with opinion makers, NGOs, and politicians. Both EEA and SEEbalance® analyses are comparative methods; the advantages and disadvantages of several alternatives are assessed according to a predefined customer benefit with a holistic approach. The SEEbalance® also considers social impacts of products and processes [6].

# 2 Eco-Efficiency Analysis for Biotechnological Processes

Industry and academia are trying to identify comprehensive product sustainability solutions. Often biotechnological processes play an important role to generate more sustainable solutions [7]. Key biotechnology indicators are often used to steer and support research activities with statistics from different perspectives [8]. To identify the most sustainable solution requires the implementation of an analytical process which follows defined procedures and assessment systems [9]. The basis for decision-making processes is often the comparison of different product or process alternatives leading to a dedicated product and a defined application [10]. To cover all relevant aspects, a life cycle perspective from "cradle-to-grave" is needed and can be used in a comprehensive assessment.

The Eco-Efficiency Analysis can be used:

- To analyze the entire life cycle of a product or production process ("cradle-to-grave," including the product utilization phase) and quantify sustainability impacts
- To compare the environmental impacts of various alternatives based on life cycle inventory data, impact assessment, and the creation of an "environmental fingerprint" presented in a spider diagram
- To contrast a product's economic value added with its impact on the environment
- To compare ecological and economic aspects with those of alternative products or processes that deliver a defined customer benefit equally well
- To evaluate future scenarios and the effects of different avenues of action

The rationale behind this assessment tool has been described in several publications. The impact assessment was done with methods mostly developed by BASF which were described in several publications. The basic information about the methodology can be found in a publication from 2002 [1]. The human toxicity impact method was developed by BASF in an approach together with internal and external toxicologists [11]. This new method called ProScale overcomes some aspects such as exposure assessment which was not included in the previous method. ProScale will create a database and might be an additional information in the Environmental Footprint of the European Commission [12]. Examples of the application of the tools for sustainability assessment show the wide range how these systems can support decision-making processes [6]. Practical examples can show how the metrics for sustainability can support decision-making processes answering different questions [13]. The analysis involves measuring the life cycle environmental impacts and life cycle costs for product alternatives for a defined level of output over the course of their life cycle.

The eco-efficiency methodology is a comparative analysis and thus does not determine the sustainability of a product in absolute terms. Thus, a product which was deemed most eco-efficient in one analysis may be a less eco-efficient alternative in another study when a different application is considered. The methodology was reviewed by external parties and was validated several times. The last review of the newly developed methodology was performed and documented in 2016 by NSF [14].

## *2.1 The Environmental Fingerprint*

The environmental fingerprint (Fig. 2) shows the environmental impact of the alternatives relative to one another in all the assessed impact categories. The alternative that lies furthest out and has the value of 1.00 is the least favorable (highest impact) alternative in the category under consideration. The closer to the origin (0, 0 coordinate) an alternative is located, the more favorable (less environmental burden) it is. In principle, it is possible to change the direction of the results to the opposite if it is clearly described in the report. The axes coordinates are calculated independently of one another, so an alternative that, for example, does well in abiotic depletion (resource depletion – mineral, fossil) can perform worse with regard to emissions.

## 2.2 *Normalization*

The normalization step uses external (study-independent) values. The environmental normalization values used are annual regional statistical values provided with standard tools on the market, e.g., from CML, ReCiPe, or other methods. For the assessment in a specific region, the regional statistics will be used. For all impact categories in this study, EU statistics were used. For the normalization of the total costs, the regional GDP (gross domestic product) is used. This value should be updated yearly and documented.

## 2.3 *Weighting*

The normalized values are aggregated using weighting factors obtained by external polling consultants. For this assessment people were selected being representative for societies and their way of thinking as important stakeholders. To be suitable for an EEA, poll respondents should cover a balance of age, gender, income, household size, and education. Even if they are no experts, they represent societies and their way of thinking. These weighting factors may be assessed for a region, country, or world.

The polls are based on the Maximum Difference Scaling approach, in which respondents are asked to select, in a series of questions, the most important environmental impact category from three options. A Maximum Difference or Best-Worst Scaling is a survey method in market research that was originally developed in the 1990s and is used to try to gain an understanding of consumers' likes and dislikes. Respondents are usually asked to select the most and least important attributes from a subset of product features [15]. The BASF EEA weighting factors have been based on polls run by the external consultant TNS, a global research agency. In the updated version, the EEA uses weighting factors generated by the European Commission in the Product Environmental Footprint program [16].

## 2.4 Eco-Efficiency Analysis

The BASF Eco-Efficiency portfolio was developed to graphically depict both economic and environmental results on a single $2 \times 2$ matrix (see Fig. 3). The normalized values from the Environmental Footprint are aggregated into a single relative environmental impact through normalization and weighting and are plotted against costs. Because environmental impacts and costs are equally important and are not weighted differently, the most eco-efficient alternative is the alternative with the largest perpendicular distance above the diagonal line in the Eco-Efficiency portfolio (Fig. 3).

The Eco-Efficiency Analysis follows some principles to enable practitioners a consistent and reproducible study design. Products or processes need to meet the same defined functional unit; the alternatives should cover at least 90% of the relevant markets. For competitors' products specific assessments based on general available information such as Best Practice Available Technology (BAT) reports (BREF) or other sources such as Environmental Product Declarations (EPD) can be used. Joint projects of different industries are another opportunity to assess the sustainability of different alternatives. For research projects, the market relevance is not a major aspect because a new development is compared to state-of-the art technologies. In the research application, most of the relevant process alternatives should be compared. Future scenarios can be assessed as well. Research goals can be checked and compared to determine if a future goal is more sustainable than an existing process. Scale-up questions are important in the modelling, and scenario analysis can support the product development as well. For the generation of significant and meaningful results, the entire life cycle is considered with specific life cycle steps. In the generation of results, the environmental and an economic assessment are based upon the ISO standards. The environmental assessment is based on ISO 14040 and ISO 14044 [17, 18] standards for Life Cycle Assessments and ISO 14045 [19] for Eco-efficiency fulfills both the required and optional phases. Aggregation and weighting for competing products in a comparative assertion by claiming the environmental superiority of a product are not in compliance with the ISO standard, and so that can be done only based on the non-aggregated and non-weighted LCA study. The summary of an EEA is shown in the Eco-Efficiency portfolio. Furthermore, ISO 14045 is a relevant standard that gives guidance for performing studies for internal and external uses [2, 3].

## 2.5 Eco-Efficiency for Astaxanthin

In aquaculture the carotenoid pigment astaxanthin is used in feed for salmon, bream, and shrimp to impart the pink hue desired by the consumer in the marketplace. Consumers regard the coloration of these fish as an important criterion of quality.

Whereas wild fish obtain carotenoids through the food chain, salmon in aquaculture depend on inclusion of these pigments in the fish feed. Without addition of these pigments, the salmon's flesh would be "yellow-gray." It is for this reason that the nature-identical pigments need to be added. As these carotenoids also occur in nature, their addition provides the same requisites for the farmed as for the wild salmon. Moreover, the deposited pigments play a vital role in the life of the fish. Astaxanthin is a vitamin A precursor for fish, and important for growth and survival, specific functions in reproduction and metabolism, and health in salmonids. The beneficial effects of the carotenoids come from their ability to protect cells from the harmful effect of oxidative compounds. The positive role of their antioxidative properties in human and animal health is widely accepted. With the astaxanthin product Lucantin® Pink, and the canthaxanthin product Lucantin® Red, BASF offers high-quality pigments for feeds, providing the necessary carotenoids for modern aquaculture [20].

### 2.5.1 System Boundaries

In the Eco-Efficiency Analysis, the chemically produced product Lucantin® Pink was compared with astaxanthin obtained from yeast and from algae. The functional unit was defined as production of 1 ton of salmon in combination with implementing the astaxanthin to the fish diet. The Eco-Efficiency Analysis underwent a critical review by a panel from Öko-Institute.V. (Freiburg).

Synthetic astaxanthin is produced in closed reaction vessels from crude-oil-derived substances. Various synthetic steps are required to obtain the final product, which is formulated to 10%. Solvents used in the various steps are recovered and recycled.

Fermentative astaxanthin is obtained from *Phaffia rhodozyma* (*Xanthophyllomyces dendrorhous*), a type of yeast that uses carbohydrate sources from agricultural production. The fermentation reactor is filled with the yeast, the growth medium, energy, and water. After the yeast cells have multiplied, they are isolated and dried. After addition of antioxidants, the yeast cells, which contain around 0.4–1% astaxanthin, are packaged and shipped to retailers.

Astaxanthin derived from algae is produced primarily with *Haematococcus pluvialis* in open ponds. Nutrients are added to the ponds, which are continuously mechanically agitated. The cultivated algae biomass is dried and ruptured before packaging. The concentration of astaxanthin is around 2%.

The system boundaries for the fermentation process are shown in Fig. 1.

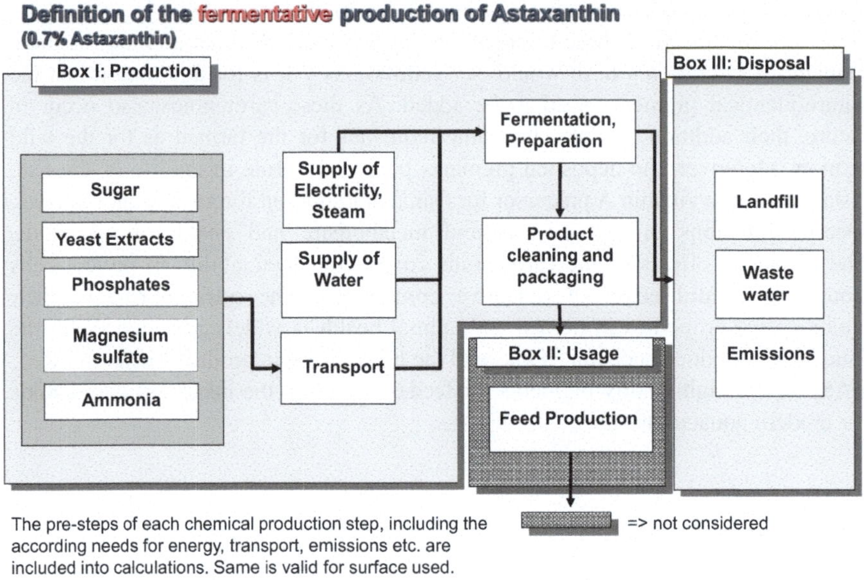


**Fig. 1** System boundary of the biotechnological astaxanthin production

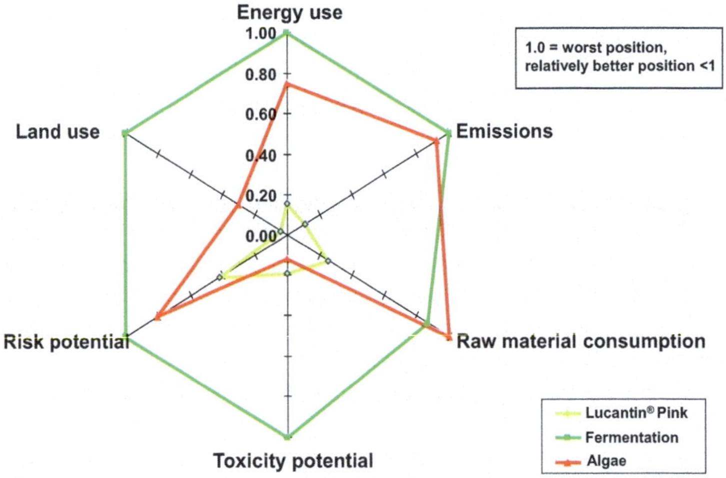


**Fig. 2** Environmental fingerprint of the biotechnological astaxanthin production and the fermentation compared with a chemical Lucantin® Pink process

## 2.5.2 Overview of the Study Results

All the life cycle steps need to be identified, calculated, and summarized in an overall environmental fingerprint (Fig. 2) which shows the normalized impact of each alternative in different environmental impact categories including all steps up to

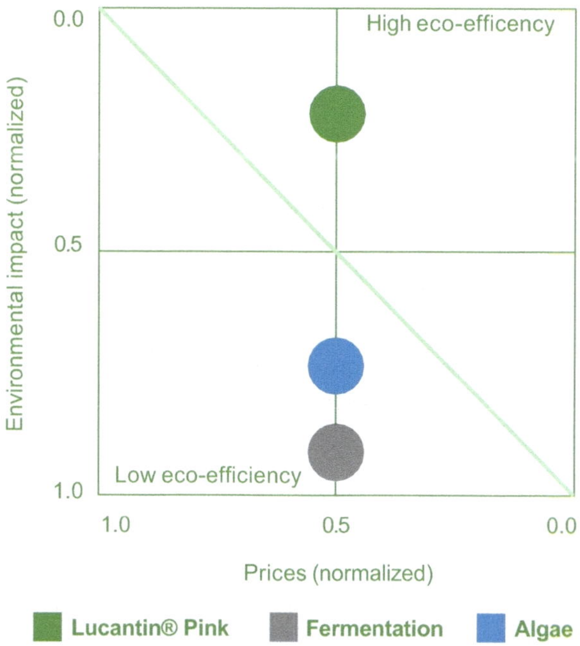


**Fig. 3** Eco-Efficiency portfolio of the biotechnological astaxanthin production and the fermentation compared with a chemical Lucantin® Pink process

point of sale. The second part of the graph shows an Eco-Efficiency portfolio (Fig. 3).

The results are calculated with an integrated approach that considers all upstream processes as well. Energies, emissions, product flows, etc. need to be calculated with background information to ensure a holistic life cycle approach. For comparable results, the quality of all these data inputs must be in the same range. Cutoff criteria need to follow the same rules to avoid imbalances. In the research field, often estimates and planning data will be implemented to model future scenarios.

### 2.5.3 Detailed Results for Raw Materials Consumption

A fermentation process looks very much different in comparison to chemical synthesis. The fermentative production of astaxanthin depends on the microbial consumption of vast amounts of sugar. Sugar as one of the important substances to run a fermentation process is taken as an example to demonstrate the way how resource utilization is calculated in this environmental assessment. The analysis covers the soil cultivation, preparation/production of seeds (sugar beets), and production and transport of fertilizers, pesticides, and herbicides. Furthermore, harvesting, processing/crystallization, and transport are considered in the calculation. All consumptions of energy and raw material for these preliminary steps are added up. In addition, yeast, phosphates, magnesium, and ammonia are necessary to run the fermentation process. The production of those materials is considered in a cradle-to-gate approach as well. Furthermore, the supply of electric power, compressed air, and steam requires considerable amounts of coal, oil, and gas.

The process of using algae in open ponds does not require direct energy for heating because it utilizes the sunlight. Nevertheless, the water in the ponds needs circulation to increase the efficiency of the system, and energy is required for the drying of biomass as well. Thus, the use of fuel and electricity is the main reason for the high consumption of raw materials linked with a specific resource impact factor. The use of methylene chloride as a solvent for the product preparation and extraction is needed as well including the energy for the recycling of the solvent.

Due to the petrochemical base of the chemical Lucantin® Pink synthesis, the use of oil and gas is predominant.

### 2.5.4 Detailed Results for Emissions

Emissions of $CO_2$ as well as all other air emissions are mainly correlated with the use of energy for chemical synthesis in case of Lucantin® Pink. Emissions which have an ozone depletion potential mainly come from the use of halogenated non-methane volatile organic compounds for the synthesis. This type of process using biogenic materials assimilates in the first step $CO_2$ which is a benefit in a cradle-to-gate approach. However, for fast consuming goods such as astaxanthin that will be consumed in the fish production, the carbon dioxide is released in short time frames, and the impact is neutral for $CO_2$ in a cradle-to-grave approach. The energy needed to produce fertilizers and all the other products needed to produce the biogenic materials have additional $CO_2$ emissions, so that these materials are in total $CO_2$ emitters.

The other air emissions are recorded separately and summed up during the life cycle. The effect of these air emissions on the environment differs according to the type of emission. The categories of effects examined in this context are the global warming potential (GWP), photochemical smog (photochemical ozone creation potential, POCP), acid rain (acidification potential, AP), and ozone layer depletion (ozone depletion potential, ODP).

Water emissions were considered as well using a distance-to-target approach for the most commonly regulated water emissions such as heavy metals, chemical oxygen demand (COD), N- and P-emissions, halogenated chemicals, and volatile organic compounds (VOC). The emissions were translated to dilution water amounts that are theoretically needed to reduce the water emissions below thresholds set by authorities. The approach was further explained in an ISO standard, for example, for different types of water footprints [21].

### 2.5.5 Detailed Results for Land Use

High impacts in the category of land use were linked in the fermentation process results from the use of arable land required for sugar production. For growing sugar beets, only high qualities of arable land are suitable. Algae for astaxanthin production are cultivated in large open ponds. Thus, the process needs a relatively large area of land. But usually the production is in desert areas of low soil productivity which means a low negative impact with regard to land use (cf. Lindner et al. this book). The chemical synthesis requires the smallest area because of the high productivity of chemical processes in the context of using land, even if sealed land with a relatively high negative impact is utilized.

### 2.5.6 Detailed Results for the Risk Potential

In the Eco-Efficiency method used for a long time, risk potentials were considered as well. In this category risks like accidents, risks for fire, or other types of risks could be considered if relevant. The risks were checked in every study if they should be considered and which impacts should be included. In this case study, it was shown that the agricultural upstream processes show higher risks due to higher accident rates than in the chemical industry. Additional risks as impurities of products by external impacts, e.g., pollution of the open ponds in the algae process or handling of high volumes of materials in the process, were identified. In the new Eco-Efficiency method, this category is not considered because of higher uncertainties in the assessment process and data gaps.

### 2.5.7 Case Study Conclusion

The *fermentative production* of astaxanthin by *Phaffia rhodozyma* performs poorly with respect to all criteria. This is due to the high-energy input and especially the large number of agricultural products necessary to generate the cell mass which contains only a small amount of astaxanthin. The substrate obtained from agricultural production – typically sugar and/or molasses – requires fertilizers, possibly chemical plant protection such as plant protection products and fuel for cultivation. In addition, land use requirements are large.

Impacts due to production of *algae-based astaxanthin* are lower than for that based on yeast but are still high compared to chemically produced carotenoid. The one exception is the toxicity potential category. Along the whole life cycle, different chemical steps are needed to produce the final product. Several compounds that are needed are linked with H-phrases from the Global Harmonized System (GHS) that lead to impacts in the human toxicity category. That does not mean that the final product has any toxic effect, but the material needs of upstream processes determine the toxicity potential. These facts should not be mixed up because the Eco-Efficiency Analysis considers the human toxicity impacts from a life cycle perspective, i.e., including production and upstream processes.

Algae production requires primary raw material such as simple mineral salts not associated with high toxicity potentials.

With the exception noted above, synthetic astaxanthin from *chemical production* process results in less than 50% environmental load in each impact category compared to algae-based product. This large reduction in environmental impact is possible due to the sophisticated production process designed to conserve material and energy and minimize emissions. Low concentration of astaxanthin in the extracted material considerably increases the expenditures of energy and raw materials to produce the purified active ingredient. Even if, e.g., the effectivity of the fermentation process is doubled, this will not change the grading of the products significantly. So, there is a need for improvement of this biotransformation process to be more eco-efficient. The Eco-Efficiency Analysis can help in this context to support the further development, can support to identify improvement potentials, and can help to set realistic goals for further developments.

Often "natural" production of astaxanthin by fermentation or algae is commonly thought to be more environmentally friendly. This might be the case, if all conditions to produce these products are optimized and based on a high efficiency of the process. To assess results accurately and to interpret them in a meaningful manner, differences between alternatives need to be analyzed carefully and need to be supported by scenario analysis and data sensitivity analysis.

In this case it was shown that the chemical synthesis among the alternative pathways to the final product has the lowest environmental impacts. The costs of alternatives are quite similar. The combination of costs on the *x*-axis and environmental results on the *y*-axis are summarized in the Eco-Efficiency portfolio (Fig. 3) to an overall result. It was shown that Lucantin® Pink was the most eco-efficient process alternative located with a largest difference from the diagonal in the direction of the upper right corner of the diagram.

## 2.6 Eco-Efficiency for Vitamin $B_2$

Vitamin $B_2$, also known as riboflavin, plays a central role in the release of energy in the human or animal body and is therefore considered as an important ingredient in feed. It helps in maintaining good digestion and good nerve function. A lack of vitamin $B_2$ results in several different symptoms, such as a higher risk of infectious diseases. Because of its significant role, there is high demand in the feed industry for industrially produced vitamin $B_2$. In the marketing but as well as for research activities, sustainability is one key aspect for the development of processes. Different process alternatives need to be considered before implementing a final process. The ingredients should be assessed as well as the whole process with a cradle-to-grave approach to avoid optimizations in the wrong direction. The question was under which conditions a biotechnological process for the vitamin is more sustainable than a chemical process? What are the major impacts, and where are the improvement potentials along the whole life cycle [22]? It was shown that in the collaboration of researchers and sustainability specialists, improvement potentials could be identified in several scenario analyses and were realized as well in the further development of the process. In this collaborative approach, the sustainability of processes could be improved significantly.

### 2.6.1 Processes and System Boundaries

The biotechnological synthesis produces vitamin $B_2$ via a one-step fermentation from vegetable oil using the fungus *Ashbya gossypii*. This fungus has enzymes that produce vitamin $B_2$ during metabolism [23]. The fungus feeds on vegetable oils, i.e., renewable raw materials. The amount of vitamin $B_2$ produced depends on the quantity of the relevant enzymes and the conditions under which the fungus is grown. In contrast, the chemical process requires no less than seven intermediate steps to obtain vitamin $B_2$ from glucose (Fig. 4).

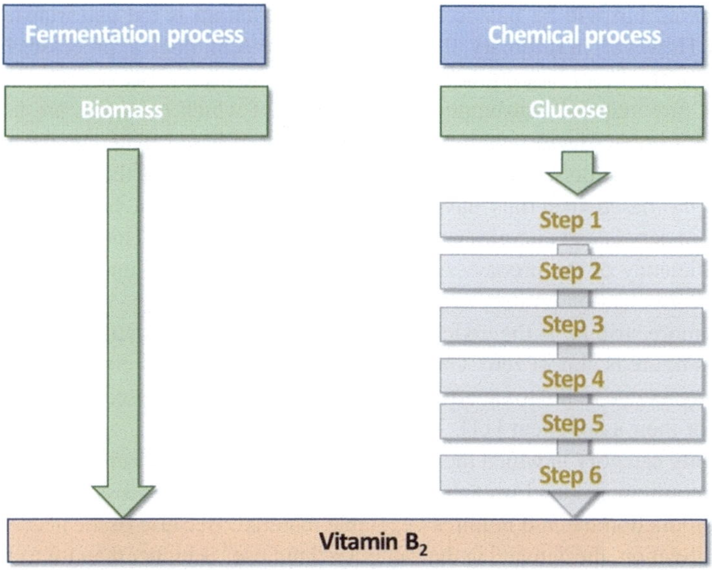


**Fig. 4** The two different pathways to derive vitamin $B_2$

### 2.6.2 Holistic Life Cycle Analysis

A holistic life cycle analysis was performed for both systems. The multistep process in the chemical synthesis used primary data from the production including all pre-steps needed to produce the intermediate compounds. Energies, effluents, and emissions were integrated based on primary data sources of the producing companies and the service providers including the end of pipe technologies needed.

For the biotechnological processes, the approach was the same including energies and emissions of the fermenter as well. The upstream materials needed to feed the fermenter and for the work-up processes step, the purification, and the waste treatment of the biomass were included as well.

### 2.6.3 Results Derived from the LCA

A glance at the ecological fingerprints of the two processes for vitamin $B_2$ clearly shows the ecological advantages and disadvantages of both. The biotechnological process has advantages over the chemical method in five of the six categories. The reductions of air emissions and consumption of abiotic resources were significant. Biotic resources are considered in our method with a factor of zero. The measures result in a 25% decrease in energy consumption and a decrease of as much as 33% in the global warming potential (GWP). The carbon uptake of the bio-based materials is considered but will be balanced out when the vitamin is fed and transferred to biomass. However, in summary, the GWP of the biotechnological processes is better than that of the fossil-based process chain for the chemical process.

In the raw material consumption, it is important which resources are used and with which impact factors they are connected. Fossil-based materials have in this impact category a disadvantage because of the short range of their reserves base, whereas bio-based materials have advantages. It is important how much of the different materials and energies are used to be more favorable. This is clearly linked to the efficiency of the processes. Scenario calculations show under which conditions the fermentation process is more eco-efficient.

Even more striking is the toxicity potential: because of the natural raw materials used, this figure is almost zero. Chemical processes have in this comparison a much higher impact due to the toxic effects of some chemicals expressed by H-phrases as a basis for their assessment [11].

The only category in which the biotechnological process performs less well is the surface requirement (land use). In contrast to the chemical process, the biotechnological production method requires vegetable material. Because of the need for land under cultivation, the demand in the category "land use" is higher than for a chemical plant. Overall, the ecological fingerprint clearly shows the environmental advantages of biotechnological production.

In Fig. 5, the most unfavorable alternative in every category is rated with value 1; the most favorable alternative is positioned closest to the center of the graph in a

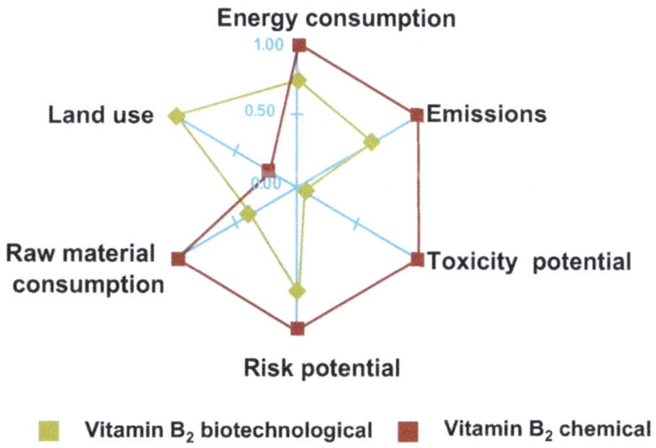


**Fig. 5** Comparison of environmental impacts of six categories (the more outside the worse)

relative manner. This step is another normalization step different to the normalization against worldwide normalization factors.

### 2.6.4 Eco-Efficiency Analysis Results

The cost evaluation is done in the same manner as the environmental approach. The life cycle is the basis for the calculation. Based on this approach, the different life cycle steps are associated with the specific cost figures. In the Eco-Efficiency Analysis, costs for materials, energy, maintenance, workers, depreciations, etc. are considered as it is done in the conventional cost calculations. Often a so-called total cost of ownership approach is applied, covering the costs the customers or the customers of the customers have. Theoretical costs like avoidance costs or approaches from willingness to pay are not included. Different depreciation models can be applied depending on the approaches and specific internal accounting methods of companies to steer decision-making processes. Country-specific assessments can be applied as well as to cover region-specific needs and conditions. The Eco-Efficiency Analysis covers costs, but depending on the system boundary defined, prices can be considered as well. In that sense, prices for products reflect the costs for the customer.

The environmental and economic results can be summarized in a matrix, the Eco-Efficiency portfolio (Fig. 6). The functional unit was defined as "production of 100 kg vitamin $B_2$ as supplement for feed" and with the reference flow of 100 kg vitamin $B_2$. The environmental impact is entered on the horizontal axis and the resulting costs on the vertical axis in a normalized approach including a weighting scheme for the environmental results as explained before. The average line (diagonal) shows points of equal eco-efficiency. It makes it easier to distinguish between alternatives if there are more than two alternatives and if they are very close together. Since costs and environmental impacts are shown, higher values are worse; lower values are better than the average value.

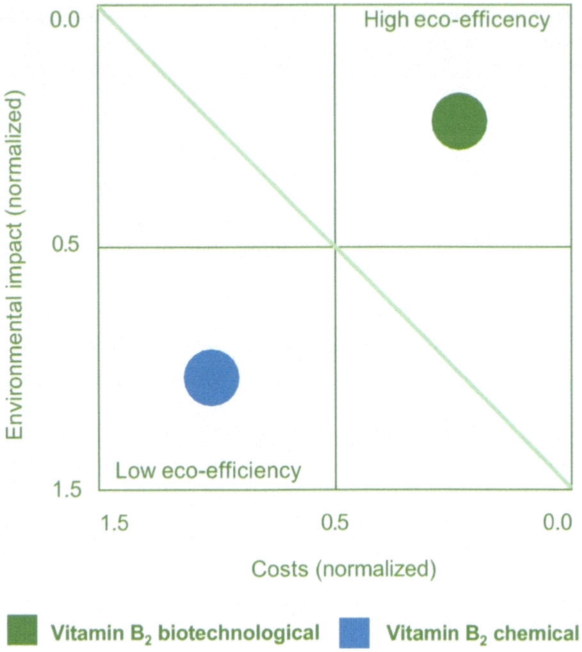


**Fig. 6** Eco-Efficiency portfolio of vitamin $B_2$ production alternatives comparing a biotechnological and a chemical process

### 2.6.5 Interpretation of Results and Its Use in Strategic Discussions

Using this form of representation, the superiority of the biotechnological method is immediately clear. It is situated in the top right-hand corner of the matrix, where the most eco-efficient process is found. The chemical process, on the other hand, is in the bottom left-hand corner because of its clearly higher environmental burden and higher costs, indicating a low eco-efficiency. The objective of the Eco-Efficiency Analysis was to demonstrate the strengths and weaknesses of the biotechnological process in comparison with the defined chemical process. The results are clear. In this case, biotechnology is the most eco-efficient process to produce vitamin $B_2$ combining economy and ecology in an advantageous way and outperforming chemical synthesis. Eco-Efficiency Analysis shows that sustainability and economic profit are not mutually exclusive but can be successfully combined [24].

The result is a basis for the decision-making process and for the evaluation of sustainable processes within the area of "industrial biotechnology." In different studies strengths and weaknesses of biotechnological processes can be compared. An additional value of the eco-efficiency tool, apart from its description of the current state, lies in the recognition of the dominant influences and in the illustration of "what if ...?" scenarios, i.e., changing one factor while leaving all others illustrates the impact of the adjusted factor on the overall assessment.

In the case of vitamin $B_2$, the operating division fine chemicals at BASF invested in a new fermentation plant, partly because of the advantages shown in the Eco-Efficiency portfolio [23].

## 2.7 *Enzymes in Chemical Synthesis*

Enzymes are proteins that regulate chemical reactions but are themselves unchanged by the reaction. Because they're often required to start or speed up a reaction, enzymes also are called catalysts. Without enzymes, many biochemical reactions would be energetically inefficient [25]. When one substance needs to be transformed into another, nature uses enzymes to speed up the process. Enzymes are used as catalysts to manufacture a variety of everyday products – like sugar, beer, bread, and ethanol. Numerous enzymes exist, and they are characterized by highly specific action and very fast reaction rates.

Like all catalysts, enzymes increase the reaction rate by lowering its activation energy. Some enzymes can make their conversion of substrate to product occur many millions of times faster [26].

Enzyme rates depend on solution conditions and substrate concentration. To find the maximum speed of an enzymatic reaction, the substrate concentration is increased until a constant rate of product formation is seen. Saturation happens because, as substrate concentration increases, more and more of the free enzyme is converted into the substrate-bound ES (enzyme-substrate) complex. At the maximum reaction rate (Vmax) of the enzyme, all the enzyme active sites are bound to substrate, and the amount of ES complex is the same as the total amount of enzyme [27].

They are also used directly in products such as laundry detergent, where they help remove stains and enable low-temperature washing [28].

There are many examples where enzymes were analyzed with LCA to find optimization potentials, identify new opportunities, and compare alternative solutions against the background of an improved sustainability. The system boundaries can vary depending on the goals of the study. One example is a cradle-to-gate assessment of environmental data for five characteristic enzyme products produced by Novozymes in Denmark. It was found that main sources of impacts of the production of enzymes are the fermentation processes and the production of ingredients for the fermentation. Environmental impacts of the considered enzyme products vary by a factor of about 10 because of differences in use of carbohydrates, applied fermentation time, formulation type, yield, and final product strength.

Contributions to global warming are to a large extent driven by energy consumption, and other energy-related impact categories (acidification and photochemical ozone formation) follow the same overall pattern (data not shown). Among the five different enzymes assessed, the GWP is in the range of 1–10 kg $CO_2$-eq/kg product, the POCP in the range of 0.5–3 kg $CO_2$-eq/kg product ethylene equivalents, and the eutrophication potential from 1 kg $PO_4$-eq/kg product to more than 20 kg $PO_4$-eq/kg product. That shows that enzymes cannot just be simplified in an average figure. The study indicates that continued improvement of yields through production strain optimization is probably the most effective single means to further environmental improvements. However, often the main impact along a value chain is linked to the processes which enzymes are catalyzing. A higher amount of environmental impacts can be balanced or outweighed, if the function in the processes is very effective and reduces the environmental impact of the process.

## 2.8 Eco-Efficiency Analysis for Laccase-Initiated Oxidative C-N Coupling

In the specific case of catalyzing a chemical reaction, enzymes can be used as well. They are probably in competition to a classical chemical process, and a sustainability assessment can help to show if the enzyme process is more sustainable and where improvement potentials for further developments are. In a case study, the laccase-mediated heterocoupling of 3-methylcatechol and n-hexylamine was investigated at the milligram to gram scale, and an early Eco-Efficiency Analysis and evaluation revealed bottlenecks in enzymatic synthesis. Eco-Efficiency Analysis was used as an optimization tool to create more sustainable solutions and improve subsequent reactions, enabling the scaled-up biocatalytic process to become more economical and competitive compared with the chemo-catalyzed procedure using sodium iodate (Scheme 1).

On the preparative scale, the water-miscible organic solvent methanol has a decisive influence on catalyst activity as well as reaction rate and selectivity. Consistent with the results of Wellington [29], the robust, stable, and commercially available Novozym® 51003 represents a native laccase that can operate at exceptionally high methanol concentrations; such high methanol concentrations have not successfully used previously.

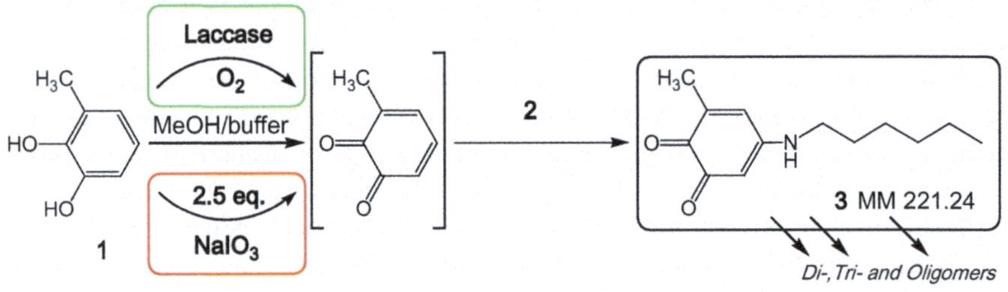


**Scheme 1** C-N coupling reaction of 3-methylcatechol (**1**) and *n*-hexylamine (**2**) to form the target product **3**

To assess, if the alternative synthesis is more sustainable than the chemical route, a holistic Eco-Efficiency Analysis was calculated. Using the systematic approach by calculating and interpreting results for different impact categories can help to identify weak points of the system as well as support decision-making if such a process should be implemented.

In the early stages of reaction and process development, this analysis helps researchers to optimize the whole system by optimizing the process but as well by probably replacing chemicals and materials by alternative materials. Often trade-offs are not easy to handle and are also findings that should be handled accurately during the process development. To get a good overview, a detailed system boundary overview needs to be generated to cover all relevant life cycle steps (Fig. 7). Red lines indicate energy flows and waste disposal. Solvent reprocessing was calculated once as evaporation to yield the crude product and once as redistillation for solvent recycling. The application or customer benefit (gray) was not considered.

Another aspect in the design of the study are cutoff criteria with the intention to reduce the workload of an LCA study to a level which on one hand can be handled and on the other hand considers all significant information. Therefore, a careful assessment is needed often combined with overview calculations.

For this example, the analysis assessed six different impact categories which are the most common indicators in LCA. The six ecological categories examined are energy consumption, resource consumption, emissions (air, water, and waste), land use, risk potential, and human toxicity potential [30].

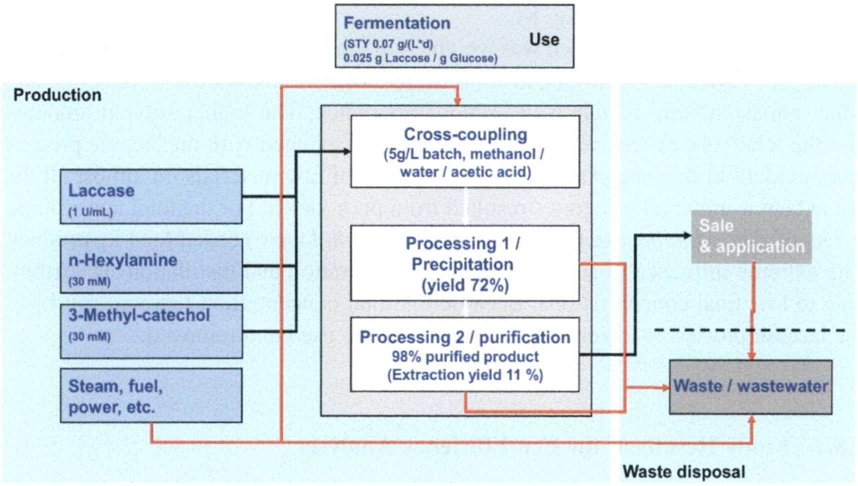


**Fig. 7** System boundary sheet of the laccase process

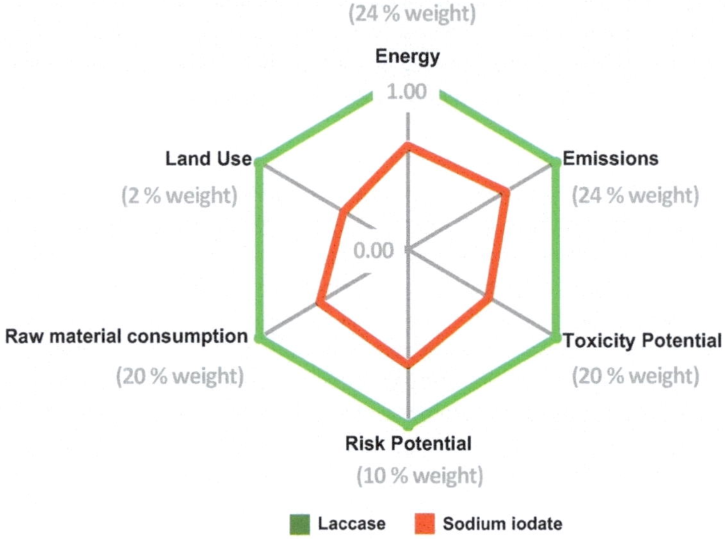


**Fig. 8** The environmental fingerprint of C-N coupling alternatives (base case)

### 2.8.1 Study Results of the Ecological Impacts

The $NaIO_3$ process turned out to have in all categories better results as shown in the environmental fingerprint (Fig. 8).

Energy consumption, which was weighted in the poll at 24%, played a major role in this Eco-Efficiency Analysis; in contrast, the roles of land use, solid material, and water emission were of relatively minor importance. The higher solvent amounts and the relatively expensive catalyst production associated with the laccase process were evident in this analysis. The high influence of raw materials on almost all the valued environmental categories resulted from poor yields. For the total application, 5.5 kg of *n-hexylamine* and 6.7 kg of *3-methylcatechol* were needed for 1 kg product. The extreme influence of solvent handling (evaporation and distillation) is attributable to low final concentrations. Because the final concentration (approx. 5 g/L) of the laccase process is lower, more solvent must be used and removed.

### 2.8.2 Study Results of the Eco-Efficiency Analysis

According to the present state of improvement, the Eco-Efficiency Analysis indicates that the laccase-initiated C-N coupling, despite higher selectivity, performs worse than the sodium iodate process (Fig. 9). This Eco-Efficiency Analysis clearly demonstrated that focusing on selectivity is not always the key to successfully improving a biocatalytic process; there might be additional factors contributing to

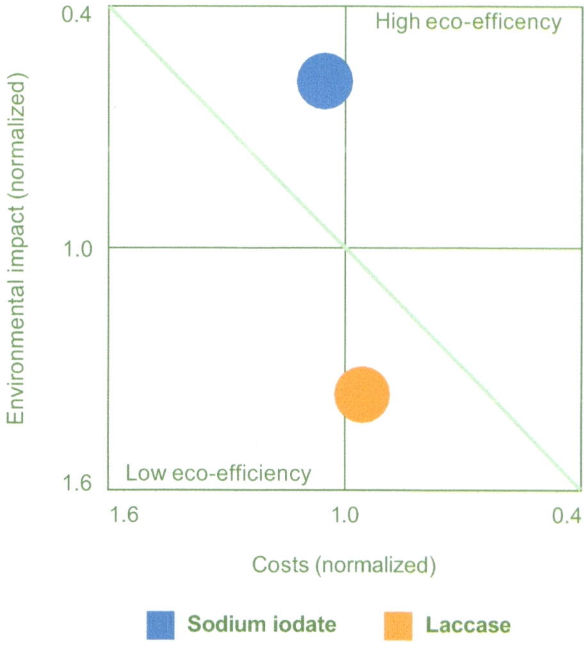


**Fig. 9** Portfolio plot for eco-efficiency. The distance from the diagonal represents total eco-efficiency. Functional unit: Production of 1,000 kg heterocoupling product of *n*-hexyl-amine and 3-methyl-catechol

the overall result. The principal reason for the superior performance of the sodium iodate process is that the product titer was two times greater (approximately 9 g/L). This key difference makes the sodium iodate procedure more efficient and more environmentally friendly compared to the biocatalytic process. The portfolio shows a clear differentiation of the alternatives so that minor changes in the processes will not change the results of the study significantly.

### 2.8.3 Scenario Analysis of the Eco-Efficiency Analysis Approach

However, the Eco-Efficiency Analysis not only includes an in-depth overview of the status quo of the compared processes but also provides recommendations to improve the ecological efficiency of biocatalysis in future scenarios. One of the future scenarios indicated that the final product concentration is critical, whereas selectivity has no significant influence on obtaining comparable eco-efficiency. Under the assumption that both processes have the same final product concentration and the same solvent consumption, the biotechnological process and the chemical alternative achieve similar eco-efficiency (Fig. 10). This result shows clearly improvement potentials and outlines that the process in general can be almost as eco-efficient as the chemical process. Linked with clear development targets and a realistic timeline, the sustainability assessment can be used for strategic planning of research processes. The Eco-Efficiency Analysis can support the development of upscaled

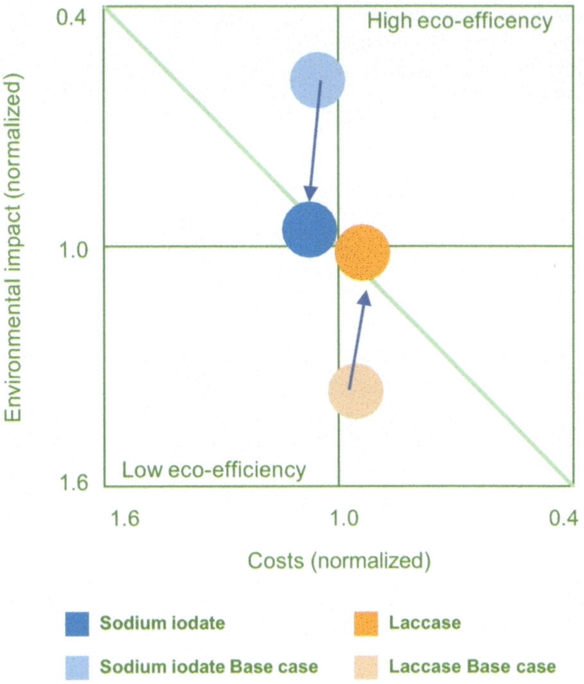


**Fig. 10** Future scenario for the C-N coupling reaction with significant improvements of the laccase process. Functional unit: production of 1,000 kg heterocoupling product of *n*-hexyl-amine and 3-methyl-catechol

processes, can identify possible risks in the upscaling, and can show improvement potentials by the generation of different scenarios.

With such a scenario analysis, the identification of major levers for the sustainability of the processes is possible. Very often, upstream impacts can also be important. Different production routes of basic materials as well as a different energy production for the processes can improve the eco-efficiency of an alternative significantly. Improvements in work-up processes, isolation of materials, and crystallization have the potential to improve systems as well.

### 2.8.4 Identification of Improvement Potentials

All previous experiments were aligned for maximal selectivity, which implies that the biotechnological process has not been fully exploited. Thus, the focus changed after the interpretation of the sustainability analysis from high selectivity to high productivity.

A new future scenario was generated where these findings were assessed; basic figures were adopted to the scenario, and a new result was created. The base case positions in the scenario are marked with light colors, and the new situation is expressed in intense colored points. It was shown that the environmental impacts of both alternatives show the same level when improving the productivity of the biocatalysis based on the volumes handled (Fig. 10). In that case, the differences on the environmental axis are negligible as well if the uncertainties of the process and the assessment are considered.

Solvent disposal and energy costs in the early process steps are triggered by the low yield of the final purification and reduce the eco-efficiency significantly. Higher yields of the final purification step in the C-N coupling mediated by laccases can reduce this impact as well resulting in better eco-efficiency results.

So different levers can be changed in more effective directions to improve the whole process. The combination of levers is possible as well and can be optimized. In other cases of the Eco-Efficiency Application, more than 100 parameters could be handled by using so-called Eco-Efficiency Managers (starting page https://www.v2.eeaman.com/), a specific Internet-based software solution that allows changing process factors in different directions and to display the results immediately. This support for researches enables them to work on technical solutions with much better performances in combination with the increase of the sustainability. Stakeholders want to see more examples like this to support their decision-making in the markets and to support products with increased sustainability impacts.

## 3 New Developments of the Eco-Efficiency Analysis Approach

New challenges in the assessment of LCA with new characterization models and systematic approaches such as the Product Environmental Footprint of the European Commission triggered a new set up of the methods applied. Furthermore, different needs of customers in different businesses were identified followed by the generation of a tool set with three different types of Eco-Efficiency Analysis. The new BASF Eco-Efficiency toolbox addresses the broad range of sustainability assessment needs by including flexible and fixed tools that cover life cycle environmental impacts and costs [31]. The underlying Eco-Efficiency methodology is based primarily on standard assessment models as implemented in the Product Environmental Footprint methodology of the EU (EU PEF). In addition, the approach avoids greenwashing by ensuring adequate environmental coverage and includes aggregation to an overall Eco-Efficiency Portfolio and Index (Fig. 11) to support decision-making based on quantified sustainability assessments. The new modular toolbox (Table 1) is based on ISO 14040, 14044, and 14045. It consists of fixed (EEA6, EEA10) and flexible tools (LCAflex) to address the range of needs.

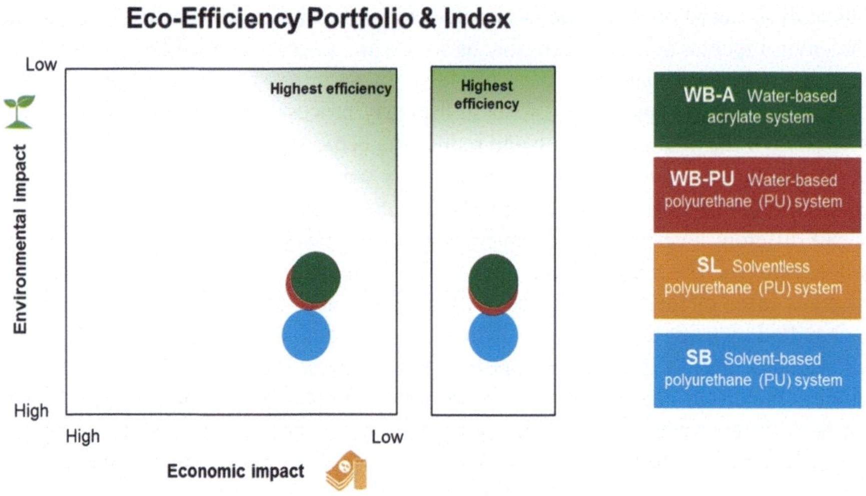


**Fig. 11** Portfolio and index aggregation of the new eco-efficiency tools

**Table 1** Overview of assessment opportunities of the toolbox

| | Tools | Sustainability dimensions (# impacts) | | | Methods based on | | | |
|---|---|---|---|---|---|---|---|---|
| | | Environmental | Economic | Social | EU PEF | ReCiPe | CML | TRACI 2 |
| Standard | Life Cycle Impact Assessment | 1-18 | | | ✓ | ✓ | ✓ | ✓ |
| | Environmental Product Declaration | 5-7 | | | ✓ | ✓ | ✓ | ✓ |
| BASF-specific | LCAflex | 1-18 | ✓ | | ✓ | ✓ | ✓ | ✓ |
| | Eco-Efficiency Analysis EEA6 | 6 | ✓ | | ✓ | | | |
| | Eco-Efficiency Analysis EEA10 | 10 | ✓ | | ✓ | | | |
| | AgBalance™ | 26 | ✓ | 28 | ✓ | | | |
| | Internet calculator possible for all tools | | | | | | | |

## 3.1 *Impact Assessment Methods of Different Types of Eco-Efficiency Analysis*

The EEA6 and EEA10 have defined environmental impact categories as well as set impact assessment methods. The EEA6 includes six environmental impact categories (GWP, POCP, AP, ADP, freshwater eutrophication, and human toxicity) expected to adequately cover environmental impact for most chemical products and processes. The EEA10 includes also impact categories which cover additional environmental aspects that become significant in, for example, assessments including bio-based materials. In this context land use, water footprint, or eco-toxicity are impact categories, which are often more significant than ADP or POCP. For GWP bio-based materials have often advantages along the whole life cycle, even if in the pre-chain there are efforts to produce different materials. All these specific needs of a biotechnological process can be considered quite well. The LCAflex is, in principle, completely flexible both in terms of environmental impact categories and impact assessment methods. However, the relevance check ensures that for each EEA6 and LCAflex, enough environmental impacts are covered, and the most relevant impact categories are included. In the relevance check, an overview calculation of the impact categories of EEA10 is done, normalized, and compared with each other. 80% of the relevant impacts need to be covered, so in an LCA flex, sometimes 3 impact categories are enough; in other cases 12 or more impact categories are needed to cover the impacts accurately. If this is not the case, then an EEA10 assessment is required.

### *3.2 New Type of Eco-Efficiency Portfolio Including the "Person-Time Approach"*

Today, we use a so-called person-time approach, where the single results are calculated compared to the specific impact an inhabitant of the Europe creates in a year and the subsequent time factor is derived. For example, a European citizen emits 9 tons $CO_2$ per year. If a study figure is 0.9 tons $CO_2$, it is equal to 10% of the total year, 36.5-person days. On this basis, results are comparable and can be aggregated with the same unit after weighting to single scores. Furthermore, the time expression can be easily understood by non-experts. Alternatives whose summed economy and environmental ratings are identical are equally eco-efficient (Fig. 11).

For biotechnological processes, this new setup enables practitioners the definition of the right tool to cover the most relevant aspects by excluding indicators which do not support the differentiation of process alternatives. More meaningful results support the acceptance of decision-making approaches in the sector of biotechnology. The experience shows that for biotechnological processes, an EEA10 is needed to cover all relevant impacts and to reach more than 80% of the impacts.

## 4 Integrating Social Assessments with the Social Analysis of SEEbalance®

Both Eco-Efficiency Analysis and SEEbalance® have been employed in the food and feed value chains as well as for bio-based processes to assess the key drivers of sustainability in various production systems [32].

New developments of different stakeholder groups pushed the development of new assessment approaches for social aspects. Different systems were used to assess social indicators which might be very important for biotechnological processes as well. In this context, the SEEbalance® was updated as well and has now a completely new Social Analysis included.

SEEbalance® is BASF's comprehensive approach to assess the social, ecologic, and economic aspects of products or processes covering the entire life cycle. Depending on the question, the four modules of the SEEbalance® can be applied individually or in combination. SEEbalance® supports decision-makers in identifying the sustainability benefits and trade-offs along the value chain of products or processes. The method does not relate the results to product mass flows because of the lack of quantitative data and to avoid discussion on "quantitative amount of child labor" for a product.

## *4.1 The Social Analysis as Module of the SEEbalance®*

The *Social Analysis* for social aspects takes new developments into account developed and introduced by different organizations and with different levels of details and indicator sets. The *Roundtable for Product Social Metrics as a cross sector initiative considers* three stakeholder groups (workers, users, local communities) and added a specific category, the small-scale entrepreneurs. Twenty-four social topics are covered in total. The new handbook published by the consortium focusses on qualitative approaches for social indicators [33] and gives a good overview of social topics linked with a defined product along its life cycle. A sector-specific approach published is the *WBCSD* guidance document [34] for the assessment of "Social metrics for chemical products" including and describing 3 stakeholder groups, 25 social topics (11 mandatory), and 168 indicators in total. The *UNEP/SETAC* guideline assesses 5 stakeholder categories, 6 impact categories, and 31 subcategories [35].

Based on these approaches, we developed a new approach that can be linked with the other methods used in the sustainability assessment of products or processes covering the entire life cycle.

Depending on the requirements and the questions to be answered in the context of decision-making, the four modules of the SEEbalance®, Life cycle costing (LCC), and LCA are integrated in the Eco-Efficiency Analysis. The Social Analysis covers the two modules *Social Life Cycle Assessment and Social Hot Spot Assessment.* The modules can be applied individually or in combination. SEEbalance® supports decision-makers in identifying the sustainability benefits and trade-offs along the value chain of products or processes [36].

### 4.1.1 The Social Life Cycle Assessment

The *Social Life Cycle Assessment* (SLCA) is a module of the Social Analysis and covers the identification of social risks along the life cycle based on existing data from external data providers.

The core of this assessment is the identification of social risks along important predetermined impact categories with a focus on the three main stakeholder groups workers, communities, and consumers.

In step 1 of the two-step approach, it is checked if information about the company performing the individual process step is available. If information is available, the respective company needs to be identified in Ecovadis™ to access company-specific evaluation forms. If no Ecovadis™ evaluation about the producing company is available, RepRisk® is used as the source for company-specific company risk information. Based on this information, the companies are rated with the scale-based approach.

If no information about the producing company can be found, step 2 is applicable. For this, information is needed to identify the relevant country or countries. In this step, Verisk Maplecroft™ is used to assess the risk of a specific sector in the respective country/ies. The results are transferred in a color scaling code from green via yellow and orange to red expressing the severity of the impacts in different stakeholder categories. Figure 12 shows an example of a generic fermentation processes with the color code results for each life cycle step. The steps can be aggregated to an overall result in different interpretation procedures.

### 4.1.2 The Social Hot Spot Assessment

The *Social Hot Spot Assessment* is another module of the Social Analysis and is therein a method for the evaluation of selected hot spots based on an in-depth-specific desktop research and link of relevant risks to Sustainability Development Goals (SDGs).

In the Social Hot Spot Assessment, a risk analysis is conducted to identify all relevant social hot spots in general. In the identification step, the possible issues are identified by first checking at the life cycle step(s) which are the most significant ones for the evaluated topic. Subsequently, specific sector issues in the relevant region need to be identified. Some basic aspects are here the definition of a unique and specific step with high importance of social indicators for the alternatives in the analysis; e.g., T-shirt production life cycle: significant steps can be working conditions in Bangladesh or cotton production in Turkey. Oil production in Venezuela as precursor for textile chemicals is less important in this context. Another criterion is the public interest, e.g., NGO which perceives specific life cycle steps as critical.

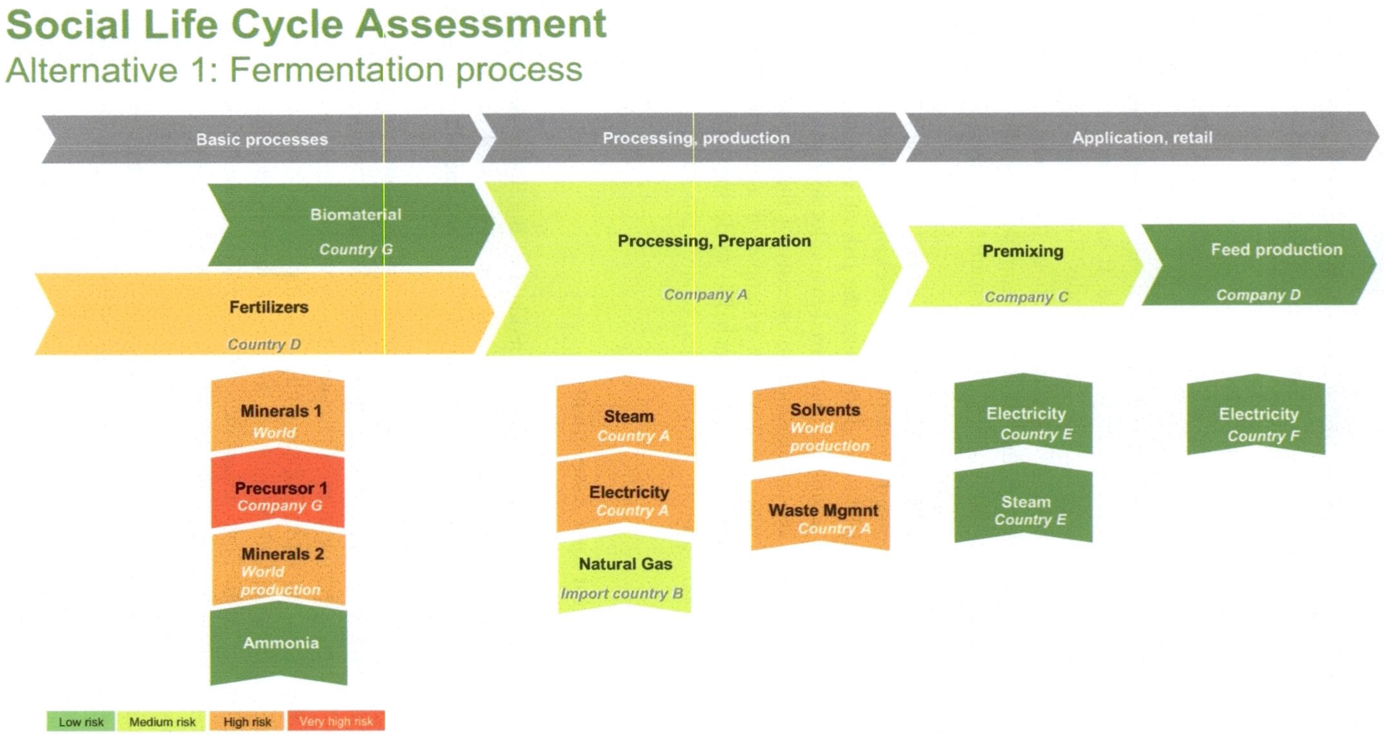


**Fig. 12** Overview of results from the Social Life Cycle Assessment along important life cycle steps of a fermentation process (illustrative)

**Table 2** Assessing social hot spots with the SDG goals and targets and identification of impacts against the improvements in the context of the SDG

**Identification and evaluation of hot spots**

Using SDG as guiding principles

| SDG | Description | Search item | Finding |
|---|---|---|---|
| 5.1 | End all forms of discrimination against all women and girls everywhere | discrimination, women | Nothing |
| 5.2 | Eliminate all forms of violence against all women and girls in the public and private spheres, including trafficking and sexual and other types of | violence, women | Nothing |
| 5.3 | Eliminate all harmful practices, such as child, early and forced marriage and female genital mutilation | Forced marriage | Forced marriage is found in this country |
| 5.4 | Recognize and value unpaid care and domestic work through the provision of public services, infrastructure and social protection policies and the promotion of shared responsibility within the household and the family as nationally appropriate | Value unpaid care | Value unpaid care is promoted |
| 5.5 | Ensure women's full and effective participation and equal opportunities for leadership at all levels of decision-making in political, economic and public life | women´s full participiation | Partially not realized |

For the assessment of the relevant risks, the Sustainability Development Goals (SDG) are used for the identification of impacts. They can be utilized to define the search topics. After identification of risks affecting the positive development of SDG, the findings are linked to the specific SDG improvement of situations (Table 2).

The findings can be linked to SDGs and shown in an overview graph. In the Social Hot Spot Assessment, the same color code is used as in the Social LCA. The Social Analysis uses the same methodological approach for the display of results. The overview can be used for the definition of improvement measures in the next step (Fig. 13).

## 4.2 *The AgBalance™ Approach for Agricultural Products*

For agricultural production which is often the basis for basic materials in biotechnological processes, some very specific indicators should be considered, for example, impacts on biodiversity, soil health, and the agri-sociological aspects. In this context, the AgBalance™ method was developed. It is a holistic method for assessing sustainability in agriculture and food production and identifying key drivers for improvement. The method was developed in an international team of different experts. It was described in detail and validated by National Sanitation Foundation (NSF), Det Norske Veritas®, and Technischer Überwachungsverein TÜV Süd [37].

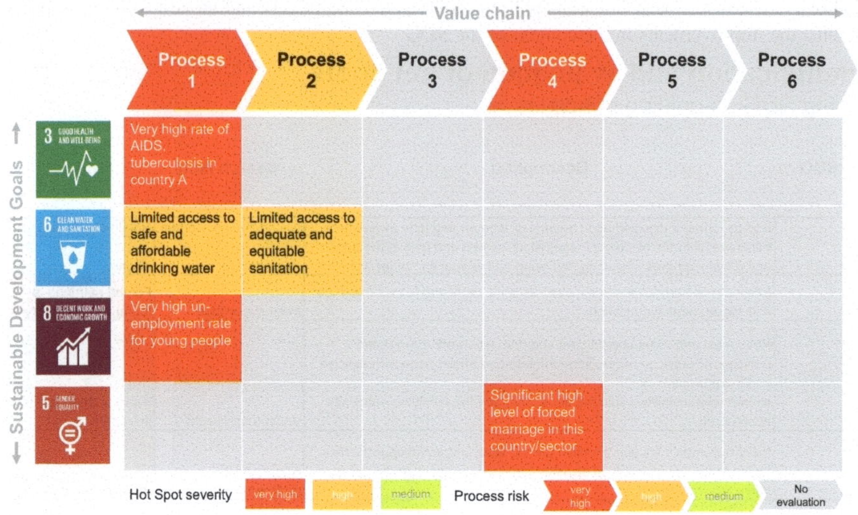


**Fig. 13** Overview of results from the Social Hot Spot Assessment showing negative impacts in the connection to SDG (illustrative)

AgBalance™ comprises a multi-criteria life cycle-based approach in combination with a defined aggregation and summary of single results into a single sustainability score [38]. AgBalance™ delivers results that enable farmers, the food industry, politicians, and society to objectively evaluate processes in terms of their sustainability profile. In doing so, a vast amount of information on individual factors can be ascertained in addition to overall statements on the sustainability of agricultural practices.

## 5 Summary and Conclusion

Biotechnology is a growing and important field of providing more sustainable solutions. To decide if a solution is more sustainable than an alternative approach, sustainability assessments are needed. They can sufficiently support decision-making processes with important information based on a holistic life cycle approach. This approach can avoid wrong developments and can reduce the time to find more sustainable solutions. Research and development departments in industry but as well in academia can benefit from modern tools that support their research activities.

To use eco-efficiency assessments more intensively, good case studies in different fields of industries and services should be available. It should be shown how it is possible to generate meaningful results in a reasonable timeframe with reasonable amount of data needed for reasonable costs of 5–10% of the project costs (depending on the size of a project) or lower. Prospective LCA will help to figure out future options as well and can support researchers to define goals and improvement potentials, further analyzed by a scenario analysis. The LCA, Eco-Efficiency Analysis (EEA) manager and SEEbalance® can support the development of new and more sustainable processes and products. These tools, their easy application and interpretation as well as the generation of different future scenarios will support the biotechnology development quite sufficiently and supports the implementation of new and more sustainable technologies.

An additional opportunity of this type of assessments is to use specific tools which can generate a high number of scenarios in a short time frame based on automatic systems which support practitioners very efficiently. Different types of such tools were developed and successfully applied and shown with some examples in this chapter.

In the example of vitamin $B_2$, e.g., the advantages of the biotechnological process versus the common chemical process could clearly be shown. In the further development, the chemical process was closed and the biotechnological process implemented. The process was improved several times and is until today an innovative, successful, and more sustainable process supporting the BASF strategy for creating chemistry for a more sustainable future.

## References


1. Saling P, Kicherer A, Dittrich-Krämer B, Wittlinger R, Zombik W, Schmidt I, Schrott W, Schmidt S (2002) Eco-efficiency analysis by BASF: the method. Int J Life Cycle Assess 7(4):203–218
2. Saling P (2016) Eco-efficiency assessment in LCA compendium – the complete world of life cycle assessment. In: Finkbeiner M (ed), Klöpffer M, Curran MA (series eds) Special types of life cycle assessment. Springer, Dordrecht, pp 115–178. https://doi.org/10.1007/978-94-017-7610-3_4
3. Saling P (2016) The BASF eco-efficiency analysis – a 20-year success story. BASF Verlag, Ludwigshafen
4. Kölsch D, Saling P, Kicherer A, Grosse-Sommer A (2008) How to measure social impacts? What is the SEEBALANCE® about? – Socio-eco-efficiency analysis: the method. Int J Sustain Dev 11(1):1–23
5. Schmidt I, Meurer M, Saling P, Kicherer A, Reuter W, Gensch C (2004) SEEbalance – managing sustainability of products and processes with the socio-eco-efficiency analysis by BASF. Greener Manag Int 45:79–94
6. Uhlman BW, Saling P (2010) Measuring and communicating sustainability through eco-efficiency analysis. Chem Eng Progr (CEPE) 106(12):17–26
7. BIOPRO Baden-Württemberg GmbH (2013) Facts and figures. Biotechnologie.de. https://www.biooekonomie-bw.de/en/articles/dossiers/industrial-biotechnology-biological-resources-for-industrial-processes/. Accessed 25 Sept 2017
8. OECD (2001) The application of biotechnology to industrial sustainability. www.oecd.org/sti/biotechnology
9. Wall-Markowski C, Kicherer A, Saling P (2004) Using eco-efficiency analysis to assess renewable-resource-based technologies. Environ Prog 23(4):329–333. https://doi.org/10.1002/ep.10051

10. Saling P (2005) Eco-efficiency analysis of biotechnological processes. Appl Microbiol Biotechnol 68:1–8. https://doi.org/10.1007/s00253-005-1951-0
11. Landsiedel R, Saling P (2002) Assessment of toxicological risks for life cycle assessment and eco-efficiency analysis. Int J Life Cycle Assess 7(5):261–268
12. Dornan P, Saling P, Coers P, de Hults Q, Downes R, Rydberg T, Tiemersma T, Weinbeck X (2017) ProScale – a life cycle oriented method to assess toxicological potentials of product systems. Guidance document. https://www.proscale.org/. Accessed 5 Aug 2018
13. Saling P, Günther C (2005) Using eco-efficiency analysis to optimize your business strategy. In: Markets meet research, Ludwigshafen, 14 Sept 2005
14. Uhlman B, Laginess T, Green D, Steinmetz D, Saling P, Grosse-Sommer A, Paczkowski N, van Gelder R (2016) NSF protocol 352, validation and verification of eco-efficiency analysis, part A. BASF's Eco-Efficiency Analysis Methodology
15. Marley A, Louviere J (2005) Some probabilistic models of best, worst, and best–worst choices. J Math Psychol 49(6):464–480. https://doi.org/10.1016/j.jmp.2005.05.003
16. European Commission (2018) European commission services towards the development of a harmonised methodology for the calculation of the environmental footprint of products and organisations (including carbon). http://ec.europa.eu/environment/eussd/smgp/dev_methods.htm, Accessed 25 Oct 2018
17. ISO 14040 (2006) Environmental management – life cycle assessment – principles and framework. Geneva
18. ISO 14044 (2006) Environmental management – life cycle assessment – requirements and guidelines. Geneva
19. ISO 14045 (2012) Environmental management – eco-efficiency assessment of product systems – principles, requirements and guidelines. Geneva
20. Baker R, Saling P (2004) Label eco-efficiency analysis astaxanthin. https://www.basf.com/documents/corp/en/sustainability/management-and-instruments/quantifying-sustainability/eco-efficiency-analysis/BASF_Label_Astaxanthin.pdf. Accessed 14 Nov 2017
21. ISO/TR 14073:2017, environmental management – water footprint – illustrative examples on how to apply ISO 14046
22. Saling P, Kicherer A (2006) Assessment of biotechnology-based chemicals. In: Dewulf J, Van Langenhove H (eds) Renewables-based technology: sustainability assessment. Wiley, West Sussex, pp 299–313
23. Saling P (2003) Eco-efficiency in vitamins at BASF. Speciality Chemicals Magazine, Hatfield, p 36
24. Saling P, Günther C (2006) Sustainable production of vitamin B2. Feed Mix 14(2):22–24
25. Alberts B, Johnson A, Lewis J, Raff M, Roberts K, Walter P (2002) The molecular biology of the cell. Garland Science, New York
26. Stryer L, Berg JM, Tymoczko JL (2002) Biochemistry5th edn. W.H. Freeman, San Francisco
27. Schomburg I, Chang A, Placzek S, Söhngen C, Rother M, Lang M, Munaretto C, Ulas S, Stelzer M, Grote A, Scheer M, Schomburg D (2013) BRENDA in 2013: integrated reactions, kinetic data, enzyme function data, improved disease classification: new options and contents in BRENDA. Nucleic Acids Res 41(Database issue):D764–D772. https://doi.org/10.1093/nar/gks1049
28. Novozymes (2017) https://www.novozymes.com/en/biology. Accessed 14 Nov 2017
29. Wellington KW, Kolesnikova NI (2012). Bioorg Med Chem 20:4472–4481
30. Illner S, Plagemann R, Saling P, Kragl U (2014) Eco-efficiency analysis as a reaction-engineering tool – case study of a laccase-initiated oxidative C–N coupling. J Mol Catal B Enzym 102:106–114
31. Grosse-Sommer A, Gruenenwald T, Paczkowski N, van Gelder R, Saling P (2017) Applied sustainability in industry: the BASF eco-efficiency toolbox for decision-making and marketing support. Poster presentation LCM, Luxembourg
32. Uhlman BW, Saling PR (2017) The BASF eco-efficiency toolbox: holistic evaluation of sustainable solutions. Reference module in earth systems and environmental sciences. Encyclopedia of Sustainable Technologies. https://doi.org/10.1016/B978-0-12-409548-9.10042-9

33. Goedkoop MJ, Indrane D, de Beer IM (2018) Product social impact assessment handbook – 2018. Amersfoort
34. WBCSD, Alvarado C, Brown A, Hallberg K, Nieuwenhuizenn P, Saling P, Chan K, Das Gupta J, Morris D, Nicole G, Wientjes F, Dierckx A, Garcia W, Combs C, Kilgore A, Satterfield B, Haver S, Jostmann T, Vornholt G, Bergman U, Feesch J, Whitaker K, Kiyoshi M, Govoni G, Mehta R, Menon A, Sen S, Upadhyayula V, Bande M, Coërs P, Debecker D, Poesch J, Viot JF (2016) Social life cycle metrics for chemical products – a guideline by the chemical sector to assess and report on the social impact of chemical products, based on a life cycle approach. www.wbcsd.org/contentwbc/download/1918/24428. Accessed 30 Nov 2017
35. United Nations Environment Programme and Society for Environmental Toxicology and Chemistry (2009) Guidelines for social life cycle assessment of products. Paris
36. Saling P, Alba Pérez A, Grünenwald T, Kölsch P (2018) Generation, calculation and interpretation of social impacts with the social analysis of SEEbalance®. In: Traverso M, Petti L, Zamagni A (eds) Perspectives on social LCA: contributions from the sixth international conference. Springer, Berlin. in print
37. Schoeneboom J, Saling P, Gipmans M (2012) AgBalance™, technical background paper. http://www.agro.basf.com/agr/AP;Internet/en/function/conversions:/publish/upload/sustainability/AgBalance/307736_BASF_Tech-E_Paper-AgBalance.pdf. Accessed 5 Apr 2013
38. Frank M, Schöneboom J, Gipmans M, Saling P (2012) Sustainable intensification in Winter Oilseed Rape production – results from an AgBalance study. In: Corson MS, van der Werf HMG (eds) Proceedings of the 8th international conference on life cycle assessment in the agri-food sector (LCA Food 2012), 1-4 October 2012, Saint Malo, France. INRA, Rennes, pp 62–67

# Index

The manufacturer's authorised representative in the EU is Springer Nature Customer Service Centre GmbH, Europaplatz 3, 69115 Heidelberg, Germany. If you have any concerns regarding our products, please contact ProductSafety@springernature.com

Printed and bound by CPI Group (UK) Ltd, Croydon, CR0 4YY
30/09/2026
02225689-0013